PROCESS
TECHNIQUES FOR

ENGINEERING
HIGH-PERFORMANCE
MATERIALS

PROCESS
TECHNIQUES FOR
ENGINEERING
HIGH-PERFORMANCE
MATERIALS

TIM OBERLE

CRC Press
Taylor & Francis Group
Boca Raton London New York

CRC Press is an imprint of the
Taylor & Francis Group, an **informa** business

CRC Press
Taylor & Francis Group
6000 Broken Sound Parkway NW, Suite 300
Boca Raton, FL 33487-2742

© 2014 by Taylor & Francis Group, LLC
CRC Press is an imprint of Taylor & Francis Group, an Informa business

No claim to original U.S. Government works

Printed on acid-free paper
Version Date: 20130516

International Standard Book Number-13: 978-1-4665-8188-3 (Paperback)

Library of Congress Cataloging-in-Publication Data

Oberle, Tim.
  Process techniques for engineering high-performance materials / author, Tim Oberle.
    pages cm
  Includes bibliographical references and index.
  ISBN 978-1-4665-8188-3 (pbk.)
  1. Manufacturing processes. 2. Materials. I. Title.

TS183.O24 2013
658.5--dc23   2013018219

Visit the Taylor & Francis Web site at
http://www.taylorandfrancis.com

and the CRC Press Web site at
http://www.crcpress.com

Contents

Acknowledgments

Thanks to Rita, Helen, Claire, and Michelle for putting up with this project for so long. I love you all. Thanks also to the many colleagues who helped me bring this book to life.

Introduction to Process Dependency

Why does one batch of product perfectly suit the needs of a particular customer while others range from barely adequate to completely useless? Why do some product types require completely different development techniques?

> The man who comes up with a means for doing or producing almost anything better, faster or more economically has his future and his fortune at his fingertips.
>
> **J. Paul Getty**

We routinely witness the introduction of miraculous new materials with amazing properties. When newly discovered, they are rare, expensive, and temperamental in their performance. Gradually, they evolve, improve, and become essential in demanding applications. These curiosities morph into multiple permutations that capture the imagination of cutting-edge designers. Still later, their price drops to commodity levels and these marvels become mundane, but they are vital to the global economy and our way of life.

Etched silicon circuitry and high-bandwidth fiber-optic cable facilitate the ongoing tech boom. The plastics and composites of the previous generation still drive our consumer culture. Steel alloys revolutionized mass production and enabled the proliferation of trains, ships, and cars. Vulcanized rubber was the wonder material of the mid-nineteenth century. The discoveries of mud brick for shelters, bronze smelting for tools, and fired clay for storage vessels lifted our ancestors into the first civilizations.

None of these advances sprang from a single inventive revelation. Each one required a production process to be developed and continuously improved. Skillful process manipulation is the key to commercializing materials with magical properties and commercial superiority. The processing techniques are always veiled in secrecy. Rivals have coveted the secrets of improved materials ever since we first transformed a lump of flint into a lethal hunting blade.

This book explores the strategies for the systematic development, improvement, and ongoing production of engineered materials. The important properties of these materials are generally dependent on the conditions that they experience during manufacturing. These products remember the forces of their own creation and they acquire distinctive characteristics from their process history.

This retained memory dictates how consistently your materials will perform in the real world. It explains why the defective machine part breaks at the moment of greatest need. This is the reason why one particular product grade works better than any other grade in the application for which it was tailored. Successful innovators understand how to adjust their processes to optimize end-user performance.

Nineteenth-century cannon shells were often hollow iron balls packed with gunpowder. When the gun was fired, a fuse in the shell was ignited. After it burned down to the gunpowder charge, the ball would usually explode, showering the vicinity with metal shards. Fuses were the engineered materials of their day.

During the American Civil War, the Confederate forces relied on color-coded paper fuses. The yellow fuse was rated to burn at 5 s/in. and the green fuse at 7 s/in. In theory, the gunners would cut the fuses to the appropriate length, based on the estimated range to their target and the mussel velocity of their particular guns. In practice, the procedures were governed by tables and rules of thumb derived from the experience of each gun crew with the materials available to them.

The nearby Richmond Armory supplied the artillerymen of Robert E. Lee's Army of Northern Virginia. However, in March 1863, this facility suffered a devastating explosion and fire, curtailing its production. For the subsequent invasion of Pennsylvania, fuses were hurriedly sourced from factories across the South. Due to wartime shortages, the gun crews had no opportunity to become familiar with their new supplies.

Lee marched his army north in June 1863, eventually engaging the Union Army at Gettysburg where they fought for 3 bloody days. At the battle's climax, Lee ordered 12,500 men under George Picket to assault the Union lines on the aptly named Cemetery Ridge. The Confederates preceded their attack with a thunderous 2-hour artillery bombardment by 172 cannons. Veteran artillery officer H. C. Cabell reported, "For over two hours the cannonading on both sides was almost continuous and incessant, far exceeding any I have ever before witnessed."*

---

* Report of Colonel H. C. Cabell, *C. S. Army, commanding Artillery Battalion*. CAMP NEAR CULPEPER COURT-HOUSE, VA, August 1, 1863.

Due to the dense gun smoke and the volume of fire, the Confederate gunners could not adjust their aim points once the barrage was underway. They could not see that many of their shells were failing to explode over the enemy's main line of defense. A percentage never exploded at all. Picket's famous charge was repulsed by the Union forces, which largely survived the bombardment. Half of his men could not return to the lines afterwards. This devastating failure made General Picket the scapegoat for Lee's defeat and the South's ultimate loss of the war.

This example dramatizes common performance problems:

- Lee's gunners did not know what to expect from the unfamiliar fuses. The unique process conditions at each factory resulted in different burn rates.
- The Confederate Army did not standardize testing across production sites to ensure consistent performance until after Gettysburg. The Union's paper fuses were produced in a single facility, to reduce product variability.
- As is often the case, the inconsistency was suspected before Gettysburg. However, nothing was proven until a report was issued the week after the battle.
- The people producing the fuses did not get timely feedback from the gunners to adjust their materials, procedures, and operating conditions to improve consistency. Date coding and lot control were not practiced. No central authority monitored and standardized production variables.
- No one tested how the fuse burn rates changed over time due to storage conditions.
- The Confederate supply chain was forced to focus on volume, rather than consistency. Wartime memoirs speak of shortages of the most basic supplies, such as shoes for both men and horses.
- Conditions in one area of the Richmond Armory were allowed to affect other operations in the same facility. Literally, there was no firewall around the dangerous workshop that recycled gunpowder from damaged weapons and shells recovered from nearby battlefields.
- The Confederate cannon-firing operation was not scalable under battlefield conditions. Individual gun crews were resourceful and accurate when their targets were in sight, adjusting their aim points and fuse lengths. However, their effectiveness declined when firing as a group in dense smoke. Without feedback, the gunners could not correct for uncontrolled material variables that influenced performance.

The history books do not indicate which process factors varied from one armory to another. They could have involved fiber orientation, drying conditions, local raw materials, storage/packaging decisions, or changes during aging. These obscure details of production technology never capture the popular imagination like battlefield heroics or the political upheaval of that period do. Since paper fuse technology was rendered obsolete by wartime innovation, there was little subsequent research.

The performance of an industrial process rarely has a drastic effect on historical events. However, the elements of the Gettysburg fuse example are found in many problems that plague products today. Small deviations in properties impact performance in subtle ways that are not immediately apparent on the factory floor. Serious business issues result from a series of minuscule deviations because performance reports are not communicated back to the people who make the adjustments. This book will discuss ways to engineer circuit breakers in the development and production schemes to detect and minimize these situations before they propagate into major problems.

Process Dependency in the Real World

> Wealth, like happiness, is never attained when sought after directly. It always comes as a byproduct of providing a useful service.

Henry Ford

There are many popular fantasies about product development. We see images of Detroit artisans sculpting concept cars out of clay that rarely match the ones at the dealers. Hip designers in black create iconic consumer goods that define our intrinsic value as individuals, but interface poorly with the previous versions. Gangs of magical elves sing a catchy jingle while turning out unhealthy snack foods. Every garage inventor expects to make a fortune with his or her big idea.

Forget the fantasies and imagine all the tangible, man-made products in the world. Ignore those things that are intangible, such as software, service, financial, and media products. Tangible goods can be grouped into three broad categories, based on their process dependency:

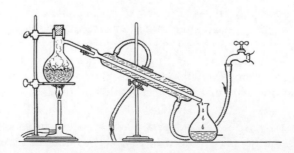

The first set is purely chemical in nature. The properties and attributes of simple chemicals are governed by their constituent matter. Their utility depends on their level of purity and the type of contaminants or by-products present and not on the production process. Generally, basic chemicals from different production sources are interchangeable commodities. Their ultimate performance has little or no process dependency.

The next group is engineered materials, which are highly dependent on the manufacturing scheme that creates them from basic chemicals. Such products remember the process under which they were produced. Retained memory will determine how effectively they perform in end-use applications. This book is devoted to the methods of developing, evaluating, and manufacturing these products to consistently ensure fitness-for-use.

The final category includes mechanical and electrical products that are assembled from components. The performance and functionality of this class depend on good design, subcomponent quality, and proper assembly procedures. However, the component parts of the product are often made from process-dependent materials.

It is difficult to classify some products into one of these three categories. Assembly techniques such as soldering, welding, and painting can be process dependent in their own right. Complex chemicals such as polymers, foods, drugs, and biochemical materials are often process dependent, unlike simple chemicals. It is difficult to pinpoint where the changeover occurs. It is easy to satisfy some end-use applications, so a wide range of similar products work equally well. Those customers will rarely experience the process dependency that satisfies or disappoints the more demanding end users.

See the "concept map of tangible products" at the end of the chapter. The connections between the important ideas in this section are graphically summarized in this map. In the chapters to follow, these maps will reveal the hierarchical association and sequential relationships between key concepts. Like Leonardo da Vinci, each one offers a visual perspective to reinforce and review the chapter content. Each reader must select the items appropriate for his or her circumstances and applications.

RULE #10 EACH PRODUCT CATEGORY REQUIRES UNIQUE DEVELOPMENT STRATEGIES

Product development procedures appropriate for software systems, electrical devices, and mechanical assemblies are not suitable for engineered materials. Process-dependent materials require specialized development techniques to consistently satisfy customers.

Note to the reader: A series of 61 illustrated rules of product development emphasize the key points of the book. Their relative importance will vary with the needs of each particular industry. In Rule #10, the important message is to analyze each project, and not just apply the same methodology as worked on the previous project or at some famously successful company.

Consider the example of the greengrocer. Peaches of a given variety often look, smell, and feel identical in the supermarket display case. When eaten, however, their taste may range from divine to disappointing because of the many factors that influence the fruit's quality.

The peach production process includes pruning, fertilizing, watering, harvesting, packing, storing, and transporting the fruit. The combination of all these steps dictates the flavor and texture. Not all of the important factors can be controlled. The balance between rain and sun is obviously critical. Market forces and transport logistics determine whether the fruit is picked at the optimal time or it is shipped to market as hard balls of fuzz.

It is not yet possible to design a succulent peach that will perform consistently under all growing and transport conditions. The farmer carefully controls the process parameters, and prays for good weather and a favorable market for the crop at harvest time. But, once the fruit is harvested, an inferior peach cannot be adjusted or corrected.

In an industrial environment, it is not acceptable to blame the weather for product shortcomings. Imagine telling construction contractors that they will have to use warped lumber because of high humidity levels at the sawmill. Try asking a computer manufacturer to accept defective semiconductors produced during the pollen season.

In most cases, suppliers are expected to filter out all the environmental factors to satisfy application requirements. The mill will dry the wood in a kiln to receive a premium price. The silicon chips are fabricated under strict clean room procedures to limit defects from airborne particulates. The process must overcome variations in raw materials, operators, and a thousand other tangible factors.

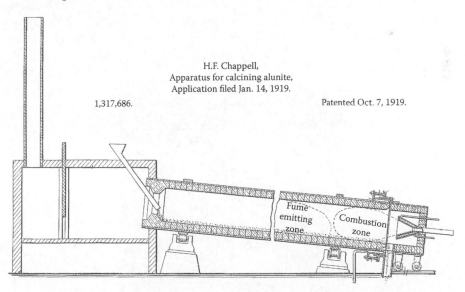

H.F. Chappell,
Apparatus for calcining alunite,
Application filed Jan. 14, 1919.

1,317,686.

Patented Oct. 7, 1919.

Portland cement is a well-understood commodity, which seems to lack technical complexity. Despite this perception, its production process must be carefully controlled to make an acceptable product. Early versions of cement were used by the Romans to build infrastructure, including their extensive network of aqueducts. After the fall of their empire, the recipe was lost until it was reinvented during the Industrial Revolution.

In the 1820s, Joseph Aspdin patented Portland cement. However, his formula was slow to cure and did not reach its full strength for weeks. Such a product would not be acceptable for most of today's construction and repair applications. Aspdin's son, William modified the process conditions to achieve a faster curing time. He kept the process improvement (higher peak operating temperature) as a trade secret.

Modern cement kilns are fed with a mixture of calcium oxide, silicon oxide, aluminum oxide, and ferric oxide. They are metered in at tightly controlled ratios and are ground to a very specific particle size. The mixture is heated in a rotating kiln where the powder progresses through a defined temperature gradient. The metallic oxides are present to form a liquid flux, acting as a solvent for the formation of tricalcium silicate at the end of a chain of chemical reactions. William Aspdin's process innovation increased the tricalcium silicate content in the finished product, improving the curing time of his product.

By adjusting feedstocks and process conditions, a variety of useful properties can be engineered into the product, to suit end-use requirements. In a mature technology sector, such as cement, the important process factors have been enshrined in procedures, industry standards, government regulations, and laws. These rules are intended to ensure a consistent performance for well-studied applications. The important factors for totally new materials can only be determined with feedback from the customers or tests that mimic the end-use conditions. This is the key to taming a complex process.

Understanding all aspects of performance is difficult when negative effects are not immediately apparent. This was the case during the mid-1960s for seven chemical companies producing 2,4,5-trichlorophenoxyacetic acid (2,4,5-T) for the U.S. military. Ideally, this class of herbicides will have a high toxicity toward plants while doing no harm to people and animals.

A steady growth of urgent orders induced the companies to drive their reactor conditions to the limits to maximize output. The process engineers were unaware that at temperatures exceeding 160°C, a side reaction tended to produce trace amounts of 2,3,7,8-tetrachlorodibenzo-para-dioxin (TCDD) by-product. The government had no specifications on the maximum acceptable level of TCDD.

Few serious problems were associated with the product at the time of use. It killed the targeted vegetation, so the customer was satisfied, but it was difficult to judge the long-term impact. Twenty years later, it was finally accepted that people were suffering from exposure to trace amounts of TCDD. In a court settlement with American military veterans, the companies paid $180 million. The troops had nicknamed the chemical Agent Orange, due to the container's color coding.

The TCDD contaminant was identified as the causal agent in a range of medical disorders that plague the soldiers and countless Vietnamese to this day. Government and industry estimates suggest that these problems were the result of only 368 lb. of dioxin present in 77 million liters of the herbicide sprayed.

The TCDD levels in the defoliant spray were found to range from 0.05 to 47 ppm. The contaminant levels were dependent on the type of production process and the conditions that were employed to process each batch. By the time that commercial production in the United Kingdom was halted, TCDD levels were less than 0.01 ppm, but even this was found to be unacceptable. The differences in the process conditions resulted in variations of over four orders of magnitude in the amount of carcinogen in the herbicide. This is an extreme example of problems that stem from a failure to recognize and control key parameters. In the company's defense, there was not even an adequate analytical test method to quantify TCDD levels in the early 1960s.

The analytical test for dioxin levels was only developed after 2,4,5-T was found to be more harmful than was expected from the original toxicity testing. Presumably, the TCDD levels were much lower in the development batches that were used for the early toxicity tests. Those samples were not synthesized under the high-rate conditions of the optimized industrial process. Thankfully, few by-products have the capacity of dioxin to affect the health of so many people. No one can accurately predict the impact of uncontrolled process parameters in a complex system.

RULE #3 UNCONTROLLED PROCESS INPUTS WILL CAUSE END-USE PERFORMANCE PROBLEMS

Process experts must know the impact of every control parameter. Nothing can be left to chance by allowing important variables to fluctuate randomly at the whim of an operator, a supplier, or an out-of-tolerance component in the production process. For the technically astute company, the manufacturing process is not a burden to be compensated for. Product developers tune production schemes to impart unique properties, which cannot be duplicated by less-skilled competitors. Discovery efforts never stop, because end-user requirements and competitive alternatives constantly evolve.

Products That Can Retain a Memory of Their Production Process

Inanimate objects are classified scientifically into three categories—those that don't work, those that break down, and those that get lost.

Russell Baker

Early in the nuclear age, atomic bomb tests were conducted in great secrecy at remote locations. The American military was shocked when Kodak scientists approached them with accurate knowledge of when those tests had been conducted.

In 1945, Kodak engineers were mystified by an unusual new category of image flaws. A careful analysis and inductive reasoning traced the source to secretive activities thousands of kilometers away from their film manufacturing lines. Radioactive fallout had contaminated raw materials, which later came into contact with their film stock. Charged particles emitted by the contamination created defects in the emulsion coating.

RULE #25 PRODUCT MEMORY IS APPLICATION DEPENDENT

In some applications, the process history causes few problems. The products function adequately despite variations in materials and processes. Or the customers expect variability and learn to deal with it. In other

applications, two seemingly identical parts will fail in completely different ways. Process dependency is in the eye of the beholder. Major issues
can occur when a company with experience in tolerant applications
acquires a new customer who is highly sensitive to process variability.

This sensitive product was changed by its exposure to a random factor during the production process. The problem's origin was unexpected and totally
unrelated to anything within Kodak's control. Only by relentlessly seeking
out the root cause(s) of an issue will a company be able to prevent reoccurrences. Kodak film was later used to detect fallout from distant atomic tests.
It was also used to protect workers in the fledgling nuclear industry by monitoring their radiation exposure. Kodak's defect investigation spawned an
unforeseen new product category.

April 23, 1963   P. M. COOK ET AL   **3,086,242**
PROCESS AND APPARATUS FOR PRODUCING MATERIALS
HAVING PLASTIC MEMORY
Filed July 15, 1960

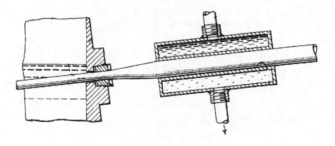

PAUL M. COOK
RICHARD W. MUCHMORE
INVENTORS

In the 1950s, Paul Cook was a chemical engineer at the Stanford Research
Institute where he studied the effect of high-energy electrons on the structure of plastics. He learned that electron bombardment can cross-link
polyethylene, significantly changing its properties. In particular, Cook discovered that cross-linking polyethylene greatly improved its ability to be
biaxially oriented. This is a process of stretching a polymer at a temperature
above its softening point, but below its melting point to form a thin film.
When the film is immediately quenched, molecular stress can be frozen into
its structure. At room temperature, the oriented material is dimensionally
stable. But, when heated, the plastic will revert to its original dimensions.
This shrink memory has a wide range of useful applications.

In 1957, Raychem was incorporated to exploit the commercial opportunities for this technology. The company pioneered shrinkable sleeves, which allowed electricians to insulate wire splices in a protective cocoon. Their oriented film could be wrapped around metal tubing and shrunk to form a plastic skin.

Metal alloys with shape memory were first developed by William Buehler at the U.S. Naval Ordnance Laboratory in the early 1960s. These properties were discovered by accident during a technical discussion on nickel–titanium blends. Buehler had formed a 0.01 in. thick sample and folded it into an accordion shape to demonstrate fatigue resistance. Curious participants stretched and bent the strip. One applied heat with his cigarette lighter. The group was surprised when the flame triggered the metal to snap back to its original shape. This particular memory was set into the material by annealing for 1 hour at 500°C. The metal reverts back to its annealed shape when it undergoes a crystalline phase transition just above room temperature. This technology is used in eyeglass frames that spring back to their original shape after deformation. Customized alloys enable mechanical actuators to be triggered with either heat or magnetic fields. The performance characteristics can be tailored to end uses by altering the composition and the process conditions.

These materials are extreme examples of material memory. Generally, process dependency is more subtle:

- Materials of biological origin
  - Paper properties change significantly with different fiber sources. Even the tree's growing conditions and soil parameters can influence product properties. Variations in moisture content can significantly change the paper's performance and dimensional stability.
  - Minor variations in microorganisms, aging time, and temperature have a huge influence on the taste and texture of yogurt, baked products, alcoholic beverages, and cheese.
  - The choice of natural fiber is very important to the final properties of textile products. One grade of animal fiber will make a scratchy sweater and another grade has a cashmere texture.

- Reportedly, humans and apes differ in only 1% of their genetic code. Keep that in mind the next time that you tolerate a 1% error rate in a complex system.
- Plastic resins and processed plastics
    - The polymer performance is sensitive to small changes in structure, molecular weight, and morphology.
    - The mechanical properties of injection-molded parts can vary greatly from one spot to another if the flow stress and cooling rates are not uniform.
    - Polymer degradation due to heat and sunlight can vary tremendously with the addition of trace amounts of antioxidants, stabilizers, and comonomers. The timing is critical when an agricultural ground cover or six-pack ring is expected to function normally for some set period and then disintegrate.
- Processed metal
    - Annealing, heat treatment, and mechanical work history are vital to the properties of metallic parts. However, their effects vary widely from one metal or alloy to another.
    - Impurities and additives can drastically change the strength, ductility, and corrosion resistance of ferrous metal. Processing conditions must be adjusted to account for the presence of these minor components. Fine stainless steel and common rust are mostly iron, but they have vastly different properties and values.
- Elastomers
    - A modest variation in the amount of carbon black and sulfur added during processing has a significant influence on the properties of natural latex rubber.
    - Synthetic rubbers, plastomers, and resilient thermoplastic alloys can be tailored to the end-user requirements in ways that are impossible for the sap of the rubber tree.
    - Changes to tire components and production techniques have brought tremendous improvements in tread life even in the last 20 years.
- Optical products
    - Fiber-optic cables are made from very high purity glass under carefully controlled conditions. Any defects or contaminants will cause signal losses in long-distance cables, reducing the distance between the amplifiers in undersea cables.
    - Flexible contact lenses are made with proprietary polymers and are highly sensitive to both bacterial and particulate contamination during manufacture. Just a few serious eye infections will shut down a global product line.

- Semiconductor products
  - A tiny microchip can have millions of component transistors and internal connections. A single defect in the wrong spot might render an entire assembly useless.
  - As chips continue to shrink, their performance is controlled by interactions at the atomic level. The lines between applied electronics and quantum physics start to blur.
- Drugs and pharmaceuticals
  - The effects of complex biomaterials in the human body are strongly influenced by subtle changes in crystallinity, isomer structure, and isotope composition.
  - The U.S. Food and Drug Administration does not just approve new medical products solely on the basis of formulation or composition. The manufacturing process must also be designated in the application. Producers of generic equivalents must often demonstrate that their production schemes will give the same result.
  - A compound that saves thousands of lives may kill a small portion of the population that is allergic to it. Such a situation may not even be discovered until the first negative interaction.

The Process Can Even Humble a Design Genius

You never are so easily mistaken as when you think you know the way.

Chinese Proverb

Leonardo da Vinci sought to create a spectacular new style of painting in a work entitled "Battle of Anghiari." The job was commissioned by the city fathers for the great hall of the Palazzo Vecchio in Florence, which still stands today. It was to be sited opposite the never finished "Battle of Cascina" by Michelangelo.

Traditional fresco techniques require the artist to paint onto a small patch of wet plaster. Each segment represents one day's work in a larger composition. Once dried, the plaster holds the brush strokes for centuries. Fresco painters are restricted in the pigments and brush techniques they can employ. Dissatisfied with these limitations, the master was anxious to pioneer an alternative wall painting style.

For this work, Leonardo scaled up an encaustic (curing with heat) technique that had shown promise in small-scale tests. The method was documented centuries earlier by the Roman author Pliny. Da Vinci selected only the ideas that he liked from the classical text, ignoring the rest. He first coated the plaster wall with a layer of pitch, and then painted a magnificent equestrian scene. Contemporaries praised his work while it was in progress. The sketch above was made as a study for this work.

After the brushwork was complete, Leonardo reportedly built a fire in front of his masterpiece, to fuse the pigments into the substrate. Unlike the small trial pieces, the wall was much harder to heat uniformly. Reportedly, portions of the work were overheated and flowed down the wall. This is an example of a process failure that is immediately obvious before the product is shipped.

Leonardo later painted "The Last Supper" on the walls of the Chiesa di Santa Maria delle Grazie in Milan. This iconic painting was one of the great achievements of the Renaissance. On this occasion, Da Vinci also employed a new technique. He first sealed the wall with a layer of pitch, gesso, and mastic before painting the scene with tempera. This allowed him to depict subtle shadowing that was not possible with the traditional fresco methods. Contemporary observers judged the piece to be spectacular. But, the customer was reportedly dissatisfied, because the process was very slow.

Within the artist's lifetime, his work had begun to fade and peel. Fifty years later, Giorgio Vasari wrote that "all that could be seen was a blur of paint-dabs." Many restoration efforts have been made with varying degrees of success, harm, and controversy. This is a situation where the process failure is not obvious until after the product is in the hands of the customer. The benefits and drawbacks of this new process could not be evaluated immediately.

Leonard employed yet another new process technique to create his most enduring masterpiece, the Mona Lisa. With the sfumato method, countless layers of transparent glaze are laid down with a thickness of only 1–2 μm. The result was an ethereal illusion of depth and volume. This masterwork was known for its subtle transitions of light and shadow. With this process, Da Vinci portrayed the human face more realistically than traditional techniques. The multitude of layers explains why da Vinci carried the painting around Italy for years, constantly tinkering with it.

**RULE #58 ALL PROCESS INTERACTIONS CANNOT
BE ANTICIPATED AHEAD OF TIME**

No one is smart enough to predict all the possible interactions for a
dynamic process situation. Only thorough experimentation and testing with the process and the end-use performance will reveal a map of
cause and effect.

By now, the reader may be confused. Cannon balls, peaches, cement, x-ray
film, herbicides, and Renaissance artwork have nothing in common. Their
only commonality is that each is process dependent in its own special way.
This is typical of the wide range of industrial applications subject to process
dependency.

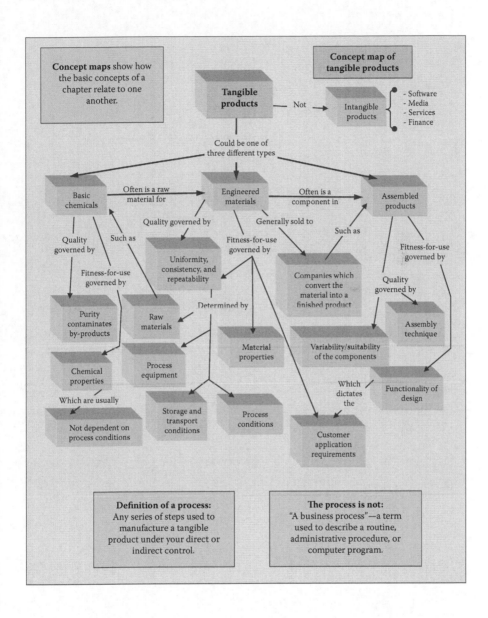

Concept maps show how the basic concepts of a chapter relate to one another.

Concept map of tangible products

Tangible products — Not → Intangible products
- Software
- Media
- Services
- Finance

Could be one of three different types

Basic chemicals — Often is a raw material for → Engineered materials — Often is a component in → Assembled products

Quality governed by

Such as

Quality governed by

Generally sold to

Such as

Fitness-for-use governed by

Fitness-for-use governed by

Fitness-for-use governed by

Uniformity, consistency, and repeatability

Companies which convert the material into a finished product

Quality governed by

Assembly technique

Purity contaminates by-products

Raw materials

Determined by

Material properties

Variability/suitability of the components

Functionality of design

Chemical properties

Process equipment

Which are usually

Storage and transport conditions

Process conditions

Which dictates the

Customer application requirements

Not dependent on process conditions

Definition of a process:
Any series of steps used to manufacture a tangible product under your direct or indirect control.

The process is not:
"A business process"—a term used to describe a routine, administrative procedure, or computer program.

Author

 Tim Oberle holds BS and MS degrees in chemical engineering from the University of Illinois. His employers have been the American Can Company, W. R. Grace, and Sealed Air Corporation. This experience includes research, manufacturing, and commercial and management roles in the United States, Europe, and Asia. Over the course of 33 years in the industry, he has performed polymer processing, converting, and fitness testing on packaging and cushioning products. These activities routinely involve the engineering of process-dependent materials. Tim has 12 issued patents and additional patents pending.

1

The Process, the Product, and Its Ultimate Life Span

What parameters actually control our process? How does process variation impact on product performance from cradle to grave? How do product properties affect the world around us?

> For a successful technology, reality must take precedence over public relations, for Nature cannot be fooled.
>
> **Richard P. Feynman**

> People mistakenly assume that their thinking is done by their head; it is actually done by the heart which first dictates the conclusion, then commands the head to provide the reasoning that will defend it.
>
> **Anthony de Mello**

In the 1970s, a large American paper towel producer was dissatisfied with its market position. Consumers thought that the brand offered little value over generic alternatives and they were becoming less inclined to pay a premium.

Like the competition, its tissue was made using the "wet-laid" Fourdrinier process. At the heart of this massive system, a wet pulp mixture is sprayed onto a fast-moving mesh belt. Much of the water is sucked away through the porous belt during a vacuum step. The resulting damp pulp then passes through an oven, where the delicate fiber matrix bonds and is dried into its finished form. The equipment is expensive to build and run but it generates tons of tissue per hour.

Since the entire paper towel industry used similar equipment, it was difficult for one company to have a significant economic or performance advantage over the others. In looking for a better business model, the company purchased the rights to a Swedish "dry-laid" paper process. This system laid down dry pulp onto the moving belt and used an adhesive technology to bind the fibers together. Energy consumption was reduced by eliminating the drying step.

The Swedish product was intriguing because it produced a fibrous matrix with excellent characteristics—strong sheets and plump rolls that could hog a lot of retail shelf space. In theory, the offering had superior performance over traditional tissue, while it was cheaper and more convenient than a

cloth rag. In focus groups, American consumers valued the look and feel of the European samples over generic alternatives.

Top management was thrilled with the upside possibilities, while neglecting to address the negative factors. Engineers hastily configured a mill in Wisconsin to produce the new offering, using the abundant softwood pulp supply in that area. An aggressive advertising campaign was employed to introduce a revolutionary new product to a skeptical public. Television commercials showed the towels surviving a cycle in a washing machine, while generic sheets disintegrated into pulp.

After the product launch, consumer reception was lukewarm, despite aggressive advertising, promotion, and generous discount coupons. People were intrigued by the hype, but were disappointed with the actual performance. They found the towels unsuited for the routine tasks of cleaning windows and picking up spills. The dry-laid tissue simply did not absorb water as well as the conventional alternatives.

An internal assessment was conducted to identify why the U.S. product was inferior to the original European samples. It was determined that the process was designed and optimized for Scandinavian wood fibers. The pulp blends from the Upper Midwest were ideal for the Fourdrinier process, but did not perform well on the dry-laid system. An ambitious management group ignored fundamental fiber science and their organization's knowledge of consumer preferences in a rush to introduce a differentiated product. They overlooked facts that were not convenient to the business plan.

A crash program was implemented to correct the problems. Formulation and process changes addressed the water absorbency issues, but the damage to the brand's image could not be repaired. Corporate management became disillusioned with the situation and wrote off their investment in the new technology, eventually divesting the entire paper division.

This product failed because its proponents assumed that a superior new process would surpass the old methods in every respect. They did not consider that a hundred years of process evolution and refinement could not be supplanted by a crash program schedule.

The end-users' paper towel expectations were based on their experience with traditional alternatives. Customers can be slow to embrace new technology if it lacks performance advantage, value, or a strong fashion/novelty appeal. The few satisfied customers could not support a massive production plant.

In this case, a pilot plant was not installed until after the production plant was under construction. The laboratory results for water absorbency were dismissed as being irrelevant, because other properties were outstanding. The marketing group instead relied on favorable focus group reports to gauge consumer acceptance. These studies indicated that end users were happy with the ability of the new towels to pick up spills. Sadly, they selected a market research company in Arizona.

The dry southwestern air skewed the outcome of the consumer tests. In that low-humidity environment, it is not necessary for a towel to absorb spilled water. Once the water spreads across a tabletop into a thin film, it will rapidly evaporate and disappear. These flawed tests confirmed false assumptions about product performance and the program proceeded toward its doom. No company can market its way around the laws of nature.

RULE #19 FULL MARKET ADOPTION OF NEW TECHNOLOGY ALWAYS TAKES LONGER THAN YOU HOPE

The market evolves on its own timetable, which rarely coincides with your needs and aspirations. Innovative new production techniques take years of fine-tuning to meet all the requirements of the existing applications. To gain a foothold, new products must have properties that offer compelling benefits in niche applications.

Critical Questions: Are you using the appropriate tools to measure product performance? Do your tests match the reality of the situation of your customers? How do all the process parameters in your production process determine the product's functionality and variability?

The critical questions scattered throughout this book are matters that you must customize to your own particular situation. If you cannot answer the aforementioned questions, then you cannot control your process to make the materials your customers want.

The Process

Life is a process. We are a process. The universe is a process.

Anne Wilson Schaef

Life is a drama, not a process.

R. Maurice Boyd

The techniques for dyeing textiles were a mysterious process art in the preindustrial world. Extracting the coloring agents from plants, animals, and mineral sources was difficult and time consuming for our ancestors. Exotic pigments were often transported thousands of miles to their point of use to obtain superior shades of color. The Tyrian purple color in the togas of Roman emperors was precious because it required the harvesting and processing of huge numbers of snails. The American plantations in the early colonial era thrived by exporting indigo to old-world textile processors. A century ago, large tracts of India were devoted to its cultivation.

Each producer had its own secret processing techniques to extract colorants from the source material and stabilize them into usable dyes. Different types of fibers required customized coloring procedures. Surface preparation treatments and fixatives are often needed to bond the color permanently to the fabric. Dye operations that did not use the optimal process techniques produced a visibly inferior product.

> *In this book, the term* process *refers strictly to a series of steps used to manufacture a tangible product under one's direct or indirect control. It does not encompass the typical business buzzword used to describe administrative procedures or routines.*

The process could be a single function performed on a small machine run by a single operator. It could be a gigantic industrial complex covering square miles of land with thousands of employees. The process might include a number of discrete steps, the last of which may be performed years after the first phase of production. The longer that you store raw materials or finished products, the greater is the impact of variability due to aging. Some process steps are performed by contract manufacturers on another continent.

The property variation in engineered materials results from processes that are dynamic by nature. They cannot be turned on and off like a light bulb. A dynamic process is not reversible. You cannot easily disassemble the material back into the original raw materials. When a product is defective, it can be difficult or impossible to transform it into an acceptable one.

U.S. Patent June 8, 1976 **3,962,153**

Fig. 1.

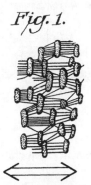

Direction of Uniaxial Expansion

All thermoplastic polymers can be stretched, expanding their surface area and reducing their thickness. However, polytetrafluoroethylene (the generic name for DuPont's Teflon) is unique in its response to the stretching process. Robert Gore was intrigued when he first observed this behavior while working for his father at W. L. Gore & Associates.

Gore discovered that when stress is applied at certain temperatures and draw rates, the polymer splits into a microstructure of tiny fibers that are joined at nodes, as shown in the diagram from his patent. At the optimal process conditions, Gore created an insulating fabric that is both breathable and water repellant. The structure of the resulting Gore-Tex fabric cannot be duplicated with traditional weaving techniques.

The polymer itself is hydrophobic, creating an excellent barrier to liquid water, which cannot pass through the tiny pores because of surface tension. However, water vapor can permeate through the open network of fibers. This unique microstructure creates a strong, waterproof matrix that combines insulation and breathability. Outside the optimal process window, the fiber matrix will tear out during the stretching process or will not have the desired insulating properties. Gore's process discovery became the basis for a line of high-performance outdoor gear.

Examples of dynamic processes that vary widely:

American Whiskey Distilleries: This process requires years of aging in charred oak casks to infuse a distinct character into the end product. The distiller must be confident that the quality and amount of raw grain alcohol going into storage will be appropriate for sale years in the future. The end product is highly process dependent, especially when graded by a connoisseur. Unsophisticated consumers may not appreciate the differences. This product is only process dependent for discriminating end users.

Plastic Molding: Many of the 3D plastic shapes in modern products are created with injection molding machines. This process is dynamic and these parts retain hidden stress, frozen in place when the molten polymer is cooled. This tension can go unnoticed for years, until the part warps, cracks, or develops cosmetic defects along the hidden fault lines.

Petrochemical Complexes: These must retool their operations when switching sources of crude oil due to differences in the properties of oils from various fields around the world. The entire plant must be reconfigured to match the fuel-mix demands of the changing seasons. Some of the products are interchangeable commodities and others are process dependent. The production economics are always highly dependent on the raw materials and the process.

Traditional Blacksmith: The traditional blacksmith converts wrought iron into finished products by repeated cycles of heating, cooling, and mechanical working of the metal. Both the process and the raw materials have a strong impact on the finished properties. Each heat/work cycle is dynamic as the part cools while being pounded with a hammer to drive out slag impurities.

Nuclear Power Plants: These have very negative fallout if the operator strays significantly from the proper process conditions. This process is dynamic because the nuclear fission reaction is not reversible. However, electrical appliances operate the same, no matter how the power has been generated, so the end product is not process dependent.

The Product

> It is the quality of our work which will please God and not the quantity.
>
> **Gandhi**

In the fifteenth century, the best European glass products were produced on the island of Murano near Venice. Their advantage came from a glass formula developed by Angelo Barovier. The artisans called it cristallo, which today is known to be a sodium–potassium glass crystal. Like most glass, it was made from sand, ash, and lime. The Venetian secret was a vegetable ash rich in potassium oxide and magnesium.

This Venetian-engineered material had a far better clarity than competitive products until the development of Ravenscroft lead crystal hundreds of years later. With this formula, the Murano craftsmen also had a considerable processing advantage. Its high melt viscosity allowed the molten glass to be blown and shaped aggressively, so that exceptionally thin, elegant finished shapes could be fabricated.

At the heart of the Venetian trading empire, the glassmakers had access to the process technology and raw materials from around the Mediterranean. With many production shops concentrated on a small island, the master craftsmen could build on the innovations of others as happens today in Silicon Valley. With this marriage of formulary knowledge and processing innovations, Murano products were a major step toward modern glass.

A product in the scope of this book could be any tangible output of a dynamic manufacturing process. Such products include complex chemicals, plastics, drugs, foods, semifinished goods, and things as small as a nanomachine. Most products are the primary output of the process, while others are less desirable by-products in the creation of something else. A process-dependent product could be an industrial material, such as a bolt of Gore-Tex fabric, that will be converted into recognizable and useful items. Or, it could be a finished shape ready for use, such as a crystal goblet. Countless process-dependent materials are components in the complex assemblies that we use every day.

Materials may be customized items of high value or low-cost, interchangeable commodities. In the first case, the primary business focus is on optimizing the quality and functional properties to the specific needs of customers who will pay for perceived value. For these applications, it is far more important and challenging to maintain consistent properties and to optimize functionality than to minimize costs.

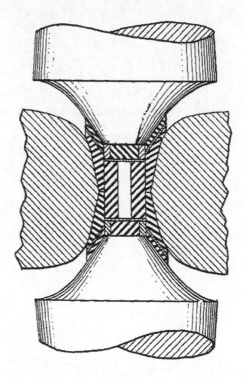

In the early 1950s, General Electric succeeded in making synthetic diamonds under tremendous heat and pressure. The best results were obtained with a belt press, at pressures in excess of 10 GPa and temperatures above 2000°C. Unlike the beautiful Murano crystal, the GE diamonds were ugly and irregular in size and shape. However, they were useful for industrial grinding and cutting applications.

By the 1980s, a second process technique was commercialized to synthesize diamonds. Chemical vapor deposition could operate at lower pressures and had much greater process flexibility. Thin diamond coatings could be applied to a wide range of substrates. Adjusting the gas feedstocks, the temperature and pressure will alter the properties of the final product. The material's hardness, surface texture, thermal conductivity, and optical properties can be tailored to specific end-use applications.

When making commodity products, the goal is to achieve the lowest cost per unit of production. This requires maximizing output while minimizing energy consumption, raw material cost and waste, and meeting industry standards of quality. The two production environments can be totally different from one another. It is difficult for one organization to run both types of production in the same operation, due to their unique demands. Both require careful attention to detail. In one case, this attention focuses on improving end-use performance. In the other case, it demands a reduction in production costs. The techniques covered in this book are applicable to the whole range of process situations.

Late in the eighteenth century, iron came in two varieties—cast and wrought. Each of the two types had its own end-use applications. They were not interchangeable for most customers, because of their property differences.

The industry learned to produce cast iron on a vast scale, depressing its price and enabling the Industrial Revolution. Its high carbon content acted as a processing aid, allowing the iron to be melted, separated from its slag, and cast into molds very efficiently.

Wrought iron was still made with process techniques that were centuries old. This low-carbon iron did not melt at the temperatures achieved with traditional methods. High processing temperatures softened it, allowing blacksmiths to gradually work out the slag impurities with laborious hammering. The most significant processing advance had been the introduction of water wheels to power trip hammers.

In 1784, the Englishman Henry Cort patented the puddling furnace as a means to convert cast iron into wrought iron. Inside his furnace, molten iron was stirred in the presence of air, allowing the carbon component to be oxidized into carbon dioxide. The resulting product had the ductility of wrought iron and could be produced cheaply in high volume. Iron producers continued to improve the process economics, later introducing the oxygen in the form of rusty scrap. The subsequent growth of the railroads was only possible because of this revolution, since high carbon iron was too brittle for use in rails.

The Knobs

Most interesting phenomena have multiple causes.

N. P. Collingwood

Every process has various means to make control adjustments and to monitor conditions. Some are clearly labeled knobs, dials, switches, or screens on a control panel. A technician with a clipboard can monitor these straightforward settings. The temperature control on an electric oven is an example of a direct control parameter. Many people consider these obvious control points to be the only kind of process adjustment. However, a variety of indirect control elements are listed as follows:

Human Factors: When baking bread, the human factors might include the technique for measuring the flour, the sequence of ingredient addition, and how frequently the oven door is opened. Such input will vary from one worker to the next and is affected by fatigue, experience, and boredom. These individual behaviors can be difficult to describe, document, and reproduce consistently. Comprehensive training, concisely documented procedures, good working conditions, periodic audits, and an engaged management team can optimize these factors. Stressful working conditions, ambiguous direction, lack of feedback, and an antagonistic supervisory climate are always counterproductive.

Equipment Components: These are discrete elements that reconfigure the process. In some cases, they are swapped in and out as needed to suit the product. In the worst cases, interchangeable parts are not identical, in the way they affect performance. The baker asks, do I use the aluminum, steel, or ceramic bread pan? Which set of beaters do I plug into the power mixer today? Is the appropriate tool clean and close at hand? What alternative might be substituted if the right one is not available? These decisions might be very consistent in production cells that always make one product. However, if the equipment is constantly being reconfigured to run an assortment of items, then the operators will constantly have opportunities to make choices that introduce variability.

Procedural Inputs: These are operational techniques, workflow decisions, and timing factors. For baking, these might include the length of time in the oven, the cooling rate prior to slicing, the preheating cycle for the oven, and the time that the oven door is open between batches. If factors are absolutely crucial to end-use performance, then they should be controlled through automation. The human work pattern is not often conducive to the precise time control of tasks that are performed intermittently. If finished properties are dependent on operator skill and they cannot accurately judge quality in real time, then performance will be variable.

Control Uncertainty: Even a labeled knob is not absolutely definitive. The baker needs to determine if the control system is properly calibrated, especially when assorted pieces of equipment are used to make interchangeable products. Whether the temperature sensor is located at the top or at the bottom of an oven will subtly change the bread properties. The temperature profile of a unit that has just come up to the set point will not be the same as one that has been running steadily all day. The display may read the same in both cases, but the bread will experience a different environment.

The human factor also has an impact on all the input types—one baker may read and set the knob differently from another, even if both are instructed to use a 375°F oven. If the temperature is critical, then use the most accurate feedback control system available to hold a tight range. Unexplainable human errors in setting conditions will also occur with some frequency. Important parameters should be engineered to prevent or highlight blunders in setting up the process.

Environmental Conditions: The temperature and humidity of the air will change the moisture content of the flour, the activity of the yeast, and the dough-drying rates. Flour mixtures will not flow and dispense the same way, depending on the weather. Opening an outside door may send a draft of cold air through the plant floor. The process conditions may require careful adjustment to accommodate the temperature changes between morning and afternoon. At worst, you could get a rude surprise when the prototype sample that was easily made in one situation cannot be successfully reproduced under different conditions. It will not be apparent if the controlling factor for this situation is actually the seasonal temperature or some other undocumented variable that has changed in the intervening time.

Second-Order Parameters: In a truly dynamic process, the temperature may not even be controlled by a knob. The heat inside a closed system might be determined by an exothermic chemical reaction, by mechanical heating due to friction, by waste heat from powered components, by the heat loss to the surrounding atmosphere, and by a variation in the cooling loop efficiency. These dynamic interactions are the most difficult to standardize and control, especially when you need to transition and stabilize the process to run different products at specific conditions.

Dynamic process control is complex when the knobs interact with one another in a nonlinear fashion. The effect of a control knob may change when unrelated adjustments are made elsewhere. Conditions will also cycle up and down at a natural frequency. Others overshoot due to lag times after adjustments and control uncertainty in the measurement system. Recording all knobs in a single snapshot will never capture the complexity in a dynamic system. A stable production process requires dampening this variability to ensure that high and low limits do not exceed the operations needs of the end-use application.

The bread-making process is complex and sensitive. However, when baking homemade bread, people are very tolerant of minor variations in appearance, taste, and texture from one loaf to another. That is the appeal of home baking. Few people would be inclined to bake for their family if every batch were subjected to rigorous scrutiny and quality control (QC). Industrial processes, on the other hand, are expected to produce uniform, repeatable batches. A loaf must yield a specific number of slices, which are free of holes. Every process control parameter is important to the outcome in terms of quality, cost, or performance.

The "concept map of the dynamic process" summarizes how all aspects of process dependency relate to one another. The raw materials, the process equipment, and the operating parameters cannot be addressed in isolation, because they are always related.

One major producer of English muffins is able to impart a unique texture to its product by skillful manipulation of the process conditions and ingredients. The techniques are so well guarded that only seven people are reported to have full knowledge of the key process parameters. Their trade secrets are split-up into the formulation, the moisture content of the dough, the process equipment design, and the process conditions. The production workers may be privy to aspects of their own operation, but not the whole picture.

For instance, the complete recipe can be concealed by the use of master-batches of additives that are prepared by a small team of specialists. The water content of the dough can be monitored with customized sensors that display only "too low", "in-specification," and "too high." The workers can use these readings to control the process without knowing the actual targets. The equipment configuration of critical contact surfaces can be concealed from casual observation behind the safety guards. The process conditions can be monitored by automated systems that do not divulge the actual numerical values.

Some people maintain that these process parameters are well known in the industry and that the unique texture is a marketing creation. Competitors in the industry have claimed to duplicate that special texture, but not repeatedly in the production environment. This is an example of dynamic process parameters being optimized and closely controlled to achieve very specific material properties.

Critical Questions: What control knobs determine the properties and performance of our product? Do all the people who directly or indirectly manipulate these knobs fully understand their impact on the value and quality of our product for the customers? Are there subtle interactions between process parameters that change the optimal settings from one day to the next?

Product Life Span

Hedgehogs are never represented on product testing panels.

The McDonald's fast-food chain offers a dessert called the McFlurry. This ice-cream product is packaged in a paper cup with a domed plastic lid. The dome has a circular opening at the top for easy access to the contents.

Consumers were satisfied, but unintended outcomes were discovered in Europe when the empty containers were discarded. Hedgehogs are attracted to rubbish bins, to lick the sweet residue from the packaging. However, the hole in the dome was just the right size to trap their little heads. Nature lovers found dead animals with the plastic covers around their necks and raised an alarm. The company hurriedly changed the container dimensions to resolve the problem and improve their environmental image.

Some people might have answered these concerns by using a degradable plastic for the domed lid. In practice, such materials fail to break down fast enough to save trapped hedgehogs.

The process ends when the product leaves the plant. However, your materials and by-products can persist for years or centuries in terms of their effect on the end users and the environment. These interactions may be hard to control and anticipate because of the time lapse between processing and the discovery of the negative impact. See the "time line of the product life cycle" flowchart at the end of the chapter for a visual depiction of this concept.

Life cycle considerations:

- Evaluate end-user interactions with the product.
  - Could fumes or odors become problematic after hours of exposure? (Retained solvent or monomer that varies from batch to batch?)
  - Will problems arise with possible off-label uses of a product? Remember that the imagination of customers is limitless. Hedgehogs cannot read the warning label.

- Will minor issues become serious problems as different batches age in storage? (Oxidation, UV degradation, partially cured adhesive, solvent migration, etc.)
  - Can product shortcomings become more severe for the customer if the process conditions or raw materials are changed even slightly? (Establish allowable ranges on the key variables through proactive process experimentation.)
- How are the customers of your customers affected by the incumbent product(s)?
  - Can you correlate those important factors with your process knob setting?
  - Are there process conditions that can optimize these factors for each subset of users?
- Consider the product's ultimate disposal, recycle, and reuse.
  - Is the product made of mixed materials (plastic + metal + paper) that interfere with recycling at the end of its functional life? Can these incompatible materials be made easier to separate?
  - Evaluate the impact of minor components on recycling and reuse. Paints, coatings, glues, and additives need to be tested. Soon after the U.S. Post Office introduced self-adhesive postage stamps in the 1980s, paper recyclers found that they gummed up their process.
- What about partially used containers that sit in a customer's storeroom for years, even against the advice on the label? Will they pose any hazard as they age?
  - Ideally, a liquid or soft material will congeal into a solid mass that can be safely thrown away as nonhazardous waste.
  - Serious problems result if the product appears normal but it reacts badly in use.
- Consider the range of environmental conditions in which the product will be used.
  - Can oil, solvent, dust, grit, or moisture in the customer's workplace affect your materials in a negative way? Do problem customers have similar environmental factors?
  - Could fumes, dust, or vapors from your product interact badly with other materials in the customer's product?
  - Will heat, radiation, sunlight, or electromagnetic interference trigger problems?
  - What if your labels fall off the container? Could the unlabeled package be confused with some other material in the workplace?
  - Are critical safety warnings printed with ink that will fade over time?
- Some products can be reused, some are recyclable, others are biodegradable, and still others can be safely disposed of in a landfill or an incinerator.

- Will changes that are intended to enhance one approach have negative effects on the others? Example: some additives that make polyethylene degradable could render it unfit for recycling. Or a plastic part may fail in use due to premature degradation.
- Packaging issues.
  - What problems will result if containers are dropped, damaged, or leaking?
  - Can the container be redesigned to make it easier to dispense the product without spilling?
  - Can it be safely stored around other materials that a customer is likely to have in the same area without harmful interactions?
  - What if your empty containers are reused to contain something else?

The "time line of the product life cycle" flowchart at the end of the chapter depicts the significant events in a product's life span. From this long-term perspective, you can appreciate the difficulty in controlling the process

No. 765,975.   PATENTED JULY 26, 1904.

E. P. HOOLEY.

APPARATUS FOR THE PREPARATION OF TAR MACADAM.

APPLICATION FILED NOV. 3, 1902.

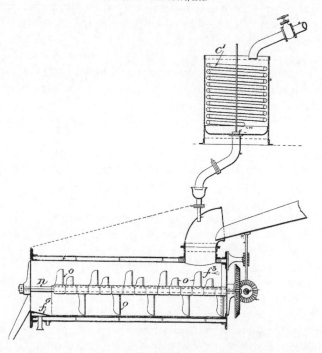

parameters to both optimize performance and minimize the product's impact on society and the environment.

The undesirable by-product of one process can become a cheap feedstock for another. In the early twentieth century, many rural roads were covered with compacted gravel. In England, this was called a macadam surface in honor of the civil engineering innovator John McAdam. Macadam became less suitable when heavier traffic loads increased the wear on these road surfaces.

Edgar Hooley improved on macadam by coating the gravel with coal tar (a by-product of coal gasification used to generate methane for street lighting). The tar was then covered with cinders or slag (a by-product of steel production). This tar-bound macadam was named tarmac.

Later, the availability of coal tar declined as electrical lighting replaced gas streetlights. However, the subsequent rise of petroleum refining created a heavy by-product called asphalt, which has replaced coal tar for road construction.

Types of Process Deficiencies

> If written directions alone would suffice, libraries wouldn't need to have the rest of the universities attached.
>
> **Judith Martin**

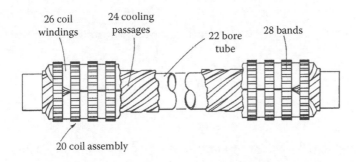

High-energy physics experiments require intense magnetic fields to accelerate and steer charged particles into a tightly focused beam. At the CERN facility near Geneva, Switzerland, the Large Hadron Collider employs electromagnets made from niobium–titanium cables. This material becomes superconductive when cooled with liquid helium. However, the cables are prone to lose this property at the current loads required for full-power operation.

As part of the production process, the magnets are "trained" in a special test fixture. The electrical current is steadily increased until the point where superconductivity is lost. As this cycle is repeated, the material gradually "learns" to remain superconductive at higher currents.

The finished assemblies were conditioned and certified for current loads designed to meet the collider's nameplate capacity and were delivered to the job site. However, construction problems delayed their installation. The magnet assemblies sat outdoors for a year. Once the work was complete, there were additional delays for testing and repairs, causing the magnets to age an additional year at higher temperatures than originally foreseen.

Tests revealed that many units "forgot" their training and were no longer able to achieve the required magnetic field density for full-power operation. While still functional, the magnets were unable to perform at their design rating after sitting idle in an environment that no one had planned for. Like many other engineered materials, the conditions of storage and shipment changed the cables in subtle ways between the time of manufacture and their ultimate use. As a result of this transformation, the materials altered the operational capabilities of the world's largest science experiment.

RULE #6 THE FUNCTIONALITY OF PROCESS KNOBS CAN CHANGE OVER TIME

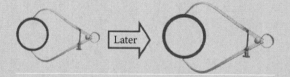

The effects that you measure today when calibrating a control knob might be different tomorrow. Materials and equipment evolve with age and use. Parameters drift with time. Secondary factors change and confound the picture. Constant evaluation is needed to compensate for these changes.

Clues that a dynamic process is not running in an optimal fashion:

- A producer is forced to pay for premium-priced raw materials. This is not because it creates a superior product, but because the process is intolerant of lesser feedstocks. Only a few specific product grades can be used, out of many potential choices:
  - Better purity, consistency, or particle size.
  - Higher octane, tight molecular weight distribution, or special stabilizer package.
  - Specialized packaging, rather than bulk delivery.

- A process runs below the nameplate capacity to make some grades, consumes more energy than it should or requires extra operators to run.
  - More frequent maintenance or a longer changeover time between products.
  - Increased QC tests to cull out a bad product.
  - Takes more time to equilibrate to the desired running conditions, resulting in a lot of scrap.
  - Operators need to constantly monitor the process to keep on target.
- Operators lose time hunting for stable running conditions rather than setting up from a menu.
  - Conditions used on the previous run do not give the same results the next time.
  - Difficult to train operators and maintain documentation.
- More time is spent arguing over the validity of the specifications than improving the process. Fitness-for-use requirements are unclear and specifications are waived on an arbitrary basis.
- The plant is experiencing too many process interruptions and the root causes are not understood.
- The process creates risk of injury or long-term health effects.
- The process is a significant source of environmental impact, especially during process excursions.
- The customers reject some deliveries as nonconforming, even though they passed the quality tests.
- The end user pays a lower price for your goods because they are inferior to the competition in ways that you cannot measure. Some end users like the "center cut" and will not accept the rest.
- The product cannot be placed with some customers at any price because it is not fit for their needs, even though they purchase a very similar product from your competitors. No one understands how to satisfy the requirements of the noncustomers.
- The customers of your customer experience more variability with goods made from your product compared to the competition. This may include an unexpectedly good performance from some batches.

None of these issues can be addressed in isolation. Subtle changes that improve one aspect of the process could cause problems elsewhere. Examples of recognized and unrecognized problems are shown in the diagram that follows.

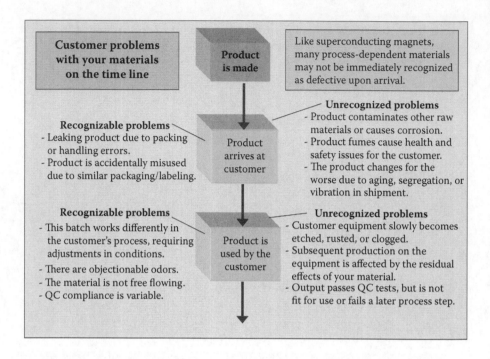

Summary

- It takes longer to adopt truly novel materials, because both the processing techniques and the end-use practices need to evolve to take full advantage of them.

- It is difficult to experiment on dynamic processes. These cannot be started and stopped at will. They generally cannot be reversed.

- Process dependency occurs in both commodities and complex engineered materials.

- Problems with process dependency may not become apparent until years after the product is produced, making it difficult to adjust the process conditions to compensate.

- The processes are controlled with a variety of knobs. Some are tangible and obvious. Some factors are inherently difficult to control directly.

- Developing products in a dynamic process environment takes more time and thought than developing production systems that are not process dependent.

- Carefully engineered product properties will gradually change in storage.
- Process dependency can manifest problems in many subtle ways.
- Process issues are interactive and cannot be addressed in isolation.

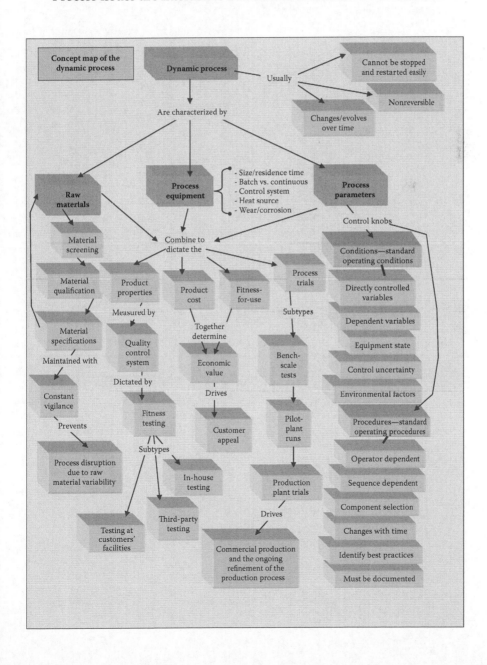

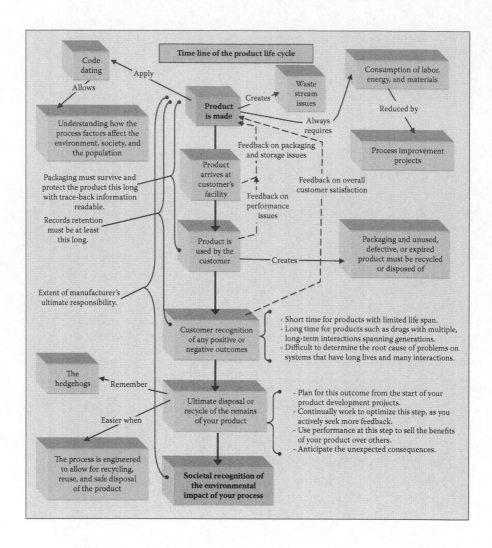

Time line of the product life cycle

Code dating

Apply

Allows

Waste stream issues

Product is made

Creates

Consumption of labor, energy, and materials

Reduced by

Always requires

Understanding how the process factors affect the environment, society, and the population

Feedback on packaging and storage issues

Process improvement projects

Product arrives at customer's facility

Feedback on overall customer satisfaction

Packaging must survive and protect the product this long with trace-back information readable.

Feedback on performance issues

Records retention must be at least this long.

Product is used by the customer

Creates

Packaging and unused, defective, or expired product must be recycled or disposed of

Extent of manufacturer's ultimate responsibility.

Customer recognition of any positive or negative outcomes

- Short time for products with limited life span.
- Long time for products such as drugs with multiple, long-term interactions spanning generations.
- Difficult to determine the root cause of problems on systems that have long lives and many interactions.

The hedgehogs

Remember

Ultimate disposal or recycle of the remains of your product

- Plan for this outcome from the start of your product development projects.
- Continually work to optimize this step, as you actively seek more feedback.
- Use performance at this step to sell the benefits of your product over others.
- Anticipate the unexpected consequences.

Easier when

The process is engineered to allow for recycling, reuse, and safe disposal of the product

Societal recognition of the environmental impact of your process

2

Fitness-for-Use Testing

How do we ensure that engineered materials will meet all of the customer needs—including the ones that the customer is not aware of? What is the optimal balance of price and performance? How do we quantify the value of our product in different applications?

> The most important word in the vocabulary of advertising is TEST. If you pretest your product with consumers, and pretest your advertising, you will do well in the marketplace.
>
> **David Ogilvy**

One large American company has been a major brand name in home appliances for decades. However, its market position was squeezed between premium product lines appealing to affluent consumers and low-cost imported equipment.

In the 1980s, the company updated its line of refrigerators. After studying the options, it chose to implement turbine compressor technology. Kitchen refrigerators commonly employed piston compressors to pressurize and circulate the refrigerant. This older technology was being replaced by turbines in larger cooling applications, such as commercial air conditioners. Turbines offer fundamental advantages in noise, vibration, maintenance, and cost of manufacturing, because they have fewer moving parts. Unfortunately, the design team focused too intently on cost reduction.

Once the design was finalized, 600 turbine units were built and subjected to an accelerated life-testing regime. This protocol was believed to simulate 5 years of typical service and had been validated during an evaluation of earlier models. The company was confident in the turbine approach because every unit passed the 2-month torture test. The new product line was subsequently introduced with a 5-year warranty and started to appear in homes across North America.

Within 8 months, the company began to receive reports of premature compressor failures. The first units to fail were located in poorly ventilated spaces. An investigation was launched to determine the root cause of a situation that was quickly becoming a crisis. Attention quickly focused on compressor bearing failures. Critical wear components were produced with an economical metal sintering process instead of the dependable and expensive

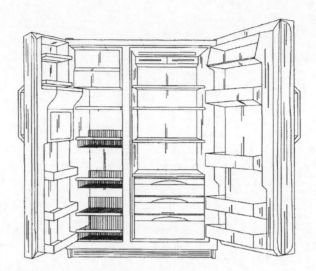

casting technique. Ironically, the powdered metal technology had previously been rejected by the company's air-conditioning division due to concerns over wear problems.

The 2-month test cycle had first hinted of the problem. The prototypes exhibited the onset of lubricant breakdown and heat damage. Under the test protocol, these distressing observations were not recorded as failures, because the systems were functional throughout the prescribed evaluation period.

The testing regime could produce failures in piston-style compressors that correlated well with the actual field performance. However, the turbine failure rate was nonlinear. They could easily survive the test period, but would experience lubricant breakdown if the ambient conditions were too warm. The low flow rate of the refrigerant in a home refrigerator was insufficient to properly cool the small compressors. This was not an issue on large equipment, where the bearings had superior lubrication and better wear resistance.

The primary testing error was in not determining that the ambient temperature was the controlling factor in the failure of the sintered bearing components. The secondary mistake was in not running the trial for a longer period, which would have revealed the inflection point where the failure rate increased. Rather, the units were tested only to an arbitrary time limit dictated by previous experience with a different technology. This test regime is similar to driving a car at night in heavy fog. No danger is visible, but you have no means to judge the safety margin between a good performance and a disaster. An appropriate fitness-for-use (FFU) test program will reveal the safe window of operation.

RULE #18 DOING INAPPROPRIATE FFU TESTS IS AN INVITATION TO DISASTER

Performing an unsuitable suite of fitness tests will create a false sense of security, masking potential problems. Dramatic material failures often have their origins in tests that seem to validate performance, while overlooking serious flaws. Test methods that are ideal for one application may be unsuitable for another, despite apparent similarities.

Jan. 21, 1947.   D. M. CHAPIN   2,414,449
DEPTH CONTROL DEVICE
Filed Aug. 18, 1943

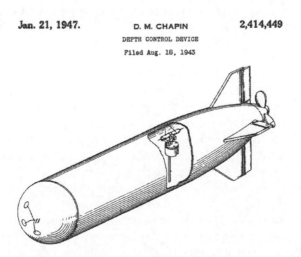

When the U.S. Navy entered World War II in 1941, their long-range submarines were armed with the technically sophisticated Mark 14 torpedo. This complex weapon was designed to explode beneath the hulls of enemy ships, striking at their weakest point. However, in the opening months of the war, the torpedo's performance was dismal, with very few successful attacks. Due to budget restrictions during the Depression, the navy had never actually performed a live fire test of this weapon system on a real vessel. The money was not available to sink obsolete ships or explode an expensive prototype. The Bureau of Ordinance had failed to discover that the sensitive detonator could not cope with variations in the magnetic field strength under actual use conditions.

When the problem was discovered, submarines switched to the traditional contact exploder, which was assumed to be more reliable. However, the performance problems persisted. Faulty prewar testing and poor assumptions were again at fault. The original trials were performed with dummy warheads that were lighter than the explosive charge. With the heavier load, the live torpedoes ran deeper than expected. Compounding the problem was a control system that used water pressure to determine its depth. The designers failed to appreciate that the measured pressure was also a function of velocity. The controllers had been calibrated during a low-speed operation on an earlier generation of weapons. When the operational speed was boosted with a stronger propulsion system, no one recalibrated the control system or verified its performance in open water.

The depth setting was adjusted and the torpedoes started to strike enemy ships. Only then did a third problem become apparent. Many warheads failed to explode on impact, allowing lightly damaged targets to sail away. A frustrated fleet commander ordered improvised tests to evaluate the detonators. The torpedoes were hoisted with a crane and dropped 30 m onto a concrete surface, to simulate actual impact conditions. In 70% of the drops, the core rod jammed under the impact force, failing to detonate the warhead. This actuator was constructed from unsuitable materials for the forces encountered in the end-use conditions. Once the nature of the problem was obvious, it was a simple matter to fabricate stronger firing pins. It took 20 months of poor performance to find and resolve all these issues.

This example highlights the need to spend time and money for a meticulous performance validation under realistic end-use conditions. Ideally, the results should tie the product performance to material properties or dimensions that can be monitored with routine quality control (QC) tests. When process changes are made, the performance of the product must be recertified with additional testing. When end-use conditions do not allow the recovery of failed products, then extraordinary measures must be taken to duplicate and analyze those failure situations. In the torpedo example, it was difficult to evaluate performance problems while the submarines were diving deep to evade counterattacks. When there are multiple failure mechanisms, problem solving is extremely tedious.

Finding the Critical Properties of Process-Dependent Products

Experience is the name everyone gives to his mistakes.

Oscar Wilde

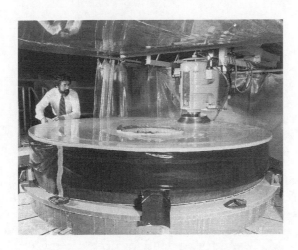

One of technology's most embarrassing FFU testing failures occurred at the Perkin-Elmer Corporation. Because of an improperly calibrated instrument, the company made a grinding error on a 2.4 m concave mirror that was produced under a government contract in 1981. The focal point of this otherwise flawless mirror was out of tolerance by a miniscule degree. One test result did suggest errors, but it was disregarded at the time. Due to a long series of program delays after production, the mirror sat unused for 9 years. During that interval, no one tested the complete optical package to ensure that it worked correctly.

Only after the mirror was in Earth's orbit did it become apparent that it was not made to specification. The Hubble Space Telescope returned blurry images, which significantly inhibited its astronomy mission. Optics experts fabricated a corrective lens, once they understood the original mistake. The cost of this component was minor compared to the money required to launch a shuttle mission to install it.

In the Hubble case, NASA acted as if fitness tests were not necessary. The physics of collecting and focusing light with concave mirrors was well understood from centuries of stargazing. In addition, Perkin-Elmer had significant experience in producing optical components for use in Earth's orbit. But, it only takes one undetected error in the development and production chain to disrupt a one-of-a-kind mission. Fitness testing must be exhaustive in such situations. With ongoing production, the performance of early products feeds back to the developers, who fine-tune the process to continuously improve performance.

Suppose that one particular shipment of your engineered material causes problems at a key account. The customer's purchasing agent calls, screaming on the telephone that the operation is shutdown because of the quality defects. Your sales force and logistics group will swing into action, getting the bad product out of the customer's facility, with few questions asked. A replacement product is rushed in. No one wants to risk offending the customer by asking too many questions during this tense moment. Maintaining the relationship by acting responsive is priority number one.

Back in the production plant, the QC area is filled to the brim with returned products. The quality staff is dumbfounded at the lack of information. A great many unopened containers were returned and no one understands exactly what was wrong and how to avoid repeating the problem. Extensive testing is performed, but reveals no insight into the problem.

Some people speculate that the batch was only returned because business was slow and the customer had too much inventory. Others assume that the complaint was a ploy to improve the customer's position for upcoming price negotiations. The plant manager proposes relabeling the containers and shipping them back to the customer as a different lot.

The marketing manager and the director of sales are fuming that the plant has screwed up another shipment and put customer relations at risk. R&D claims that manufacturing has changed the process parameters to improve output. No one understands why the product has been returned and which specific containers gave the customer problems. In fact, few people understand the customer's application needs.

This hypothetical company focuses its development efforts on production optimization. "How can we make the product as cheaply as possible, while complying with the specifications?" This approach is necessary when making a commodity material. However, as you move up the food chain to value-added products and advanced materials, then a different strategy is needed. FFU testing must be employed to gauge product performance. This ensures that the cannon balls and torpedoes explode correctly. FFU tests guide development, process scale-up, and ongoing operations. To do so, it is essential to understand the possible failure modes for your product in each application.

Failure modes in decreasing order of importance:

- *Catastrophic failures that occur without warning*: These are product defects that will cause harm to humans or equipment. Unpredictability means that the material must be removed from its end use well before it reaches the point of failure. These situations can be very costly when they require the replacement of entire product assemblies that incorporate the defective batches of your material. Imagine the need to replace or service orbiting telescopes, implanted medical devices, or undersea cables because of process variability during the original manufacturing step.

- *Nonconforming performance that is not immediately evident*: In this case, the customer's product or service is compromised, but the customer is not aware of the problems until some later time. These issues are hard to resolve when they are intermittent or involve aging effects. Faulty torpedoes, defective magnets, and fading paintings are examples. These failures are not evident to your customers when they convert your materials into finished products. Detecting the problems may require considerable data collection and analysis. Significant expense will be required to recall, inspect, and repair assemblies that contain your defective component.

- *Product variability that degrades end-use performance*: The customer has to do more rework, sorting, or inspection when batches of your product have variable properties. Some shipments work very well, while others perform poorly, even though they are all "in-specification."

- *Esthetic/cosmetic factors*: These soft issues are harder to quantify and measure. Purely visual defects generally suggest to customers that your product is inferior in other ways. The worst situation is when your material looks okay but blemishes or imperfections appear on the customer's product at the end of their process.

The importance of each of these factors needs to be understood for each end-use application. Not every customer will have the same values or tolerance for risk.

FFU Tests

Test fast, fail fast, adjust fast.

Tom Peters

In 1792, Yale graduate Eli Whitney left his home in Massachusetts to find employment in the southern states. While in Savannah, his hosts complained about the incumbent process of refining raw cotton. Cotton seeds were laboriously hand separated from the freshly picked boles. This tedious procedure limited the region's ability to export its crop. Seeing this situation firsthand inspired Whitney to devise a ginning process to remove the seeds.

Whitney was aided by the local abundance of raw cotton and his experience with factory methods. He soon had a prototype process that he could test and modify. He received feedback from the local farmers to ensure that the equipment could be used and maintained by their workforce. Once satisfied with the solution, Whitney traveled to the nation's capital, demonstrated his invention to the secretary of state, Thomas Jefferson, and was granted the 72nd patent issued by the newly established patent office. The FFU tests were so persuasive that the process was put into commercial use immediately.

In this case, the refined cotton fiber was not process dependent. Its functionality and downstream performance were independent of the cleaning technique. The process was not dynamic—it could be stopped or started at any time without affecting performance. This development was historically important because it significantly reduced the cost of cotton production, enabling an exponential increase in exports. By 1861, U.S. plantation owners were producing the majority of the world's cotton. Tragically, this profitable trade encouraged the planters to accelerate African slave importation to tend the crop. This increased the political and economic tensions that initiated the American Civil War.

Whitney was unable to derive significant revenue from his invention. He hoped to charge plantation owners a royalty on each pound of cotton processed. But, the equipment was easily copied by local craftsmen, allowing widespread infringement on the patent. The local courts were not inclined to protect intellectual property rights at the time.

In-house testing does not relieve the need to evaluate and qualify the product at actual customers' facilities. It does weed out the unsuitable candidates

RULE #27 CONSTANTLY REEVALUATE THE FFU PERFORMANCE AND REFINE YOUR TESTING TOOLKIT

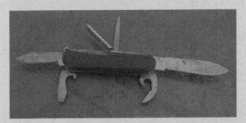

The most important tool for process development is a battery of internal tests to accurately predict product performance. By screening candidate samples, you can eliminate those that are least fit for the application. The survivors can be ranked by their likelihood of success and compared to other options. Such insight allows process variables to be correlated with the end-user performance. This usually leads to the creation and testing of additional samples to confirm these links and optimize performance. The findings from subsequent customer trials of the best samples will be used to further refine the internal test suite. Once the product is in production, new raw materials and process changes can be validated to maintain customer satisfaction.

and narrows the choice of materials for the field-testing, but on-site trials are the ultimate confirmation of product value.

FFU testing should not be confused with routine production QC tests. Generally, the factory QC tests and the specifications are validated with FFU tests. In some cases, the plant might adapt versions of fitness tests to monitor production.

See the distinctions and similarities between the two types of product tests in the "concept map of product testing techniques" at the end of this chapter. FFU tests identify the essential product properties that control the performance in the end-use applications. QC tests monitor these properties on an ongoing basis as a surrogate for the customer needs.

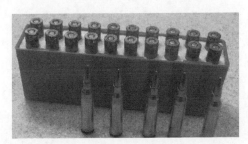

Imagine a producer of small-arms ammunition for hunting, police, and defense applications. The quality assurance program requires frequent dimensional measurements on the brass cartridges at the forming process. Inspections monitor the weight of the propellant dispensed by automated fillers and the coating uniformity. A robotic test fixture ensures that each finished unit has the proper overall dimensions, weight, and balance, rejecting those exceeding an acceptable range.

These measurements give feedback to the operators on the aspects of quality that they can influence. The readings will be tracked and charted to ensure that average values do not drift over time or abruptly shift, due to process factors. Other factors such as the particle size and burn rate of the propellant have been precertified by the vendor before the materials arrive.

Obviously, the FFU testing will involve loading ammunition into a gun and firing it under controlled conditions. The gun might be anchored in a fixture and activated remotely to eliminate human factors. The test cell could be instrumented to monitor all aspects of performance—sound, muzzle flash, accuracy, frequency of misfires, projectile velocity, and spin rate. The distribution of bullet holes around the target is recorded to monitor the precision and variation within the batch. The particular battery of tests would be chosen based on the end-user requirements for the specific product. Conventional hand firing by an expert might evaluate the intangible factors of user experience.

FFU tests on experimental samples give direction to the development team for the next round of trials. The team will adjust the parameters and create additional batches of samples until the perceived customer requirements are met or exceeded.

Tests must evolve to better simulate the changing end-use environment. The development team refines test procedures in response to customer feedback, and compares prototypes and production samples with ammunition from competitive sources. The team tries to improve upon the performance of competitive offerings, looking for ways to exceed customer expectations.

With this information, the company will be prepared to approach the customer with its best candidates to close the sale. Test data support advertising and product data sheets. The reports are understandable because they speak the end-user's language. They are a persuasive means to demonstrate advantage in a market driven by technical requirements.

FFU testing during development:

- Use the tests to eliminate unsuitable candidates. Identify which process options cause the negative outcomes. Design your production QC tests to screen out these undesirable results.
- FFU tests will drive cost/benefit analysis, determining which process parameters provide value.
- Rank the best options by how well they suit the market needs. Compare the costs of the potential process paths and calculate the return on investment of each to determine which is most attractive.

- Generate more samples based on the optimal conditions and raw materials from the first-round tests. Determine if the FFU tests agree. If not, look for uncontrolled process factors or test conditions that affect the results. If they do agree, adjust conditions to determine the size of the process window.
- Use the FFU test results as examples in patent applications. Once you have filed an application, field-testing can be undertaken to validate these findings. Unproven, valueless, or contradictory claims can be dropped from the patent application later.
- Sell the value that your prototypes bring to the customers with test reports. Use that value proposition versus commercial alternatives to appropriately price the offering.
- In a totally new product space, there will not be customers to evaluate your new products. You may need to create the product, the supporting equipment, and the application technology.
- Make sure that all the people who perform the tests use very consistent techniques. Use blind tests of identical control samples to verify that all the individuals obtain comparable results. Repeat these validation tests to ensure that the results remain consistent as personnel and equipment change over time.
- Always employ one or more control samples for subjective tests. The evaluators should be blind to the sample identity. When human bias becomes an important factor, then extreme measures should be taken to conceal sample identities.
  - Use a large population of judges to randomize differences.
  - Question the standards if there is a wide variation between the judges.
- Any unexpectedly good or bad test results must be replicated. Repeat them with samples from the same batch and also from new batches made under similar conditions. An unusually bad test can provide as much guidance as a terrific one. Resist hiding or ignoring embarrassing negative outcomes. Do not pick and choose the results. Listen to what the data are telling you about the material and the test method's validity.

FFU tests in ongoing operations:

- Alternate raw materials can be evaluated with FFU testing to understand if they alter performance and properties. Consider the effect on the product value and the overall production operation. Do not focus exclusively on the price of each feedstock alternative.
- When changing specifications, perform FFU tests on product samples made at the extreme limits of the allowable range. The desired situation is to have a uniform end-use performance across the range.

- Your finished product specification limits should be periodically explored with FFU tests just to verify that nothing has changed in the process. This confirms that you have identified the critical process knobs and that you are maintaining control over the end-user experience.
- When evaluating new conditions or procedures, conduct FFU tests to ensure that there is no adverse impact. Monitor for process equipment deterioration that occurs over time, causing performance drift.
- Perform FFU tests on the output of newly commissioned production lines to verify how the new process compares to the older ones. Determine if the output of all lines can be used interchangeably. Initially, new equipment is often restricted to specific applications.
- As new customers qualify your existing products, ensure that the FFU tests adequately predict their requirements. Create new tests, if the needs differ from the existing applications.

Creating new FFU tests:

- Validating a reliable battery of tests can take as much time as developing a product. An ideal test can be tuned to discriminate between small differences in process variables or material properties.
- Track which tests are appropriate to each end-use application and significant customer. This is the first step toward creating customized products that become irreplaceable to your customer.
- Record and retain exhaustive data on all trivial details. Nothing should be "understood," assumed, or "taken for granted" in notebook entries.
  - Suppose the customer poses a question about the report? It is more credible if the author can re-sort or re-search the data to get the answer immediately, rather than repeating the tests.
  - Record which piece of test equipment is used in each evaluation. In the ammunition example, record the serial number of the specific gun. A notebook full of data becomes worthless if there are important variations between guns that were used interchangeably.
- Once a given test is developed and deemed to be useful, create a detailed procedure. Refine and update these instructions with revision control to compare the test's results over long periods of time. Photograph typical examples of failure modes or retain the samples to maintain a standard for defining defects.
- Apply statistical methods to ensure that conclusions are valid. People become emotionally encouraged or discouraged during a testing campaign, depending on their expectations and anxiety to get a certain result. Resist jumping to conclusions before sufficient data are accumulated and analyzed.

- Simulations of some customer applications may have very low rates of failure. This necessitates a huge number of tests to confidently measure differences. Employ more extreme test conditions, increasing the failure rate to show differences between the various options.
  - The ammunition might be stored in a hot, humid condition with a controlled amount of saltwater or corrosive vapor. Shaking boxes on a vibration table will simulate extreme transport conditions. A carefully engineered "defective" firing pin might be employed to increase the number of misfired cartridges in all batches.
  - Verify that the stress factors give a valid multiplication of failure across all sample types.
- Test at both ends of the performance spectrum. The bullet must fire when struck by a firing pin even after stressful conditioning. On the other hand, it must not discharge while inside the box of cartridges. Verifying the safety margin requires a very specific family of tests.
- Rate the consistency of the performance across the spectrum of end-use environments. A rifle bullet for hunting or match competition must exit the gun at the same speed, regardless of age and storage conditions. Variations in the muzzle velocity will change the trajectory when firing at long distance. Both fresh shells and samples subjected to forced aging conditions should be fired at different air temperatures. Look for the process conditions that produce a very tight distribution of test results across all conditions.
- Test with a wide range of customer equipment, unless designing for a very specific end use. In the ammunition tests, both cheap, unreliable weapons and expensive, well-made guns are evaluated. The benefits in one market segment may be drawbacks in another.
- Consider who will be using the product and under what conditions. Imagine plastic vials will be filled with vaccines in a refrigerated pharmaceutical clean room and chilled with dry ice. The tests must simulate conditions that the product will be exposed to during its entire life span. The empty vials might experience vibration, abrasion, and impacts at temperatures up to 50°C during shipment. They must resist dimensional changes, cracking, and discoloration. After filling, the vial might experience temperatures as low as −78°C during shipment to the end user.
- Keep customer relationships in mind when formatting your results. Suppose that you are trying to sell tires to the Ford Motor Company. All the road test data in your presentation should be conducted on Ford vehicles. How the product performs well on a Toyota may not impress the Ford engineers. In a second round of testing, you will use prototype vehicles provided by Ford that will be in production when your tires are purchased, several years down the road.

- Make sure that the test conditions are appropriate and consistent with the customer requirements. The road tests may have been conducted at temperatures ranging from 1°C to 30°C, due to the needs of other customers or industry standard guidelines. However, if Ford requires data from −20°C to +35°C, then additional testing will be needed.
- FFU testing should be treated as serious experimentation. Randomize the sequence of samples, because test fixtures change over time due to wear, temperature fluctuation, and fatigue. Repeat the control samples during the campaign to check for test result shifts.
- Record data in a consistent, neutral terminology if the product is prone to lawsuits or governmental investigation. Few managers are aware of the poorly chosen or colorful terms written in notebooks by bored technicians. Avoid judgmental terms such as *catastrophic failure* or *unacceptable defects*, especially when testing to failure or under exaggerated conditions. Consider whether to retain or destroy the original notes after the data are summarized in a report or transferred to a database. Keep original notes that are necessary to document patent applications or regulatory requirements.
- Most FFU tests predict failure (or the lack of it) rather than commercial success. Testing on-site is the only positive proof of concept.

FFU tests will evolve as the project progresses through each stage. Testing regimes that give insight during the initial screening phase may not be appropriate for the final qualification and approval of the finished product. Stages of internal fitness testing:

- *Screening Tests.* Early in the development project, it may be appropriate to screen for one particular attribute on mock-up products. Such samples may not be appealing to customers, because they are not complete products. They identify promising process options and materials with a singular aspect of performance.
- *Comparative Tests.* Later in the development effort, rank alternatives to narrow the choices toward the optimum. This can be challenging when the competing options are very different or they have not been refined to the same stage of performance. Replicate results to account for signal noise, testing bias, and experimental errors.
- *Product Mock-up.* Simulate the customer's finished products using your materials in preparation to select the optimal prototype(s) for feedback.
- *Total Fitness Evaluations.* Toward the end of the project, the product must be given a reality check to ensure that it will consistently satisfy customer needs. This is the engineering equivalent of a dress rehearsal, where every aspect of the product must be scrutinized. Large sample populations are needed to evaluate the impact of low percentage failures.

- *Historical Consistency.* Once a product is commercial, ongoing output should be compared to a retained control, to ensure that its performance matches the original development samples. There is often a lag time between a subtle process shift and the customers' realization that your product no longer meets their requirements.

Getting Creative with FFU

I hear and I forget. I see and I remember. I do and I understand.

Confucius

The de Havilland Comet, the first jet airliner, was a showcase of new technology. Its performance far surpassed the piston-engine competition when introduced for passenger service in 1952. Because the plane was pressurized and flew faster and at higher altitudes, the fitness requirements were not well understood.

One engineering prototype was subjected to FFU testing as part of its airframe certification. This torture test consisted of repeated cycles of pressurization and depressurization. This strain eventually ruptured the airframe at the point of highest stress—the corner of one of the square windows. However, based on the high number of cycles before failure, engineers concluded that the planes could expect a long service life.

Several early crashes were blamed on pilot error or bad weather. However, within the span of a few months in 1954, two planes broke up during commercial flights, with the loss of all on board. In one case, the accident was so abrupt that a routine radio transmission was cut off in midsentence. The fleet was grounded and an investigation was launched.

Stress concentrations at the square window corners were found to initiate cracks, which spread catastrophically across the aluminum skin. This would occur after metal fatigue slowly weakened the plane's skin during many cycles of takeoff and landing. In a more rigorous FFU test, a second aircraft body was cycled only 3057 times before suffering failure. In this test, pressure cycling was done in tandem with wing flexing to better simulate flight conditions. Tragically, this trial was still underway when the second plane took off for the last time.

The investigation determined that both the production process and the design were faulty. During assembly, rivets were punched through multiple layers of aluminum sheet in a single step. This was faster and easier than drilling a clean hole and inserting the rivet in a second operation. The technique produced irregular holes, which initiated fatigue cracks. The postcrash design change also eliminated the square window frames, which amplified the stress on the fasteners. The picture above shows the redesigned passenger windows of the later models. To this day, airliners have rounded windows to minimize airframe stress.

In this case, the FFU conclusions derived from earlier applications did not scale up to the more stressful end-use conditions. The designers may have relied too much on their successful wartime experience with materials and production techniques. They did not consider that the occasional mysterious loss of a single warplane over enemy territory was not scrutinized as closely as a peacetime airliner crash. Today, prototype airframes are routinely stressed to the point of catastrophic failure. The test results will dictate the service schedule for the replacement of key components during mandatory maintenance programs.

RULE #47 PERFORMANCE PREDICTIONS ARE DIFFICULT IN NEW APPLICATIONS

It is hard to predict the service life and failure modes for novel materials in end-use situations for which you have no standard of performance. Proprietary applications are the most difficult, because you cannot see everything that goes on at secretive end users. Some targeted applications are so large and expensive that they cannot be accurately replicated in the laboratory.

New application development:

- Determine if the customer has internal simulations or QC tests to screen promising materials. Learn the procedures and duplicate them in your FFU laboratory. Are tests a firm requirement or a guideline?
- Is it possible to buy small, laboratory versions of the customer's unit operations? Check the equipment in the customer's product development laboratories. Concentrate on the portions of the operation that cause the most problems for your materials. Simulate the critical features of the customer's equipment at the interaction points with your product. This is especially revealing if those components are normally shielded or hidden inside a production machine while it is running.
  - Outfit a simulator with instrumentation to evaluate the critical interface.
  - Screen the full performance envelope encountered on such machines.
  - Mount the simulator inside a climate chamber to evaluate a range of ambient conditions.
  - Document the results with high-speed cameras, strain gauges, or pressure transducers.
- Consider contracting experts who have worked in the customer's industry and understand the end use of these products. This can be very tricky if the customer perceives that you are buying their trade secrets with an eye toward empowering new competitors. The relationship can become difficult when hiring a disgruntled employee from a key customer. It is invaluable to have someone who really understands how the products are used and evaluated in specific market segments.
- Industry groups, trade schools, private testing companies, and equipment vendors will have similar equipment. They may offer an independent perspective on how to solve problems. They may also have insights into future trends in the market.

- In rapidly evolving fields, university researchers may be at the cutting edge of the technology.

- Seek out industry standard tests as a starting point for your work. Ensure that the tests are appropriate for the application and are recognized by the customer. Use an independent laboratory or certification authority to validate your internal results and to build credibility. Standard tests are generally supplemented with FFU tests specific to the customer application or market niche.

- Place your experts on committees that write specifications and standards for the customer's industry. These groups provide a wealth of information if you have the time and expertise to sift through it.

- Use control samples that mean something to each customer, such as competitive materials.

- Your presentations may show where competitive products exceed yours in some aspects of their performance. This can build your credibility and reveal customer priorities in subsequent discussion.

- "Cherry-picked" results that only disclose results favorable to your material are always suspect.

- Rarely will a single FFU test give a complete picture of product performance. A series of different tests will combine to simulate all aspects of product performance and failure.

- Determine what product certifications are required by law, industry practice, or the customer's insurance company. These might include compliance to Conformance European (CE) or Underwriters Laboratories Inc. (UL) guidelines.

- Duplicate the customer's QC tests on incumbent raw materials to benchmark your product against the competition. Consider adjusting the process conditions to score higher than the competition on these tests.

- The customer's product can be purchased and rebuilt with your materials or components substituted for the existing part(s). Subject them to a battery of customer tests to compare performance.

Paved roads around the world are delineated with painted lines to guide motorists. These paint stripes separate our cars from the oncoming traffic and advise us on lane changes. Ideally, they will be resistant to the constant wear of tires and weather. The functional life span of paint is a critical factor in the cost of road maintenance. So, governments are keenly interested in buying the most durable paint formulations.

To evaluate the longevity of a potential alternative, highway departments apply candidate paint grades in test strips on a busy road and monitor how well these swatches stand up over time. In addition, they evaluate how quickly they dry, whether they show up well both day and night, and how easily they can be applied with the existing equipment. The durability tests take the longest time, often lasting for months or years.

The time needed to perform these tests became an issue in 2010. The U.S. government appropriated vast sums of money for road construction and repair, in an effort to stimulate the economy. However, the producers of methyl methacrylate had scaled back production due to a slow demand for their product. Full production levels could not be restarted instantly. This raw material shortage inhibited the production of paint formulations that were approved for highway use. Qualifying alternatives for the 2010 construction season was difficult, due to the time needed for the road testing.

Know Thy Customer

> This is a world in which reasons are made up because reality is too painful.
>
> **Barry Diller**

Nov. 21, 1939.   N. A. CHRISTENSEN   **2,180,795**

PACKING

Filed Oct. 2, 1937   2 Sheets—Sheet 1

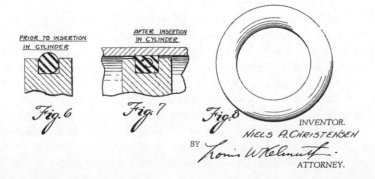

PRIOR TO INSERTION IN CYLINDER

AFTER INSERTION IN CYLINDER

Fig. 6   *Fig. 7*   *Fig. 8*

INVENTOR.
NIELS A. CHRISTENSEN
BY
ATTORNEY.

In the 1930s, hydraulic systems started to replace pneumatics to power actuators and braking systems in heavy-duty applications such as trains. However, it was problematic to get a good seal around piston shafts, resulting in a constant leakage of the oily hydraulic fluid. Niels Christensen studied this problem extensively and experimented with a wide assortment of gasket materials and configurations in an effort to seal these surfaces.

He would mount candidate samples in a test fixture and cycle the piston a number of times, monitoring their effectiveness at containing the fluids. Afterwards, he would examine the stressed materials under magnification for signs of damage. The best results were obtained with rubber rings that had a circular cross section that were seated in grooves in the piston. Trial and error suggested that these grooves should be 50% larger than the O-ring. Christensen was issued a patent in 1937, but his solution was not widely accepted.

At the onset of World War II, the aviation industry drastically increased production. Hydraulic systems were needed to actuate control surfaces, because human muscle power was insufficient as planes grew in size. Christensen had little credibility with aircraft designers, so a persuasive demonstration was needed for the army procurement team at Wright Field in Dayton, Ohio. O-ring grooves were cut into the hydraulic system on the rusty, worn-out landing gear of an A-17a. The U.S. Army Air Corp was very pleased when the improvised gaskets performed well for 86 consecutive landings. The government commandeered Christensen's patent in exchange for a lump-sum payment, because it was critical to a wide range of military needs during the war.

Critical Questions: How well does your product perform for the customers? How does it compare to the alternatives? What are each customer's unique requirements and what do they all need in common? What hypothetical material properties would provide improved performance in the application? What are these improvements worth to the customers?

No one person can answer all these questions. The sales staff meet with their purchasing manager and hear that the prices are too high. They talk to the warehouse manager and listen to complaints about that late shipment and some damaged containers last year. They take upper management out to dinner and the golf course, getting a lecture about how tough the economy is. Sometimes, these people do not know what could be improved from the perspective of the people who handle the product.

Salespeople rarely spend hours on the customer's plant floor with the operators who use your materials. They are never present on the night shift when the least experienced workers are misusing the product. Someone needs to spend significant time in the end-user plants to interview employees and observe the actual situation. Ask questions. Search out their wants, needs, opportunities, and constraints. Every interface between the end users and your product should be studied and understood.

Development team members must understand how similar products are used in a range of different applications and what features/properties need improvement. It is also important to see how the noncustomers perform the same functions. The deficiencies and strengths of industry standard offerings should be understood. Opportunities to add value should be sought out. This might be as simple as altering the package or the instructions. It could be as complex as designing a proprietary system of equipment, products, and services that fundamentally change the customer's operation.

The team members should gather enough information to devise FFU tests and interpret the results. They should bring back critical components that interact with the product. Such components may be incorporated into the test to duplicate the application environment. Think holistically about the product's use. Consider if altering the end-use conditions would improve the customer's operation, along with the functionality of your product.

Rohm and Haas revolutionized house paint in the 1950s when they introduced the technology to enable water-based acrylic paints. The company developed an acrylic emulsion called Rhoplex AC-33, which served as the binder for house paints sold by a range of companies. Rohm and Haas' expertise was in the chemistry of acrylic resins and their ability to tailor the polymer properties to the end-user's performance requirements. The needs differed with each application, such as interior versus exterior locations.

In order to fully understand the various applications, Rohm and Haas' experts had to spend extensive time learning how different experimental samples worked for the painters and the formulators. Once they identified the optimal formulation, these same people had to convince paint companies and individual painting contractors that their product offered superior performance in terms of resilience and consistency of application. Years of diligent effort were required to achieve industry trust.

Types of customer tests:

- *Focus Groups*: These are traditionally used for gathering impressions of consumer products, but they are used in industrial settings as well. The biggest concern (and strength) is that the participants are outside of their normal product-using environment. So, the results may not match the commercial reality. On the positive side, this tool can discover feelings about aspects of a product, in isolation from other factors, such as price, barrier to entry, or an existing solution.

- *Home Visits*: These often require more preparation and resource commitment than a focus group. They allow you to watch users evaluate a product in their normal environment. With a sufficient sample size, new revelations will generally appear. However, personal attention may lead the participants to be more enthusiastic about the product than would normally be the case.

- *The Audit*: In this case, you travel to customers' locations and spend time watching them use the product in the course of their normal operation. While doing so, you may be measuring efficiency and effectiveness—time to perform an operation, percentage of failures/rework, and failure modes. You may also do operator training, make suggestions for operational improvements, and generally offer your expertise to help the customer in any way possible. The product that you are watching may be your standard offering, or competitive materials. This is a good means to verify that your process is operating consistently and that the product is still optimal for the changing end-use environment.

- *The Customer Trial*: This situation involves a FFU test of an improved or modified product. Your objective is to do a complete head-to-head evaluation of this new version versus the customer's incumbent solution. You provide a large amount of free product to convince the customer to do so. A sufficient number of your employees should be on hand to collect data and ensure an accurate recording of the positive and negative results.

- *The Blind Trial*: Some customers are not willing to let you into their operation to experiment with their operations. They insist on testing your product themselves, free of outside observation. The results of such trials are subject to information filtering based on the customer's commercial agenda. Compare the impressions of as many individuals and customers as possible to widen your perspective.

Considerations for any customer trial:

- Keep safety as your first priority. The worst possible outcome is to cause injury or damage to your customer. Evaluate as many possibilities as you can prior to exposing customers to your product. Limit the impact by starting the trial on a single machine, work cell, or other discrete unit, where any problems can be observed and contained. Begin with test conditions that are the least risky and watch carefully as the customer increases the production rate or the severity of the test.

- Exposure of your ideas to the marketplace will affect your ability to obtain patent protection. Consult a lawyer about the need for the customer to sign agreements to protect confidentiality.

- Treat the trial as an experiment—compare the results to a control and follow a plan. Record all results, conditions, and procedures.

- Consider the relationship with the customer from a commercial perspective. What will they expect in return for allowing the experiment? What is the benefit for them?

- Will the customer report details of the test to your competitors? Even if the customer is bound by confidentiality agreements, individual employees will gossip about the details if they have friendly relationships with your competitor's salespeople. No secret is safe in field trial situations.

- Be careful what you promise to a customer during a test. Sometimes, it is necessary to evaluate a prototype that cannot be produced economically today. The test can buy information to justify a capital improvement of your process. Make sure that the customer fully understands the situation.

- All your participants should be well briefed as to the procedures and goals prior to the test. They must understand what they can and cannot say. They should come equipped with the required safety equipment and training for the work environment.

- Any technical activity at a customer location should be well coordinated with sales and marketing functions. Trials should enhance commercial relationships, not stress them.

- Limit the number of variables in the test. Use internal FFU testing to screen down to the vital few samples. The logistics of doing trials at customer locations make large sample sets very awkward.

- Look for creative ways to reward the customer's employees for putting up with your experiments. Just because the trial is beneficial to their management does not mean that the extra work is pleasant for them. Food, tee shirts, hats, and other complementary thank-you gifts often go a long way toward getting cooperation and candid comments. Clear with the customer in advance.

- Consider who will best operate, learn, and communicate in the customer's workplace. Your PhD researchers might not interact well with the blue-collar workers. Recruit participants with the social, language, and listening skills to gain the trust of the people whose knowledge you seek.

- Customer trials should provide feedback to improve internal tests. Differences in test results between the two places must be minimized through continuous improvement.

RULE #60 YOU NEED NOT MAKE FLAWLESS MATERIALS, BUT THEY MUST PERFORM FLAWLESSLY FOR CUSTOMERS

Flawless product performance is accomplished with exhaustive FFU testing that guides development toward the best compromise between process efficiency and end-user satisfaction. These two goals are not mutually exclusive. Customers will not be well served in the long run if your process is not stable.

Summary

- FFU tests determine the degree to which your product satisfies customer needs. They define the performance window and the consequences of failure.

- Testing to failure will reveal the margin of safety in an application. Compare operating window sizes for prototype products to the incumbent solutions. Test all failure modes.

- FFU testing can be done at any stage of the product's life. Tests are created and performed most often during the course of product development.

- FFU tests can be performed inside your organization or outside, but they must be conducted to rigorous, repeatable, and controlled standards.

- FFU tests are not static. They can be constantly refined to match changing end uses.
- FFU requirements dictate the contents of product data sheets, material safety data sheets, specifications, and QC test procedures for production operations.
- Product attributes that are important to one customer may be insignificant to another. FFU test suites must be flexible enough to evaluate a range of performance situations, conditions, and requirements.
- Creating and validating totally new FFU test methods is time consuming and expensive when trying to satisfy unfamiliar applications.
- Field-testing should be carefully planned and executed to gain insight into product performance while maintaining the customer relationship.
- Safety is the first priority of customer trials. Do no harm to your client.

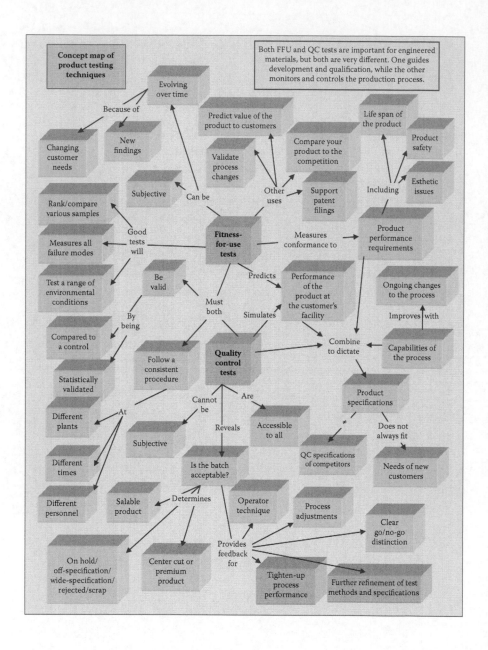

Concept map of product testing techniques

Both FFU and QC tests are important for engineered materials, but both are very different. One guides development and qualification, while the other monitors and controls the production process.

3

Technical Creativity and Idea Generation

How do we foster an innovative culture to capture ideas and then refine, evaluate, retain, and implement them effectively? How should we reward people for their contributions?

> We can't solve problems by using the same kind of thinking we used when we created them.
>
> **Albert Einstein**

In 1878, a young Thomas Edison was experimenting with a system to receive, record, and replay telegraph messages at high speed. This work had applications for the Western Union Company, which paid Edison to pursue incremental improvements.

While working around the equipment, Edison noticed that the compressed strings of dots and dashes sounded like faint human voices. The ghostly sounds were not unlike the voice transmissions over Bell's newly invented telephone. This observation triggered a divergent idea in the inventor's fertile imagination. His inspiration had no immediate commercial application, but it was very compelling.

Edison sketched a design and asked John Kruesi, his best machinist, to fabricate a model. The laboratory staff was perplexed when told that this simple machine would record and play back human speech. They expected a joke or a magic trick. The group was stunned when the prototype actually worked as predicted.

The U.S. Patent Office approved Edison's application in a matter of weeks, because nothing remotely like this concept had ever been proposed to them. The invention was so novel that years passed before anyone developed a business model for selling equipment or prerecorded music. The public flocked to exhibitions of the technology, generating more ticket revenue than Edison had spent to develop, patent, and build the first models. People were amazed at the utter simplicity of the device itself.

During this period, the population was routinely dazzled by new products that sprang from electrical technology. However, the early phonographs required no electrical components. The proof of the concept was a simple mechanical device. Four centuries earlier, Leonardo da Vinci could have entertained his patrons with it. Even the ancient Greek Antikythera mechanism of 2000 years ago was far more intricate and scientifically grounded than Edison's first phonograph. No other genius conceived of this device,

because no one had asked for it, or even dreamed of such a thing. (We should remember that the Frenchman Charles Cros proposed a phonographic theory in 1877. However, he did not build a working model.)

Edison had multiple advantages when pioneering audio recording technology. He had the full range of microphones, speakers, and other gadgets close at hand for experimentation. He spent long hours working around the equipment in his Menlo Park laboratory. His experimental schemes could be designed, tested, and redesigned quickly while he watched, learned, and innovated. He was not tied up with endless telephone calls, PowerPoint presentations, and e-mails. And finally, his experiments were not limited to customer requests.

RULE #23 WATCH PRODUCT PERFORMANCE AND LEARN

The inspiration for material and process improvements comes from watching products in use, asking questions, and spending quality time with many different process operations. Do not limit these experiences to specialists tasked with innovation or market research. Reading concise reports by others is not as beneficial, nor is speculation without follow-up experimentation.

Creativity in an Industrial Environment

Most innovations come without epiphanies, and when powerful moments
do happen, little knowledge is granted for how to find the next one.

Scott Berkun

Sept. 13, 1955   G. DE MESTRAL   2,717,437

VELVET TYPE FABRIC AND METHOD OF PRODUCING SAME

Filed Oct. 15, 1952

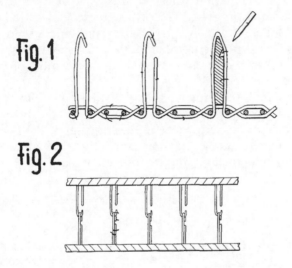

Fig. 1

Fig. 2

INVENTOR

George de Mestral.

One of the classic tales of inventive creativity was the 1940s development
of Velcro closures. George de Mestral famously had his "Eureka!" moment
while looking at the hooked microstructure of burrs, which had attached
themselves to his dog's fur, during a hike in the Swiss countryside. Many
accounts stop at that point, assuming that Mestral's inventive flash was sufficient to create a product.

In fact, it took years to refine the invention into a patented, commercial
product. Early prototypes were laboriously made by hand. It was easy to
make loops, but hooks were tedious and expensive until the inspiration to
form them by cutting open loops. Mestral discovered that heating nylon fiber
loops under an infrared light would permanently set their shape, retaining a
hooked profile through repeated use.

While the product was tremendously novel, it was not very profitable during the term of Mestral's patent. Outside of NASA, designers did not view Velcro as a fashionable replacement for zippers, buttons, and tie closures. Industrial customers were slow to implement high-volume products, initially using Velcro in specialty applications. Wide-scale use only came after the patent expired, allowing low-cost manufacturers to innovate across a wide range of applications.

Evaluating numerous ideas is the key to achieving market advantage. Most innovations are mundane improvements in raw material selection, processing techniques, test methods, sales opportunities, fitness enhancements, business practices, and marketing campaigns. With diligent evaluation and implementation, these small steps make the difference between stagnation and progress. The "Wow!" discoveries are only a tiny portion of industrial development. Depending on them for your growth is like using your investment portfolio to purchase lottery tickets. Even after you hit upon a winning combination, the bright idea must be developed into end-user solutions.

In a business environment, creative thoughts might pop into any employee's right brain as frequently as happens in an artist's loft or a revolutionary's coffee shop. To some, these new ideas represent change, risk, uncertainty, and inefficiency. Others recognize that competitive advantage requires evaluation, experimentation, and implementation of dark and scary visions.

RULE #44 PROCESS-DEPENDENT INNOVATION REQUIRES LONG, HARD WORK

Innovating with a dynamic process technology demands persistence to demonstrate and implement customer solutions. Brilliant ideas will not instantly be recognized as the final answer. Proposals must be compared with competing ideas before fitness-for-use (FFU) tests reveal the best option. In response to findings along the way, the original idea may morph into a completely new form before it is commercialized. The process path that yields the optimal solution can never be predicted at the outset.

New ideas in isolation are not good enough. Concepts must be communicated, tested, weighed, and eventually implemented. Charles Cros found that simply postulating a grand idea is not sufficient.

In the midst of this confusion, the management laments the limited imagination of its development staff. The products lack that edgy buzz, which wins awards. The trade press ignores our new ideas. Nothing revolutionary has been introduced for years. "Our engineers are just too dull and uninspired!"

This misconception comes from assuming that creativity cannot be nurtured, because it is an innate ability that a person either has or does not have. Novel ideas are not pulled from a hat, fully formed and completely obvious in their utility and advantage. In a process-driven development environment, many ideas must be cultivated and evaluated before the optimal solution emerges. Developers settle for the first favorable answer when under time pressure. There are usually better solutions waiting for patient competitors.

People who are perceived to be creative are often those who are motivated to package and present their ideas in a way that best communicates utility to the organization. People who are judged to be less innovative are those who try to sell raw suggestions and get discouraged when they receive little support or constructive feedback.

In a process-driven company, it is expensive and time consuming to test ideas that change the output of a complex production process. Imagine that significant testing is needed on multiple customers to judge whether an adjustment makes an incremental improvement. The management needs great confidence in a complex proposal before it spends money to authorize such trials.

This innovation roadblock can be overcome when the organization has a wide span of process knowledge and capability. Development is easier when diverse process resources can rapidly create an assortment of prototypes. Internal FFU tests are needed to evaluate the alternatives. Experts should be available to explain the underlying science, suggest alternative technologies from other fields, and interpret confusing test results. With these advantages, improvements can be refined and patented before your customers and competitors see the course of your work. There is never enough expertise within the company to fully understand and exploit all of the market needs. Extending the span of knowledge with partnerships to other organizations is vital to developing in totally new situations.

The "span of process knowledge and capability" chart at the end of the chapter describes the essential elements of product understanding. The breadth of this insight will differ from one company to the next. All organizations have unique perspectives on what should constitute their spans of knowledge. The larger the company is, the more challenging it is to find where the required information or capability resides.

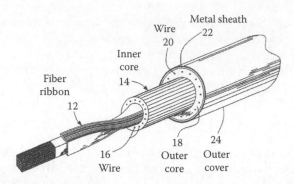

Fiber-optic cable has a diverse spectrum of technology applications. The product has been evolving for years to deliver better performance in specific end-use areas. For instance, improved glass purity has increased the distance between the amplifier facilities on undersea communication cables. Modern fibers still carry 80% of their signal strength after a kilometer of transmission. The materials used in eyeglasses are considerably less pure, losing the same amount of light in only 3 ft. of thickness.

These advances are not in the hands of any one inventor or one company. The improvements in fiber properties come after years of research on raw materials and processing techniques. To deliver a better performance in an application, complementary upgrades must occur in the amplifiers, switches, and software systems. Network operators must marry the new cables and equipment with systems designed to eliminate constraints in order to improve the overall system performance. The advances in one aspect of the larger network must be matched and coordinated with the others.

Innovation in this environment is not a matter of a lone genius making a breakthrough. It demands the communication of the needs, capabilities, and compromises between many partners. The benefits of one individual idea might take years of work to be realized. This environment requires innovative organizations, where a web of creative individuals share and build on each other's ideas toward a common goal. This situation puts a premium on understanding which parts of your knowledge base are free to share and which sectors are tied up with secrecy agreements.

Reality Check on the Innovation Climate

It appears that those managers who support individual's research freedom in pursuit of the core mission, encouraging interdisciplinary teams and managing with a 'soft touch' run the most productive outfits.

Arthur Molella

2,130,948

SYNTHETIC FIBER

Wallace Hume Carothers, Wilmington, Del., assignor to E. I. du Pont de Nemours & Company, Wilmington, Del., a corporation of Delaware

No Drawing. Application April 9, 1937, Serial No. 136,031

Nylon fiber was developed by a DuPont research group headed by the brilliant, but emotionally troubled chemist Wallace Carothers. His team was seeking synthetic versions of natural products such as silk and rubber. In the course of fiber research, they synthesized several polymers in the polyamide family (later trade named Nylon), but they were not impressed with the properties. Being chemists, rather than process engineers, they did not explore the processing capabilities of the new materials. The focus of their work shifted toward polyester chemistry.

In 1934, a member of Carothers' group, chemist Julian Hill, noticed that he could draw strands of fiber by prodding a moist lump of polymer with a glass rod. These fibers were intriguing because they were extremely stretchy. Carothers was away from the laboratory that day and his staff was free to experiment in an entertaining fashion. Hill and his coworkers ran down the hall with clumps of polymer to see how far they would stretch. Instead of breaking, the filaments were transformed by room-temperature drawing. This was especially beneficial for polyamide, which acquires exceptional fiber properties from this dynamic process step.

The cold-drawing process orients the polyamide chains, greatly increasing their modulus and tensile strength. An otherwise useless polymer was transformed into a fiber with silk-like properties. Considerable development work was needed to refine the product and create a suitable production process. The finished material had a myriad of uses and became DuPont's most successful product. Within 5 years, it was a huge hit in the production of women's stockings. The key patents listed Carothers as the sole inventor, overlooking the contributions of his team. The moody and temperamental scientist committed suicide before his new material was commercialized.

Is your company conducive to novel ideas and ways of looking at problems? How would the management react if it saw laboratory technicians running down the hallway with invisible strands of fiber? Are your employees rewarded for pursuing risky approaches? Is your organization resistant to change? A simple survey can reveal the innovation climate in a technical organization. Ask these questions of both the decision makers in management and the technical workforce to compare their perspectives.

Rate each statement on a 1–5 scale: 1 = strongly disagree to 5 = strongly agree:

1. We are anxious to explore and implement new ideas and fairly reward all those responsible.
2. The company actively listens to all employee suggestions and pursues the most promising ones.
3. Producing innovative new products is more important to the organization than any other activity.
4. Employees are empowered to run their jobs the best way they know to advance innovation.
5. People who question and test the existing paradigms and practices will be rewarded.

Add the five answers together to calculate an average score for all people in management and another average for those who perform development work. The key metric of this survey is not whether the company scores high or low on the scale, but the difference between the average management answer and the people doing the work.

If both groups give low scores, the organization is obviously happy filling a static niche market where execution is more important than innovation. Everyone is on message that new practices are not as important as following procedures. The best strategy is for the management to be consistent with that philosophy and hire people who are comfortable in a static environment.

If both groups give high scores, then a culture of innovation exists within the organization. Such an organization creates a fertile environment for creativity. However, even companies that are passionate about innovation will always look for improvements to their development practices.

If the management has a significantly higher opinion of its flexibility and openness, then the creativity problem originates at the top. Those in charge want to maintain control over imagination, which is inherently unmanageable. Inspirations from respected experts, consultants, and other successful companies might be embraced and held up as examples. Those that come from deep within the workforce are not given the same consideration. In order to improve, the organization must identify and remove the barriers (both real and imagined).

If the management has a significantly lower score than the workforce, then it is running the organization like a high-security prison. The inmates are so cynical that questions are answered in a totally sarcastic manner. Flags of concern should be raised if some particular employee levels or entire work groups have much lower scores than the others. This is an indication that the organization has preconceived notions as to who should innovate and who should stick to the basics. Creative companies are not so selective.

Fostering Technical Innovation

Politeness is the poison of collaboration.

Edwin Land

Jan. 24, 1928.   1,657,411
A. SCHERBIUS
CIPHERING MACHINE
Filed Feb. 6, 1923

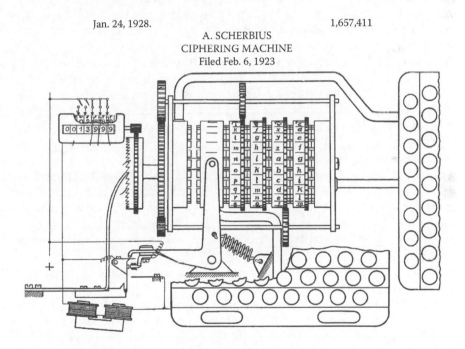

During World War II, the Germans used the ingenious Enigma code machine to protect their most important communications. This compact electromechanical device could both encrypt and decipher text messages using preassigned machine settings. The U-boat fleet was equipped with the most sophisticated version of this system. Its keys were changed on a daily basis and messages were kept short to deter code breakers. The system had so many possible permutations that the Allied intelligence services could not decipher them in a timely manner with mathematical analysis and mechanical calculating systems.

Only an intensive, multidisciplinary effort provided the clues to unlock the constantly changing code settings. The weather conditions, tactical situations, ship-sighting reports, radio triangulation intelligence, captured documents, and knowledge of German procedures and abbreviations all provided clues to crack each day's machine settings.

The correlation of such diverse knowledge was far beyond the capacity of any one individual, no matter how intelligent or lucky. Only the cross-functional collaboration of diverse experts could bring it all together. Mathematicians, naval officers, meteorologists, linguists, and technical experts exchanged ideas that unraveled the operational secrets. This was done under a blanket of secrecy and intense pressure.

The wartime code breakers could not hold back ideas to get exclusive personal credit for a breakthrough. They had to share facts, hunches, and insights. There was no way to predict which subtle clue would be the means to break each day's U-boat code settings.

Adjusting a dynamic production process to alter the material properties and provide new solutions is a challenging task that cannot be forecast or budgeted. No one can predict who will conceive the optimal idea and who will identify the process path to turn it into a practical application. Out-of-the-box thinking does not spring from a methodical system or adhere to a timetable. Rarely are the best ideas recognized immediately as being superior to the alternatives. They never pop into someone's head fully developed and ready for production.

Potential solutions may be hammered out in a series of interactive debates to separate the good ideas from the bad. The dialogue might occur in cubicles, on the plant floor, or at the coffee machine. Partial solutions spring to life in the course of passionate discussions between two or three people who understand the technology and are willing to argue abstract viewpoints.

The ultimate solution starts as a germ of an idea that captivates one or more people with its potential. At the embryonic stage, an expert will recognize that the concept represents one of many ways to solve the problem. The light bulb moment does not constitute a commercially viable solution until it is roughed out and tested against the alternatives. This evolution will require time and money.

Once infected with the basic principle, the developers will not let a day go by without pondering and discussing it with others. During that time, versions will be evaluated on paper, by computer simulations, or in bench-scale experiments. Seeing these models and tests will spark additional suggestions and observations from coworkers. This birthing experience can be brought to a halt by the urgent reality of the daily job. Most good ideas die a slow death of neglect and apathy because there are not enough hours in the day to breathe life into them. The best ideas are reincarnated elsewhere again and again until someone finds a way to keep them alive.

Successful process trials, FFU tests, or the construction of a working prototype will prove which ideas are viable. At this point, the concept has not yet been proven economically feasible or compatible with the company's business objectives. Only a program of development and testing will answer those questions.

Ehrenfried Walther von Tschirnhaus was an eighteenth-century expert on glass and ceramics, who was interested in reproducing the Chinese porcelain process. Under the patronage of Augustus the Strong in Saxony, considerable resources were applied to this task. Tschirnhaus recruited a cross-functional team, which included Johann Böttger, an experimental chemist (alchemist), mineralogist Gottfried Pabst von Ohain, and Dr. Bartholomaei, a physician and naturalist. This eclectic group brought more scientific skill and diverse experience to the task than was available within the pottery guild.

Tschirnhaus had a parabolic mirror to generate high temperatures for small-scale laboratory experimentation, allowing samples to be screened quickly. In September 1707, their operations were scaled up to a pilot plant located in three cellar chambers of the Dresden fortress. This facility provided ample space and security, where they reverse-engineered China's porcelain secrets before the rest of Europe. Initially, the team produced a redware with a high iron content, but was soon able to make white porcelain as better kaolin deposits were found. By 1710, a factory was opened in Meissen and significant sales were recorded in 1713.

Tschirnhaus died in 1708, during the scale-up phase, before the team succeeded in making a pure white product. The Saxony porcelain was named for Böttger, but historians disagree over who deserves the credit. In fact, no single person was responsible. This complex technology required multidisciplinary skills and the extensive resources of a wealthy patron.

RULE #54 PROCESS INNOVATION REQUIRES OPEN NETWORK ACCESS TO ALL APPROPRIATE RESOURCES

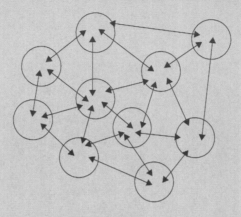

Innovators must have access to diverse technology and feedback on their work to develop process options. Ideas need to bounce back and forth between knowledgeable people, changing and growing with each iteration. Weak options must be eliminated with solid data, not opinion and speculation. Effective industrial development requires that many promising process alternatives are fully explored and documented.

Creating a culture of industrial creativity:

- People must have some unallocated time to discuss ideas, sometimes for hours on end. People who are paged, telephoned, and interrupted every 5 minutes are less effective.

- Those looking for ideas must have access to diverse resources, inspiration, and feedback. The spark that illuminates the concept might come from anywhere. It could be a new hire who has not been blinded by the company's paradigms or a factory worker with a unique insight. Interacting only with a homogeneous research staff is not an asset for out-of-the-box thinking. Process innovation does not happen in a vacuum or while doing paperwork.

  - Metcalfe's law: The value of a network is proportional to the square of the number of its users. Think of your employees as a network of resources. The more links and idea sharing between individuals, the greater the available innovation resources.

- The participants in a creative discovery must be comfortable to say anything. Coworkers cannot assist in the birth of a new concept if they are afraid of ridicule or are reluctant to shoot down bad ideas. Innovators must fight to keep their proposal alive and be flexible to learn from rejection.

- Structured creativity sessions discourage negative responses with rules enforced by a facilitator. Sarcasm and ridicule cannot kill the spark of a great concept. Bold new solutions only take hold after the weak alternatives are stamped out. There is never a shortage of bad ideas, because they all sound good in theory.

- There can be no suspicion that participants are hoarding their best ideas to gain sole credit. "Credit hogging" causes the death of collaboration.

- When the group voluntarily selects itself for informal brainstorming, the participants tend to be comfortable and self-motivated by a passion for the topic. When the boss collects a team and proposes the topic, there are no guarantees of comfort, trust, and motivation.

- Competition between functional and operational groups can poison the creative process. The free exchange of ideas among people of diverse technical disciplines is a vital ingredient. Creativity is stifled when people bring a political agenda rather than an open mind. The participants need a collaborative attitude.

- Many sound ideas are suboptimal or impractical in any given situation. However, removing constraints and limitations is a key to implementing innovative solutions, which, in turn, may trigger more options. Tomorrow's situation will be different and that nonconforming idea might be brilliant in a new setting. Ideas are "good" or "bad" only in the narrow context of one situation.

- If the concept is not reduced to practice and patented, it could be used against you by a competitor.

- Passing raw ideas up the chain of command is generally not productive. A busy boss cannot be expected to seize the optimal concept and assign a team to pursue it. A skeptical subordinate will often resent having to flesh out other people's wacky ideas. The person carrying the idea is advised to share it so that it may be applied to another problem. Others may supply a key observation or build upon it. Valid models will gather an enthusiastic constituency when others recognize that they represent a useful tool. With each contribution, the good idea grows toward a practical application. Impractical ideas will fade away unless they are kept on life support by a proud sponsor.

- Innovators must be willing to share everything, even at the risk of losing credit for their contribution. Management must ensure that people understand that they will be rewarded for freely sharing their ideas and assisting others to realize theirs. The organization can easily kill this spirit with an ill-conceived rewards program with a limited number of heroes. People must be recognized and compensated for their attitude, work ethic, and behavior.

- The availability of tools (toys) allows inventors to play with their ideas and prove them out. These may be customized software, rapid prototyping equipment, or diverse raw materials that can be manipulated into interesting and useful forms. They might include competitive products, customer goods, and things that no one can anticipate.

  - Companies need a budget or a slush fund for pointless purchases.
  - Sufficient tools are rarely all available. Empower people to acquire, borrow, or rent them.
  - The information systems (ISs) department should enable non-standard software tools, rather than obstructing them.
  - Be flexible on which tools should and should not be employed by your developers. Even if you are 100% correct, the constraints will annoy and distract them.

- A picture is worth a thousand words. Having someone who can draw, model, or manipulate ideas is invaluable for the team to visualize and compare concepts. Once an idea becomes tangible, it can be modified, copied, manipulated, documented, and communicated. Excited, engaged people may have one picture in their mind, but the rest of the group might not be getting it, due to the ambiguity of the language. This is important as teams become more culturally diverse.

- Innovation happens when the appropriate individuals passionately discuss problems among themselves. Intrusive management rarely assists their task.

  - Busy people often lack the patience to listen to technical specialists talk their way in and out of blind alleys. A batting average of one great idea out of a hundred discarded proposals could well be a major league performance.
  - Never try to dictate what a productive discussion should look or sound like.
  - Innovation springs from dissatisfaction with the incumbent solution. The supervisor can mistake a creative session for a time-wasting complaint session.

- Secrecy and isolation are counterproductive. Interested people overhear the conversation, join in and contribute. Alternately, the discussion may spark a totally unrelated idea for another problem.

- Some teams of people will generate ideas spontaneously at a high rate. Other combinations of individuals have no such chemistry. There may be no way to predict the outcome.

- Access to relevant information is vital to creativity. In a process-dependent development, conclusions are reached by correlating details from widely different sources. A wide span of process knowledge can speed the evaluation of new ideas and the entire project.

- Employees should be constantly making suggestions for improving all aspects of the business. These may be formal proposals or offhand comments. All suggestions should be evaluated for their utility. This important function is not in anyone's job description if the idea does not apply to an ongoing project. Useful feedback is vital. Silence and indecision will kill innovation.

  - Evaluate proposals analytically. Look for both weak points and potential. Are they detailed or just a rough idea? Are they understandable to experienced technical professionals or are they so narrow and abstract that only a few specialists can appreciate them? Do they address a specific problem/need or are they a vague idea?

  - Have alternatives been considered?

  - Are you really listening and understanding the submission or slotting it into existing paradigms? Ask questions and suggest additional work to support the idea with models, experiments, calculations, or additional documentation. Identify people who can help the applicants improve their proposal.

  - Consider appointing one or more experienced people to screen proposals and mentor people with unsolicited ideas. Avoid employees who need structured settings.

  - Monitor the effectiveness of the idea screeners. Are concepts being properly fleshed out and vetted before being passed up the line? Rotate the responsibility to identify those with the special skills to do this important job.

  - Follow up on promising submissions, especially if circumstances prevent an immediate evaluation. Losing track of ideas in the queue will demoralize employees.

Formalized Brainstorming

The only difference between a problem and a solution is that people understand the solution.

Charles Kettering

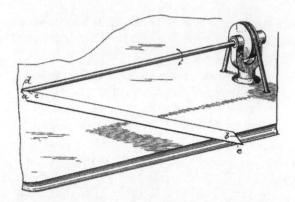

In 1888, the inventor Marvin Stone patented a process for making paper drinking straws. His vision created a totally new means of forming a tube, winding a flat strip of paper in a spiral wrap on a rotating shaft. A strip of glue along one edge of the paper prevented it from uncoiling when released. Coating the tube with paraffin wax provided water resistance.

The early drinking straws that Stone improved upon were stalks of natural straw, which came in limited sizes. Although wax-coated paper had existed for years, no one used it to make waterproof products in three-dimensional shapes. The paper cup was invented a decade after Stone's drinking straw entered mass production. The spiral-winding process is still used to make paper cores and cylindrical containers for dry products.

A generation after Stone's insight, Alex Osborne codified an innovation system for the generation of novel ideas. Osborne worked in advertising, which was expanding rapidly in the 1930s with the growth of radio broadcasting. His technique is still an important tool for idea generation and has many strengths and benefits. However, dynamic organizations will not limit themselves to this one innovation tool.

Traditional brainstorming guidelines:

- Brainstorming sessions should be focused on a clear and specific question. Participants should be satisfied with the selected topic and understand the problem that they are trying to solve.

- Groups of up to 12 people are asked to propose raw ideas, which are recorded for all to see.

- Participants ponder and jot down proposals in advance, prior to hearing other's ideas.

- Each person proposes one inspiration at a time in rotation around the room. These are not necessarily solutions to the problem, but descriptions of causes, effects, and alternatives.

- During the first rounds, ideas are best expressed in a few words, preferably single sentences.

- Everything is recorded exactly as stated, regardless of quality, validity, or utility.
- The goal is to generate a large quantity of raw, unfiltered information.
- Wild, silly, and impractical suggestions get people thinking out of the box.
- Input that modifies, restructures, or builds on earlier entries is encouraged.
- Proposals need not be the original, novel ideas of the person who submits them.
- A moderator may advance alternate perspectives when the pace slows, to jog additional input. They might also nudge the group away from a single concept that is consuming their energy.
- Criticism is forbidden in the first stage. Even praise and self-criticism are discouraged.
- When all the ideas are on the table, clarifications and amplifications are given. The participants might take to the pen, putting their thoughts into sketches, diagrams, and connections.
- Individual ideas are summarized on sticky notes, which are moved around to group them.
- The list can be cleaned-up, consolidated, and fleshed out into complete sentences. The primary concepts are identified and the related ideas are collected beneath these.
- Tree structures or mind maps can be created to show how ideas fit into a hierarchy.
- The team votes to rank the ideas or concepts based on their ability to solve the problem.
- Narrow the list to a few action items to pursue immediately. Retain the full list.

During World War II, radar developers were tasked with an impossible assignment. They needed to develop a microwave receiver that could withstand the stress of being fired from cannons. The goal was to create a smart shell that could detect targets and explode in their vicinity, rather than requiring a direct hit. The incumbent radar systems were notoriously delicate and could not survive 20,000 g of acceleration.

The developers used clever solutions that employed the conditions of use, rather than trying to mitigate them. The liquid electrolyte in the battery was isolated in a glass tube, separating it from the metal anodes and cathodes. This extended the battery's storage life and eliminated the need for an on/off switch. When the shell was fired, the glass would shatter, releasing the electrolyte. The rapid spin of the shell in flight distributed the liquid into the surrounding disks, bringing the battery to life. This sudden burst of power activated the shell's radar sensor, allowing it to detect the presence of a target.

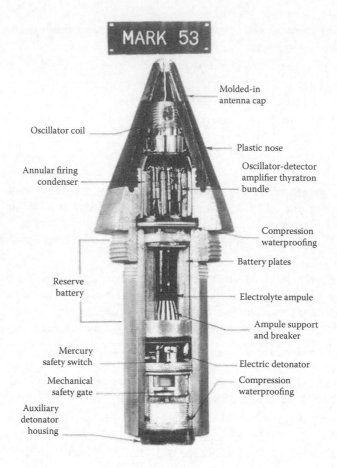

MARK 53

Molded-in antenna cap

Oscillator coil

Plastic nose

Annular firing condenser

Oscillator-detector amplifier thyratron bundle

Compression waterproofing

Battery plates

Reserve battery

Electrolyte ampule

Ampule support and breaker

Mercury safety switch

Electric detonator

Mechanical safety gate

Compression waterproofing

Auxiliary detonator housing

Mechanics of Creative Development

There are five ways in which to become wise: be silent, listen, remember, grow older and study.

Arabian Proverb

In 1903, French scientist Edouard Benedictus accidentally bumped a glass flask, sending it crashing to the laboratory floor. While cleaning the mess, he was surprised to see that the cracked glass held together in its original shape. Benedictus discovered that the flask's contents had evaporated away, leaving a clear film of cellulose nitrate on the inside surface. This unintended plastic coating held the glass fragments in place, preventing them from flying in all directions. The phenomenon was curious, but not immediately useful.

Later, Benedictus read reports that motor car drivers were being seriously injured by shards of broken glass during accidents. Recalling the flask, he set to work to reproduce the same effect with flat panels. Reportedly, 24 straight hours of experimentation with various coating options produced the first pane of safety glass. The prototype was a sandwich of cellulose nitrate between two sheets of glass, bonded together with heat and pressure. The product was patented in 1909, following years of work to refine the product and a process to make it.

The fledgling auto industry did not embrace safety glass for a number of years. The first practical use for the product was in World War I gas mask lenses, an application Benedictus never imagined.

There are always more product proposals available than can be used by the marketplace. Even the best ones must be rigorously tested to evaluate their impact on all aspects of the process and the end users. This evaluation can be broken down into a systematic method that many people follow intuitively.

Six steps of innovation:

- Fixate on a specific situation or problem that needs to be improved, eliminated, or realized.
- Accumulate and record information pertinent to the issue.
- Generate numerous raw ideas on how to approach the problem.
- Select the most promising idea(s) to investigate and an action plan to do so.
- Experiment to prove/disprove/compare the value of various alternatives.
- Struggle to realize the practical solution or application.

Fixation: Progress is seldom made by jumping impatiently from one problem to another. A person or a small group needs to focus on an important goal and pursue it doggedly. This step is not optional. It must be clear to everyone what the problem is and what the ground rules are/are not for a successful solution.

Accumulating information is often skipped in the haste of getting to the fun parts of the exercise. This step requires you to find out what is already known about the problem, the required properties, and the potential process options. When the work of previous developers is well documented, gathering this information is easy. Organizations with a wide span of process knowledge have an advantage in this step. If nothing else, the gaps in the written record will provide direction on where to experiment. Conventional wisdom, urban legends, and old assumptions should be noted for what they are.

Generating ideas is a popular activity. People love to offer their observations, speculate, and make suggestions. Brainstorming meetings are a pleasant diversion from work. Better yet, you may be able to claim partial credit for the eventual solution just by being present. Beware of idea ownership, when influential individuals bend the evaluation so that their own suggestion wins out.

Avoid creating short lists of ideas at this stage. The number of suggestions should be large until the following step evaluates and pares the choices down to a manageable number. The more ideas you have, the better is the opportunity to combine and build on them or to spark completely new revelations that open hidden paths. Reach out far and wide to include many different points of view.

Selecting ideas for experimentation requires more work than generating them in the first place. The most common choice is to first pursue the quick options or ones with the best chance of success. The alternative is to first eliminate the ideas or groups of ideas that are the easiest to discredit or prove false. Evaluate as many ideas, rather than locking on the first one that shows promise.

Truly innovative companies find ways to test ideas and evaluate their utility. Less innovative organizations pare the list because the daring ideas are incompatible with their existing systems. Novelty entails considerable work and risk to implement. The challenge is to decide which ideas to kill and which ones to promote.

Experimentation is never an end in itself. The test that proves everything to the experimenter might be pointless to the rest of us. Ideally, the work will narrow down the choices until one option remains. In the worst case, all options are eliminated, forcing you to loop back to the selecting step. Negative outcomes can spark idea generation through failure mode analysis and process problem solving. A wide span of capability within the organization provides useful feedback and alternative options.

Experimentation is more important for process-dependent materials than for assembled products. It is easier to change a product design than to manipulate a production process to consistently achieve new properties and improved performance in an engineered material. There will be dead ends and answers to questions that were never asked in the first place. Process innovation requires comprehensive work—raw material screening, equipment modification, new procedures, and the validation of FFU tests.

Struggling to sell a new approach may be the hardest phase. For the entrepreneur, this means finding capital and convincing customers to take a chance. For the academic, it requires getting grant money, publishing results, and selling the solution to skeptical peers. For a cog in the engine of industry, it means getting face time with management to explain the benefits and carving out a share of the budget. Edouard Benedictus raced through the first five steps, but spent years struggling to sell his invention. The struggle demoralizes idealistic people who expect a victory parade. Ideas are stolen, misunderstood, undermined, discredited, reverse engineered, or lost in the in-box. Their time may pass before they catch on.

The six steps may result in a validated idea, ready for development into a commercial product. Development projects have a higher probability of success in an industrial environment when they are preceded by an intense effort to gather background information and establish fundamental interactions. At the end of step six, Thomas Edison had a working prototype of the light bulb and the phonograph. Such things are created at great cost, can only be operated by a specialist, and have limited features. The working model demonstrates feasibility, but it is not yet a market-ready product.

Prototypes demonstrate the potential to a wide range of interested parties—management, the sales force, patent examiners, potential customers, and investors. A viable process path allows the estimation of costs, time lines, and market potential. On many occasions, the six-step process will determine that the concept is not viable. These innovation steps will suggest new concepts that no one expected.

In the 1930s, a team led by Otto Röhm was trying to polymerize methyl methacrylate between two sheets of glass. Theoretical research by German

chemists some 50 years earlier had documented this synthesis route; however, the material had never been developed into a commercial product. The team's goal was to create an improved safety glass. Their experiments were a dismal failure, because the glass sheets easily separated from the polymer coating.

However, the researchers were intrigued by the plastic sheet that was left behind. They had created a crystal clear cast acrylic plastic, which had excellent physical properties. After several years of development, their accidental discovery was commercialized as Plexiglas. Ironically, this material replaced the original safety glass in many applications. As was the case for Benedictus, Röhm's product was initially successful in military applications, such as aircraft windows and turrets. Later, the durable polymer found many civilian markets when it was pigmented for use in automotive taillights and decorative signs.

Organizational Hostility

Even as corporate leaders chase the vital, elusive spark of creativity, their organizations' structures, processes and norms extinguish it wherever it flares up. Their cultures and routines privilege analysis over intuition and mastery over originality.

Roger Martin

In the United States, it is becoming more difficult to produce cured sausage using time-honored European techniques. Products such as traditional salami, soppressata, and saucisson sec start as raw meat mixtures that are stuffed into natural casings. They should be hung in the open air at room temperature, picking up airborne yeasts to initiate the fermentation process. A skin of white mold often covers the rind.

Safe procedures have evolved in each region with centuries of experience. These artisanal products rely on naturally occurring acids, seasonings, and salt to inhibit harmful organisms. The rich, distinctive flavors and texture are dependent on a dynamic process.

Today, the traditional process technology is in decline. The U.S. Department of Agriculture and local boards of health have cracked down on procedures that do not comply with their food processing guidelines. In particular, they require "kill steps" such as heat treatment to ensure that the meat is free of pathogens. Curing is mandated to take place in controlled environments. The flavor and variety of the time-honored meat products are becoming difficult to achieve.

The consumers of these distinctive products are greatly disappointed with the output of the safe process. The traditional producers feel that they are unfairly restrained from using the full range of processing options because of unnecessary government regulation. They complain that beer and wine producers use a similar biology to mature products at room

RULE #15 CHANGE IS SCARY

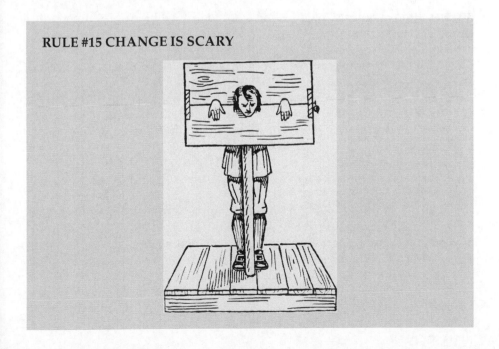

> Every organization will generate objections to an innovative idea. The source of the opposition will be honest, passionate, and supported with reasonable arguments. There will always be some reason why you should never change anything.

temperature. However, beverages are governed by different regulations than meat. Development professionals feel the same way when their proposals are rejected or ignored. Rigid thinking and snap decisions seem to unfairly limit their options and inhibit innovation.

When innovators discover a simple solution to a life-threatening problem, their organization should swiftly adopt it. This was not the case for the most powerful navy in the eighteenth century when an effective cure for scurvy was first discovered. This disease was common on long sea voyages, but is now recognized to result from vitamin C deficiency. A navy surgeon, James Lind, felt compelled to research this topic after learning of the deaths of 380 sailors out of a 510-man crew who had sailed only a few years earlier on a circumnavigation of the globe. His groundbreaking experiments involved separating patients into groups to directly compare one treatment to another. He published the results in 1753, showing that a daily ration of fruit juice could reverse the onset of the disease and keep sailors healthy and productive.

The British Admiralty was unwilling to implement his solution or follow up on the findings for the next 40 years. Many sailors died because an entire generation of naval officers was unwilling to adopt new methods. The slow action was partially Lind's fault. His experimental methods and results were first class. However, he recommended a distilled fruit juice concentrate, to overcome the logistical concerns of preserving raw fruit on long voyages. Lind never actually tested this option. It turns out that the heat-treatment process destroys the juice's vitamin content. Lind skipped the struggling step completely.

This pattern of bureaucratic behavior was repeated during the American Civil War in the 1860s. Both armies were armed with single-shot rifles that could be fired only one to three times per minute, depending on the skill of the individual soldier. Multiple entrepreneurs came forward with repeating rifle prototypes that offered more firepower. Despite urgent field requests for better weapons, the Union Army Ordnance Department was slow to move away from the outdated rifles that it knew and understood. Its leaders argued that the fast-firing arms would waste ammunition, overtaxing the supply lines. Only the direct intervention of President Abraham Lincoln eventually prodded them into reluctant action. By the end of the war, some units were equipped with innovations like the Spencer carbine.

Learning from this war, the Europeans eagerly upgraded their killing technology. Many armies adopted fast-firing rifles for their wars of the late nineteenth century. The American army was not so progressive. The Ordnance Department scrapped the proven Spencer design after the war and equipped the troops with modified single-shot rifles and carbines.

In 1876, George Armstrong Custer's Seventh Cavalry Regiment was armed with the slow-firing weapons when it recklessly attacked a large Indian force along the Little Big Horn River. The Indians were more enthusiastic about modern weapons than the ordinance bureau, purchasing surplus Spencer carbines and other repeating rifles on the black market. Not only was Custer outnumbered at his last stand, but he was also outgunned. Custer should have recognized his firepower deficiency and acted accordingly. Despite this embarrassing defeat, the army retained outdated weapons for the rest of the century.

Is your organization hostile toward new ideas? It need not passionately embrace every new idea, just not reject them immediately. Most companies insist that they support and nurture innovation. Despite good intentions, creativity is discouraged in many ways.

Cultural barriers to innovation:

- Cranky support systems block new approaches, especially those that are truly novel. This includes ISs that cannot account for items outside the existing data structure.

- Overburdened sales and customer service functions are reluctant to call on a completely new class of customers to gather leads. Incentive plans may discourage salespeople from spending time away from current customers or those that are not immediate prospects. Sales management will not dedicate a specialized sales team to a new product until its volume justifies the move.

- What does the compensation system encourage?
  - Are people rewarded for thinking outside the box and expanding their process knowledge? Are they recognized for helping coworkers explore new concepts? Or, are they only judged on getting assigned projects done quickly with the available technology?
  - By definition, specific novel ideas will not be listed on an annual list of expectations.
  - What are the rewards for spending time on risky hunches that do not succeed?
  - Creative ideas may not fit within the scope, budget, and timetable of existing projects. Technical professionals need some flexibility to pursue things outside of their daily chores.

- Manufacturing plants are hostile toward new products that require nonstandard equipment configurations and operating schemes. They would rather see the plodding engineer come up with an unimaginative solution that runs on the existing process. This gives the flexibility to run the product on any line and avoids the need for new training modules and evaluation procedures.

- The more revolutionary the idea is, the more frightening it is. A conservative organization will empower numerous people to kill or delay uncomfortable ideas.

- The research organization may be the only place to incubate nonstandard concepts. If there is a disconnect between the ideas that it selects and the solutions that the wider company need, then the implementation rate will be very low.

- Management, sales staff, and marketing personnel may react negatively when they see rough, unfinished prototypes. People are drawn to polished working models with a flawless performance. Perfection is generally only possible in the late stages of a project.
  - Failure is a recurring experience when evaluating new ideas.
  - When screening raw proposals, a different mindset is required, compared with deciding when a product is ready for field-testing on customers. In the first case, you try to judge future potential and in the second, you are concerned with influencing end users with a good first impression.
  - Blind alleys should be documented as fully as those that appear to succeed.
  - The absence of failures at the early stages suggests that the development is following a narrow, predictable path and is not thoroughly exploring the landscape.
- Different functional areas may be given a contradictory direction from top management. The R&D organization is told to pursue an out-of-the-box innovation. However, if manufacturing is focused on cost cutting, then it will be difficult to get those new products into production.
- Innovators are discouraged from refining new ideas when they perceive that the corporation will not pursue and implement the innovation. If the capital budget is tight, then the engineers will assume that no money will be available to pursue their ideas. If an alternate idea is being backed by politically astute managers, then researchers will discount their ability to overcome the opposition.
- There is only one thing more destructive to innovation than a risk-averse lawyer—an organization where everyone thinks like lawyers and no one commits to making a decision. The management must set a strategic direction and make decisions after listening to legal advice on a risk analysis.
- Sometimes, development work will reveal that the plan is flawed. The management must be willing to cancel or revise discredited projects and adjust its strategy.

Overcoming Systematic Barriers to Innovation

When great innovation appears, it will almost certainly be in a muddled, incomplete and confusing form.

Freeman Dyson

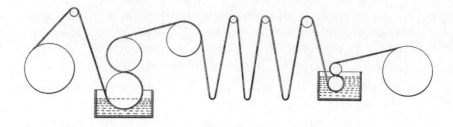

In 1923, Richard Drew was working on sandpaper development for 3M. While visiting auto body shops to test prototypes, he observed problems with masking surfaces during painting. Two-tone paint jobs for cars were very popular at the time. These required sharp lines of demarcation between the different colors. The adhesive strength of the existing tape products was generally too strong for this application. The variable adhesion strength often damaged the surface that the painter was trying to protect. This gave Drew the inspiration for a consistent product with a less aggressive adhesive, specifically for paint applications.

Following some initial failures, the 3M president lost faith in the tape project and ordered Drew to focus exclusively on sandpaper. Despite this directive, he continued to experiment secretly with the tape idea. Eventually, he built a pilot production process by making small equipment purchases that were below the approval limit. Once the prototype product had found success with test customers, his idea was embraced. Drew expanded his research from paper masking tape to clear cellophane tape for packaging and household applications. 3M went on to become the first name in tape products.

RULE #4 CREATIVE THINKING AND EFFICIENT PROJECT MANAGEMENT WILL ALWAYS CONFLICT

Divergent ideas never fit the critical path time line of a tightly focused engineering project. Their utility is variable and each one has a low probability of success. Innovation is an annoying distraction to the demanding job of running your process in the systematic way that delights your customers. However, without it, every company will inevitably stagnate. Cultivating a balance of revolutionary ideas with efficient implementation is the ultimate challenge for every process-dependent organization.

Techniques for balancing innovative ideas with operational efficiency:

- When proposing a totally novel approach, research the idea and its ramifications before formally presenting it. Take the time to create a clear explanation of what the idea is and is not. Know where it fits in the existing structure and what the conflicts are.

- Reveal the brainstorm to one person at a time. Use models, experiments, and cost estimates to demonstrate it. Be clear on what you want from the person—advice, support, or resources? Seek their participation by allowing the person to build upon the idea. Be patient, no one should be pressed to make a hasty decision, to abandon their comfort zone.

- Try to focus the discussion on how to test and prove the idea. Demonstrate how much damage the idea could do if a competitor were to commercialize it ahead of you. Look for synergies and needs that complement the existing infrastructure. Do not continue to add interesting features that render it increasingly incompatible. Do not oversell the idea by proposing it as the solution to every problem.

- When demonstrating the idea, show things that have been perfected to work multiple times in a row. Never allow the decision makers watch the failures during the learning process. Avoid making speculative predictions that will discredit the entire concept when one is proven false.

- When pitching the idea, be mindful of people's natural fear of uncertainty. Pose process improvements as an evolution that builds on the existing business, rather than a revolution that overthrows the existing order of things. Avoid using terminology that has bad associations with earlier product failures.

- Do not position the idea as an all-or-nothing proposition. Propose some calm niche of the business for an initial trial. Use that first application to demonstrate its benefits. Set your creation free by enabling its use by others, instead of trying to control it and taking credit.

- Carefully consider the format of your presentation when trying to sell ideas to people with nontechnical backgrounds. Avoid displaying equations, dense tables of data, and scientific jargon in the naive belief that it will bolster credibility. Stress the potential benefits in terms that your audience can relate to. Demonstrate the ability to solve customer problems or gain access to new market segments. Propose a tangible action plan to do so.

- Develop your idea into a toolkit that can be tested as the solution to a range of problems.

T. DAVENPORT.

Electric Motor.

Patented Feb. 25, 1837.

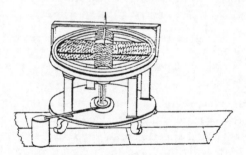

In the 1830s, American blacksmith Thomas Davenport was inspired to give up the ironworking profession and branch out into electrical engineering. He subsequently built and demonstrated one of the world's first rotary electric motors. After much struggle, he patented this revolutionary concept. However, in a time when the only electric power came from crude batteries, his devices had few practical applications. Industrialists were focused on the steam-power revolution instead.

Davenport sought advice from the leading electrical authority of the time, Joseph Henry, who had developed industrial electromagnets. Henry reportedly advised him to build his models on a smaller scale, because miniature units would be cheaper to construct. He felt that potential investors would expect full-sized demonstration units to work perfectly, but would have lower expectations for tiny versions.

Davenport's invention failed to attract commercial interest because it was too far ahead of its time. Due to the lack of a reliable power supply, motors were not ready for industrial or transportation applications. Had Davenport thought to run his invention backwards, using a water wheel or a steam engine, the situation would have been different. He would have invented the electric dynamo and solved the problem of obtaining reliable electric power.

In the early 1950s, both sides of the Cold War were seeking to develop guided missiles to defend against enemy aircraft. The most difficult application was for lightweight units that could be carried on fighter planes. The U.S. Air Force was focusing on radar as its targeting system. However, a small development group at the China Lake Naval Ordnance Test Station felt that infrared radiation was more appropriate. This organization struggled to sell its ideas to the weapons bureaucracy to get enough funding for a working prototype.

It tinkered with the individual elements that controlled the performance of its system. The best components were put together to form a "dog and pony show" for visitors, to sell the viability of its concept. An infrared detector was mounted on a radar antenna and was wired into the actuators that controlled its movement. The guests at the base were amazed that the entire assembly would track their movement by the emissions of a glowing cigarette ember. Other elements of the system were mounted on bench racks for demonstrations. Over time, the decision makers were impressed by the performance of these prototypes. Because the team had answers for the critical aspects of guided missile functionality, the Sidewinder program was funded.

Developers in industry must address all the important concerns of their targeted applications. Performance must count for more than presentation skills or the beauty of the graphics.

Encouraging innovation:

- Constantly evaluate alternatives to existing practices in light of the commercial environment, competitive pressure, and technical trends.

  - Radical change is often triggered by technology that is well known, but was previously constrained in this application. Make everyone aware of the needs.

  - Resource a portfolio of investigations ranging from comfortable, modest improvements to scary, paradigm-shifting leaps of faith.

  - A large investment in incumbent process technology discourages the consideration of alternative options. Be clear on your willingness to embrace alternatives.

- Not every project runs on the fast track with an urgent timetable. Use a different mindset for exploratory projects. Consider at the outset whether they need to build on the existing system or be separate from it. Staff and manage the projects accordingly.

- Avoid hasty negative decisions. If there is any potential, let the advocate have some time and resources to flesh out the idea. Even if it does not demonstrate utility for your current situation, file it away for future use. Reward people who take the initiative to propose and pursue ideas outside your comfort zone.

- Never compel employees to do exploratory research or to brainstorm ideas that they dislike. Sometimes they will become believers. More often, they will go through the motions and generate a list of negative observations. Rather than giving the assignment to an unmotivated genius, you are better off giving it to a passionate plodder.

- Never let innovation be the first casualty of budget tightening. It is very easy to devote 100% of your technical resources to short-term goals at the expense of long-term thinking. By their very nature, creativity and unstructured exploration rarely address immediate needs for cash flow. However, this operating mode will only initiate a downward spiral when the development pipeline is emptied of new proposals. Difficult financial situations call for different thinking in all operational aspects, not just tighter controls.

- Create a separate organization to implement commercially useful proposals that are incompatible with the larger organization. Your competition will have different constraints. Patent misfit concepts, even if you cannot employ them.

- When an organization discovers a promising new tool, there is a strong temptation to overemphasize it, at the expense of other options. Successful ideas generate overconfidence when people extrapolate them into untested areas. This practice can backfire and create a negative impression when misguided applications fail to live up to expectations.

- Few good ideas will suit today's priorities. No one manager can know the needs of every corner of the company. Never permanently delete any idea. Post it to a list of suggestions for solving problems. Include commentary on why these proposals were not implemented. Make the list widely accessible. Develop a toolkit of solutions that can be tested where needed.

- One person's creative ideas have no more validity than anyone else's. Managers and experts who push their own ideas over all others will demoralize employees and discourage creativity.

When lithography was invented at the end of the eighteenth century, it was the biggest advance in printing technology since Gutenberg. The process is still used today and continues to evolve. The original technology employed a porous limestone printing surface. By applying a greasy pattern to the stone, its surface properties could be selectively altered to create hydrophilic and hydrophobic areas. Parts of the surface accepted an ink coating and others repelled it. The inked areas transferred an image to paper that was pressed against the stone.

This process was very flexible because an artist could apply a graphic design directly to the stone, which could be used to make a print job shortly after. This innovation was not the product of specialists in the printing or scientific fields. Alois Senefelder was an actor and playwright who had difficulty getting his work published, due to the high cost of printing. Despite his lack of a technical background, he was able to perfect the process, patent it across Europe, and publish a practical guide for using the technology.

> *Critical Questions*: How do we focus the company's development effort to best deliver profitable products? Should we concentrate on incremental improvement, radical change, or a leap to a completely different market? At what point should we modify a successful strategy to adapt to a new commercial reality? Should I listen to that prophet crying out in the wilderness or today's bottom-line profit?

Needs Bank and Opportunity List

Chance favors the prepared mind.

Louis Pasteur

$$\underset{F}{\overset{F}{C}}=\underset{F}{\overset{F}{C}} \longrightarrow \left[\begin{array}{c} F \ F \\ | \ \ | \\ C\text{--}C \\ | \ \ | \\ F \ F \end{array}\right]_n$$

In 1938, researchers Roy Plunkett and Jack Rebok were working with fluorocarbon gases in a series of refrigeration experiments. The pair was employed by Kinetic Chemicals, a joint venture between DuPont and General Motors that was formed to develop and commercialize chlorofluorocarbon refrigerants.

In the course of their work, Plunkett and Rebok filled a pressure cylinder with tetrafluoroethylene and HCl. Later, they were puzzled to find that it had mysteriously "gone bad." No gas would come out when the valve was opened. Upon investigation, it was discovered that the contents had polymerized into a waxy white solid. The new plastic was inert and insoluble in all available reagents. Plunkett's 1939 patent application speculated that the material would be useful because of its acid resistance combined with its ability to be molded into useful shapes. The nonstick properties of Teflon coatings were still in the future.

At the time, there were no commercial applications because the production cost was prohibitively high. In a chance conversation, the polymer came to the attention of a secretive government research project during the war years. This group needed a gasket material that could resist exposure to a corrosive uranium hexafluoride gas. These gaskets were vital to the process of enriching uranium 235 for the Manhattan Nuclear Bomb Project. In this application, material costs were irrelevant.

Development work on the material passed to DuPont, where Robert Joyce patented improvements on the process, increasing the production rates and yields. His 1942 patent application was not published until 1946 because the material was declared vital to national security. Joyce's work suggested that Plunkett's accidental polymerization of one tank of tetrafluoroethylene was initiated by oxygen contamination in that cylinder.

RULE #39 PEOPLE RARELY SOLVE PROBLEMS THEY ARE NOT AWARE OF

Things to look for:
- --------------
- --------------
- --------------
- --------------
- --------------
- --------------
- --------------

An organization should keep a list of needs, such as process issues, desired material properties, unmet performance requirements, and customer requests. Employees across the company should be watchful for potential answers. These insights might come unexpectedly in vendor meetings, trade shows, laboratory experiments, and customer interactions.

Some items on the list will be hypothetical future products requested by a business group. These opportunities have not been accepted as projects because the enabling technology is not available. Other items may describe process shortcomings, product deficiencies, or customer problems. Entries on such lists must be as detailed as possible. Constraints need to be listed along with failed approaches. Potential contributors must know who to contact. They should also know what the resolution is worth to the organization. Not all the entries will be detailed. Some could be casual questions that might help employees understand the business and its technology. These might be similar to the frequently asked question (FAQ) section of a website. One idle question may get an expert thinking in a new way.

Some entries ask for an innovation that will turn the whole industry on its ear. Employees should be free to post answers to the questions. Some will explain why the request is physically impossible or how it violates a basic law of nature. A more flexible mind might reject the conventional answer and propose a wild-card approach that employs a novel approach to remove current constraints.

Some companies use specialized commercial websites to make their requirements known to the world at large. Monetary rewards are offered to anyone who can propose a remedy. Potentially, millions of technologists are willing to ponder the predicament in their spare time for a potential payback of a few tens of thousands of dollars. Such offers must be posed in such a way that the business strategies are not revealed to competitors. Avoid associating the company with requests that appear to undermine the market positions of your customer base.

Just as important as the needs list is a parallel database of answers that are looking for a question. Clever technical tricks and shortcuts appeal to the intellect. It is important to share these keys to the process within the company. One old idea may be combined with a new insight to solve a problem that did not exist when the original brainstorm was banked in the database.

The late Peter Drucker recommended that every professional write a monthly memo listing every surprising thing that they had encountered during the previous month. Scientists are always looking for unexpected results that violate the rules. These exceptions create the next breakthrough.

Maintaining and updating the database are tedious. However, lists of needs, industry contacts, capabilities, and surprising findings are critical elements of your span of knowledge. The challenge is not in creating lists, but in indexing, referencing, and maintaining them to allow ideas to be found quickly. It can be a challenge for developers to access all the diverse needs and knowledge in global organizations.

Summary

- Industrial innovation requires more than generating a good idea. Most concepts need prototyping and testing before the best option is selected from among many options.

- Engineered products require cross-functional cooperation to innovate because the necessary information and capabilities are spread across the company. Separating employees and resources with physical distance or departmental politics inhibits innovation.

- Wide exposure to diverse technologies prepares developers to solve complex problems.

- Time, discussion, and experimentation are required to weigh options and alternatives when trying to solve customer problems with a dynamic material production process.
- Traditional brainstorming is a useful tool, but it should be tailored to fit your situation. Informal discussion between passionate collaborators is often a more effective technique. Rigid procedures and filtering discourage idea generation and refinement.
- Organizations inhibit creativity and risk taking with a culture that does not tolerate failure. Anticipating what success should look like is not helpful.
- The more revolutionary the idea is, the less comfortable people are. A conservative organization will empower people to kill or delay ideas that scare them.
- Proximity to the action—the customer or the process—creates better decisions.
- Question the assumptions behind project objectives, if only to better understand them.
- Efficiency and creativity will inevitably conflict, but both are essential elements for developing and producing process-dependent products.
- Maintain a list of needs/problems looking for solutions so that the entire organization is aware of them. Keep another list with solutions in need of a problem.

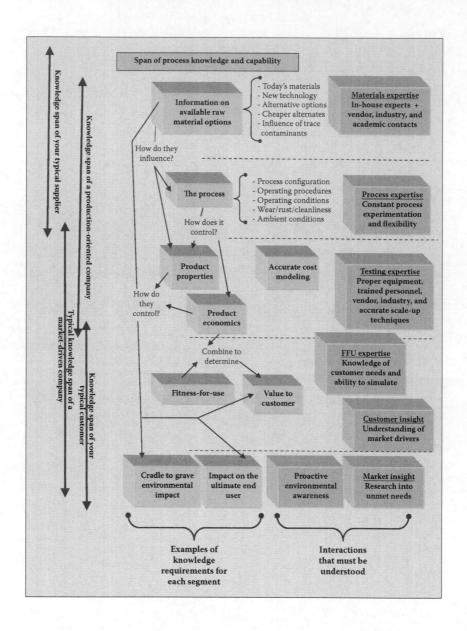

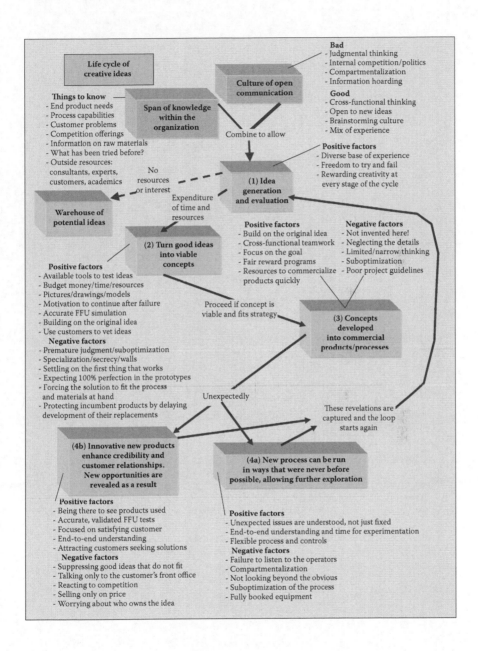

Life cycle of creative ideas

Things to know
- End product needs
- Process capabilities
- Customer problems
- Competition offerings
- Information on raw materials
- What has been tried before?
- Outside resources: consultants, experts, customers, academics

Span of knowledge within the organization

Culture of open communication

Bad
- Judgmental thinking
- Internal competition/politics
- Compartmentalization
- Information hoarding

Good
- Cross-functional thinking
- Open to new ideas
- Brainstorming culture
- Mix of experience

Combine to allow

Positive factors
- Diverse base of experience
- Freedom to try and fail
- Rewarding creativity at every stage of the cycle

No resources or interest

Expenditure of time and resources

(1) Idea generation and evaluation

Warehouse of potential ideas

(2) Turn good ideas into viable concepts

Positive factors
- Available tools to test ideas
- Budget money/time/resources
- Pictures/drawings/models
- Motivation to continue after failure
- Accurate FFU simulation
- Building on the original idea
- Use customers to vet ideas

Negative factors
- Premature judgment/suboptimization
- Specialization/secrecy/walls
- Settling on the first thing that works
- Expecting 100% perfection in the prototypes
- Forcing the solution to fit the process and materials at hand
- Protecting incumbent products by delaying development of their replacements

Positive factors
- Build on the original idea
- Cross-functional teamwork
- Focus on the goal
- Fair reward programs
- Resources to commercialize products quickly

Negative factors
- Not invented here!
- Neglecting the details
- Limited/narrow thinking
- Suboptimization
- Poor project guidelines

Proceed if concept is viable and fits strategy

(3) Concepts developed into commercial products/processes

Unexpectedly

These revelations are captured and the loop starts again

(4b) Innovative new products enhance credibility and customer relationships. New opportunities are revealed as a result

(4a) New process can be run in ways that were never before possible, allowing further exploration

Positive factors
- Being there to see products used
- Accurate, validated FFU tests
- Focused on satisfying customer
- End-to-end understanding
- Attracting customers seeking solutions

Negative factors
- Suppressing good ideas that do not fit
- Talking only to the customer's front office
- Reacting to competition
- Selling only on price
- Worrying about who owns the idea

Positive factors
- Unexpected issues are understood, not just fixed
- End-to-end understanding and time for experimentation
- Flexible process and controls

Negative factors
- Failure to listen to the operators
- Compartmentalization
- Not looking beyond the obvious
- Suboptimization of the process
- Fully booked equipment

4

Finding Product Opportunity

Which concepts should we develop into products? And, more importantly, which product opportunities can be ignored or deferred? Why must the character of the company dictate the R&D organization's structure and choice of projects?

> You can't just ask customers what they want and then try to give that to them. By the time you get it built, they'll want something new.

Steve Jobs

In the early 1970s, Philips Electronics led the race to create the digital storage of music and video. Laser technology allowed rapid access to vast amounts of digital data on rotating surfaces, compared to the incumbent analog methods. Commercialization of a consumer product was initially inhibited by the size, power requirements, and cost of the laser equipment. As their size and price dropped, digital audio systems became commercially feasible for consumers.

The Philips researchers worked with 300 mm diameter glass disks, to store enough data for full-length movies. In the music arena, stores were configured to display analog 300 mm LP record albums. Marketers were accustomed to the ample graphics display space on the cardboard sleeve. Stereo system cabinets were designed to accommodate a top-loading turntable and to store the large albums. In short, the entire music industry infrastructure was geared around large disk media.

Due to the dense storage capacity of the digital format, a 300 mm disk could hold hundreds of songs, rather than the 10 to 20 songs on a typical record album. From a technical perspective, this was quite an impressive achievement. The music industry was less enthusiastic about that approach. They could not imagine a business model to sell prerecorded music in such large units.

When Sony and Philips collaborated on a standard format in 1980, the compact disk (CD) that we know today became a reality. The 300 mm prototype shrank to 115 mm (later adjusted to 120 mm). The music recording companies could sell albums with the traditional number of songs. Seventy-four minutes of capacity was the benchmark, based on the length of Beethoven's 9th Symphony. Retail stores were compelled to redesign their display space and find new ways to prevent shoplifting. The designers of album cover art downsized their work to accommodate the new packaging. Stereo systems shrank in size and no longer required special furniture.

The pioneers of revolutionary new concepts cannot allow paradigms to dictate all of the product features. Carefully consider which product attributes

exist because they have value and which are artifacts of incumbent technology. Never ignore valuable new approaches just because there are compatibility objections. This is especially true when those issues are internal to your organization. Competitors are not constrained.

Radical change within a market segment is often blocked by technical barriers, such as the early laser equipment. As these limits are removed, a stagnant concept suddenly becomes viable, allowing the entire industry to change quickly. It pays to keep track of these constraints so that advances

RULE #49 PREDICTING THE OPTIMAL NEW PRODUCT IS IMPOSSIBLE

The questions of which new products to develop and what properties to optimize are critical issues for engineered materials companies. The picture is complicated by long development cycles, the sizable investment in production equipment, and the uncertainty of what the market constraints will be in the future. It is important for early production processes to be flexible to accommodate future situations. As the market grows, equipment becomes more specialized to serve niche applications.

in one field will enable innovation in an unrelated market. Include them on your opportunity list database.

Development strategy questions:

- Should the company develop new products by systematically studying the market needs and then seeking the appropriate materials and process technology to provide a solution?
  - Is the search limited to the needs of existing customers?
  - Which market segment should we focus on?
  - What are the market's unmet wants and needs?
- Should the company restrict the search only to materials that the existing production equipment can produce with optimal efficiency?
  - Will those new products interfere with the existing grade slate?
  - What resources will be needed to market those products into new outlets?
- Should the company invest in new or modified process equipment to commercialize interesting materials that the R&D department has discovered?
  - What if the market is very small at the outset because there is no incumbent solution?
  - Is it feasible to service only selected customers with limited production from a pilot plant line until the opportunity increases?
  - Can the modified process make our current products too?
  - Will the market select another option while we try to decide?
- Is it wise to move into similar end uses that require a higher or lower value than the incumbent market?
  - Low-margin products require minimal overhead and direct manufacturing cost.
  - Higher-value products may require significant overhead expense to service the market.
  - Can the company be a supplier of both high-value and low-cost products at the same time?

No two companies are the same. A brilliant strategic opportunity for one organization may be disastrous for its direct competitors. Take a close look at what sort of company you have (or want to become) before considering what product strategy to follow.

Market-Driven Organizations

> Spend a lot of time talking to customers face to face. You'd be amazed how many companies don't listen to their customers.

> **Ross Perot**

By the late 1970s, 3M saw office supplies as one of their key market segments, anchored by the tape products originally developed by Richard Drew. The office supply division was intently focused on meeting the needs of their customers. For several years, they had poured resources into the traditional marketing of their Post-it Note concept, but found little interest. Many free samples were sent to businesses, but the office workers did not see much use for the product in their daily job; they were not seeking such a solution to their paperwork needs.

The 3M marketing executives felt that the product had many unique advantages, if only they could communicate them to potential customers. To do so, the entire office division sales force and a hoard of temporary workers were deployed on an intensive campaign in Boise, Idaho. They gave away samples, demonstrated the product's many uses, and got customers fired up about sticky notes that did not deface important documents. In doing so, sales mentally shifted gears from defending their new product to evangelizing it. Unlike the variations in adhesive tape, this product could not be explained in a brochure. The prospective users needed to see it work before the light bulb went off. Market-focused organizations must be prepared to employ every means at their disposal to sell their value to their designated market.

RULE #7 ALL PROJECT DECISIONS ADDRESS CUSTOMER NEEDS AT MARKET-FOCUSED COMPANIES

Market-focused companies seek projects that satisfy the perceived needs of their selected end users. If they lack the technology or expertise to address those opportunities, then such companies acquire whatever is required. A full range of products, services, and advice are available for the target clients. Ideally, a portfolio of potential solutions is maintained to service future customer needs.

The product and process development decisions of a market-focused company should always be dictated by the needs of the existing and potential customer base. These marketing considerations must drive all the other operational functions within the company, including the choices of which projects to pursue and which process technology to acquire. This is not to say that product maintenance or manufacturing cost reduction ideas are not worthwhile. However, when evaluating them, the market manager should agree that the efficiency improvements in the process justify the possible impact on the end user.

Market-focused companies cannot tolerate the second-guessing of management's strategic direction by big egos in research or manufacturing. These groups must be searching for ways to tailor the process toward the market needs identified by the product managers. If this is not feasible, then they should be attempting to acquire more appropriate technology.

Marketing must completely understand the market requirements and the long-term trends that are driving the customer's business. They cannot just react to the whims of a customer's purchasing agent. This knowledge will drive financial decision making, which, in turn, will dictate where the development resources are focused.

When companies are market focused, they must constantly add new skillsets and core competencies to the R&D and manufacturing organizations. These companies cannot assume that the current products of their process equipment will be sufficient to meet customer needs. New capabilities must constantly be found to satisfy fluid market requirements. By perfecting new solutions before customers know they need them, as with Post-it Notes, they will gain market advantage. By exploring and patenting alternative process technologies, they can preclude competitive threats.

In 1893, Frederick and Louis Rueckheim devised a popular new snack food for the World's Fair in Chicago. It consisted of peanuts and popcorn coated with molasses. Process refinements improved the product over the next several years. The brothers developed a tumble-blending process to coat the product with a small quantity of food-grade oil. As a result, the product was free flowing, rather than sticky and clumpy, making the confection easier for consumers to dispense, share, and eat.

To facilitate distribution, Henry Eckstein developed a moisture-barrier package to act as a protective liner for the Rueckheims' product. In doing so, he created the flexible food packaging industry. The "Eckstein Triple Proof Package" was dustproof, germproof, and moistureproof. This wax-impregnated paper and the associated packaging system allowed the company to distribute Cracker Jack packages on a national scale. Without protective packaging, the snack would spoil quickly.

This new approach created the first shelf-stable, packaged snack food that could be nationally branded and advertised. Their Cracker Jack caramel-coated popcorn and peanut mixture is still sold today, as a small facet in a huge industry. At every stage, the company's developments were aimed toward servicing a market segment that had not existed until they pioneered it. A customized product formulation, specialized process techniques, and appropriate packaging were all critical to meeting the market needs.

Manufacturing/Process-Centric Approaches

Never worry about theory as long as the machinery does what it's supposed to do.

Robert A. Heinlein

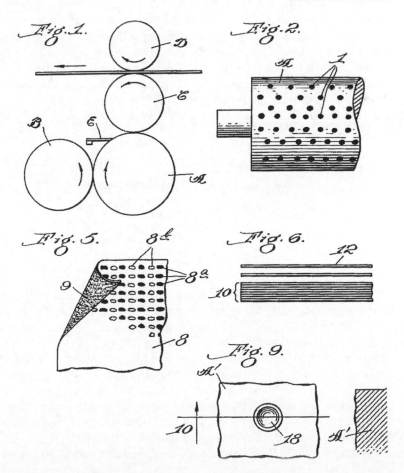

In 1913, Herbert Faber and Daniel O'Conor founded the Formica Corporation, after quitting their jobs at Westinghouse. Their only asset was an idea for making a laminate of paper and phenolic thermoplastic polymer. Their process involved saturating kraft paper with a liquid mixture of phenols and formaldehyde. Layers of the coated sheets were then compressed between polished sheets of heated steel. Time, heat, and pressure partially cured the resin and fused the composite together. Afterwards, the sheets were baked to drive off moisture and complete the curing process.

The end result was a thin, rigid sheet with a smooth, hard-surface finish. The company sold this product into a wide variety of different applications for the next 10 years. The 1913 patent application primarily focused on insulation for the electrical industry. At this stage, the Formica Corporation was a process-focused company, which would sell to any and all end uses that needed their product.

By the mid-1920s, the product found wide success in countertop applications. Formica was easy to clean and it resisted cigarette burns. This end use was so compelling that the business transformed into a market-driven company. Surface designs such as wood grain were incorporated into the surfaces to boost bar and restaurant installations. The need to satisfy the market drove a material shift to melamine resins in 1937. This allowed the firm to offer bright white counters, which appealed to consumers and were cheaper than alternatives such as marble. In 1957, the company hired avant-garde designer Raymond Loewy to create decorative surface patterns that were stylish and fashionable.

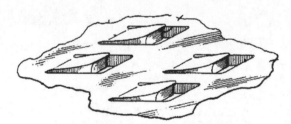

In the 1980s, Grace Manufacturing was a small metal-finishing operation that made components for computer printers. The market for that class of printers was in decline and the company was seeking new customers to stay in business. Their expertise was a metal-etching process that produced very sharp edges as an unintended consequence. This suggested that the company could reconfigure its plant to make wood-shaping tools such as planers and blades.

It patented this approach and went into competition with companies that created the same effect by stamping a raised cutting edge with a mechanical process. Grace's etching process resulted in recessed cutting edges, which were sharper and smaller than those made by the competition. The company managers were surprised when end-user innovation started to bring them applications in food preparation functions. Chefs found that Grace's plane created tiny shavings of hard cheese, rather than the larger chunks produced by stamped blade technology. Soon, more kitchen applications were discovered for these versatile tools.

As a process-based operation, it was willing to adjust its manufacturing operation to serve any market that found a benefit from its technology. Medical applications soon followed when orthopedic surgeons discovered that Grace's cutting surfaces could delicately adjust bone surfaces inside the human body.

RULE #8 PROCESS-FOCUSED COMPANIES DO NOT CATER TO ANY ONE MARKET

Process first

Companies driven by process capabilities and technology have the opposite mentality to those that are market focused. They seek end users to fill their production capacity. New products must fit either the existing facility or a new process that springs from their technical core competency. Customer requests must be rejected if they do not suit the process. Development looks inward at the process, rather than outward at the marketplace. The sales force must constantly be looking across many markets for unmet needs where the properties of your materials offer solutions.

Early in the twentieth century, Corning Glass Works was focused on glass production for a wide range of end-use applications. In 1912, they introduced a specialty grade of borosilicate glass. This high-tech material was used in abusive lighting applications, which required the product to be transparent, heat resistant, and virtually shatterproof. The primary market for this technology was in transportation applications, such as headlights for locomotives and assorted signaling devices. The new product

was so durable that the market for replacement railroad lanterns almost disappeared.

The following year, the wife of a Corning executive complained about the poor performance of conventional glass bowls in cooking applications. Within two years, the company had modified the borosilicate formulation (reportedly to remove heavy metals) and launched Pyrex brand cookware. Initially, consumers were skeptical, because glass had such a poor reputation for temperature resistance. Slow customer acceptance could be tolerated, because the company had a diverse product portfolio. By 1927, Corning had sold 30 million pieces of Pyrex. As with the Formica example, Corning adapted their material with attractive designs to make it more appealing to consumers as the market matured. The division became market focused in later years, while still part of the glass company.

Capital-intensive manufacturing facilities can be either a burden or a benefit depending on your outlook. A process-focused company maximizes the value of its installed capital, rather than just serving the needs of a particular market segment. The cost of duplicating the production capacity is a barrier to entry for competition. The workforce's process knowledge is a vital core competency. These companies will modify a process to better meet the market's needs. However, any such change must be weighed against other alternative uses for those resources and their ability to run the process wheel.

Sales/marketing in a process-driven organization: Be opportunistic with pricing when trying to sell out the capacity of a line or a plant. The costs may fluctuate considerably if plant utilization levels are not stable. Low-margin business may be sought to fill today's production schedule, and then rejected the next month to make room for profitable orders from other customers. Be prepared to reschedule delivery dates for orders while the production schedule is optimized in response to changing order patterns and process upsets. Find out early in the selling cycle whether the customer expects product customization and intensive service. The sales compensation package should incent salespeople to concentrate on finding high-margin applications. Consider partnering with market-focused companies that can sell your material as part of their portfolio of solutions.

Be prepared to provide samples of products that the process can make, but for which no orders currently exist. Keep a database of properties, availability, and potential sales prices for such offerings, when making presentations to prospective customers. Market research involves watching for these new applications, which dictates how the process needs to evolve to serve the future order mix.

R&D focus: Debottlenecking, alternate raw material qualification, and process flexibility enhancements may be the most important goals for the development group in a process-driven environment. Product innovation and application development are often delegated to the customers, who better understand their own markets. When evaluating new applications, R&D needs to determine how to reconfigure the process to service the need. Projects should be rejected if they fail to suit the process. New applications should not be seriously entertained until it is determined that their fitness-for-use (FFU) range lies within the process capabilities. Market-driven competitors may approach that decision from the opposite direction.

Balanced Approach

> I just invent, and then wait until man comes around to needing what I've invented.
>
> **R. Buckminster Fuller**

Four decades after the Pyrex success, Corning was still a company focused on glass technology. However, their kitchenware division was a market-focused organization, which tailored their Pyrex products to the needs of their consumer base. In their research laboratory, Dr. Donald Stookey was experimenting with a photosensitive glass. He was tasked with improving material properties that might be useful for computer and television applications.

In the course of one experiment, the oven overshot its temperature setting and ruined his test. He was cleaning up the resulting puddle of scrap material when it slipped from his tongs and fell to the concrete floor. Unexpectedly, this normally brittle ceramic did not break. This accident revealed that high process temperatures greatly improved the durability of what became known as pyroceramic glass. Stookey had the expertise to determine that the high level of nucleation in his sample created an unexpected crystalline structure during the temperature excursion. This finding was exploited to refine the material properties to suit a variety of potential end-user needs.

Once again, Corning found a consumer use for a material that was originally intended for an industrial application. In this case, commercialization was much easier, because there was a market-focused group in the company ready to evaluate and exploit the opportunity. The broader company found a variety of niche applications for Stookey's discovery. However, it is best known as CorningWare, and is still widely used as bakeware in kitchens worldwide, often in conjunction with Pyrex lids.

Corning was pursuing basic research into ceramic and glass technology while also marketing their earlier discoveries to the appropriate market segments. This is a balanced or hybrid model employed by some large companies that seek to diversify. But, unlike a conglomerate, the chance discoveries in one area can have useful applications in another division.

In 1953, Dr. Daniel Fox was working in a General Electric (GE) research laboratory. He was assigned the task of finding an insulating material for magnetic wire that would resist high temperatures under humid conditions (a material improvement project to support GE's electrical equipment business). While experimenting toward this goal, he accidentally rediscovered polycarbonate plastic, which had previously been synthesized in Germany in 1902.

Unlike the German chemists, Fox saw that this polymer had unique properties that were sure to have applications, although not in the power generation field. The GE management saw the potential to create a plastics division and follow DuPont's business model. A new business development group took the product and started seeking customers. Its job was not quick or easy because polycarbonate processing techniques and end-user applications had to be developed along with the various polymer grades. Eventually, however, the material's exceptional attributes guaranteed that end uses would emerge.

GE followed a hybrid approach that did not rely on the process/manufacturing approach of trying to find markets for its existing capacity. Neither did it choose a market segment and devote its energy to servicing those customers. Instead, it grew its business from a new technology, which required it both to cater to customer needs and to build an efficient process operation.

RULE #14 THE BALANCED APPROACH IS HARDEST

An organization with a balanced approach must master a very broad span of knowledge. It marries an appreciation of customer needs with cutting-edge research. Forming a strategy in this middle ground is more complex than forming one at either of the two extremes. New products can spring either from a market need or from a new research discovery. This creates a conflict because the approach for commercializing each situation can be very different. Overhead costs are generally higher.

Marketing in the balanced model:

- Isolate existing, market-driven groups within stand-alone business units. Create customized policies and procedures tailored to their needs, instead of corporate guidelines.

- Seek customers who can obtain the best value from your technology. These are rarely all concentrated in one convenient market. No one sales strategy will suffice.

- Some new customers will be located in places not served by your distribution capability.

- Build relationships with partner companies to sell integrated systems of equipment, software, services, or related products with yours to serve individual market segments.

- Hire sales and marketing people with skills and backgrounds appropriate for the markets you need to pursue. Do not assume that they should be clones of those who succeeded in the past.

- Organizations in constant flux must restructure their sales organizations in accordance with the changing technology landscape. Do not allow the established infrastructure to anchor your company to an inappropriate customer relationship strategy.

  - Create separate sales teams to focus on the new markets. These individuals may require a different compensation arrangement and management structure from those required by the current organization.

  - Insulate long-term customers from your changing focus on new opportunities. Dedicate salespeople to serve their needs and act as a long-term point of contact with the organization. People do not like constant change to their supplier interface.

Technical goals for balanced companies:

- Stay ahead of the industry and be the technology leader. Offer compelling product attributes and constantly enhance your capabilities. It is not necessary to offer the most elegant idea—just be first to market with new materials that define expectations.

- Constantly refine and improve your products to build a cutting-edge reputation. Customers will not jump onto a custom platform unless it promises compelling benefits now and in the future.

- Structure for a wide span of knowledge and capability. Communicate and cross-train people to understand those they could interface with. Each individual discipline cannot be an island of self-directed activity. Minimize the degrees of separation between nodes across the enterprise.

- Protect technology with patents, copyrights, and customized features for individual customers. Do not allow low-cost competitors to quickly copy and undercut your position.

- Be prepared to acquire new production facilities that are appropriate to the new products, technology, and markets. Close, upgrade, or sell facilities as your product mix evolves over time. Resist building large, integrated complexes that limit your flexibility to react to technology progression. These are the hallmarks of a process focus.

Balanced organizations share many common elements with both market-focused and process-centric companies, as seen in the R&D focus diagram below. A company's strategy will shift as its profit margins change over time. No one approach is right for every situation. R&D functions must suit the character of the organization that they serve.

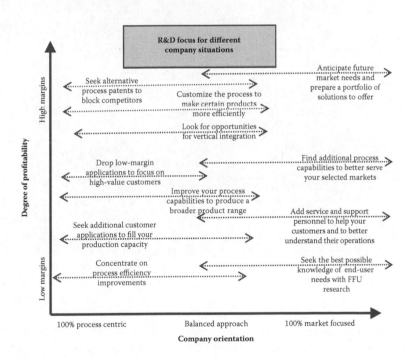

Paradox of Product Evolution

It wasn't raining when Noah built the ark.

Howard Ruff

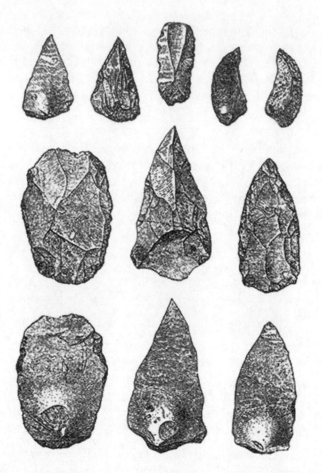

The evolution of new tools or concepts throughout human history often follows a common pattern. In the beginning, there is a huge diversity of designs or concepts as developers struggle to find effective solutions. Progress is constrained by material scarcity or process limitations. No one option is acceptable to all end users.

Early humans had numerous different concepts for spearpoints as they chipped rocks, shaped bone fragments, and sharpened wooden shafts. These early solutions were sufficient to keep our species from extinction. Eventually, one dominant technique for flaking glassy rocks such as flint into razor-sharp points was discovered, perhaps multiple times. This Paleolithic process created the optimal spearpoints, especially for hunting big game during the Ice Ages. Flint became precious, often being transported long distances from outcrops to the point of use.

Archeologists believe that a process to further refine rocks was developed by our ancestors as long as 164,000 years ago. Innovative cavemen discovered that heating certain rocks to 350°C for 5–10 hours altered their crystalline structure. Otherwise useless rocks, such as silcrete, could be worked to a sharp edge after such treatment. The resulting engineered material could be more precisely worked into tools than untreated rocks. These spearpoints were so effective for hunting big-game animals that the technology spread swiftly to widely different geographical sites.

Later, proficient toolmakers evolved the successful design into a wide variety of new variations. These were customized for hunting different animals, cutting vegetation, and for the unique conditions of each location. Spearpoints were downsized into arrowheads and upsized into specialized cutting tools.

The pattern was repeated in the automobile evolution. In the early days, there was a diversity of incompatible models powered by steam, electricity, and internal combustion engines. None of these designs was optimal or affordable to the mass market. Quite suddenly, the roads were dominated by a very few mass-produced models around 1920, because the assembly-line process was so efficient. During the decades that followed, the variety of vehicles increased steadily as different manufacturers sought to differentiate their products and create market niches. Production evolved to incorporate brand customization. Despite the differentiation, all these vehicles retained the same fundamental elements as the original Model T.

The nature of product development changes as the product platform matures. In the early stage of a product, many different technologies and processes may be employed to satisfy the same application needs. Different companies will have widely varying market size estimates depending on what fraction of the problems can be addressed by their solutions. The wide acceptance of one solution by a large number of end users will suddenly define a need that was previously described in many different ways. Few consumers felt that they had home computing needs in the early 1980s. However, within 15 years, most of them owned desktop computers. Within another decade, the market had proliferated microprocessors and their functions into a range of specialized formats, such as pads, smart phones, embedded controllers, and entertainment systems.

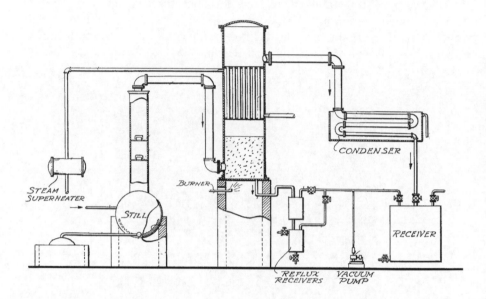

Traditionally, the production of high alcohol content beverages was only accomplished with a distillation process at atmospheric pressure. There is a limit on the alcohol content that can be achieved by fermentation, because the alcohol is toxic toward the organisms that produce it. With distillation the alcohol content can be increased beyond the 12%–15% level. However, this technique tends to destroy heat-sensitive flavor components.

Recently, small vacuum distillation operations have started to boost the ethanol content at reduced temperatures compared with the old methods. This offers the opportunity to retain tastes and aromas in liquors that had previously not been feasible. As a downside, this process will produce a different tasting product than the traditional techniques. While this method opens up a world of possibilities, duplicating the original product is not one of them.

There are numerous historical examples of a business leader forecasting a very limited market for an innovation, which later became widespread. The error occurs when only today's unmet needs are considered. Once a solution drops in price, new applications are uncovered by end users who are constantly looking for better ways to solve their problems.

All innovators need to be mindful of the stage of maturity in their target marketplace. Those who create a niche position in an immature market risk becoming obsolete if a mass solution sweeps the industry. This may be a disaster if they have made a large investment in a specialized production process that has few uses. However, niche applications may be very appropriate when some customers seek to diversify away from a mass-market solution that is adequate, but not optimal for their situation.

What to Look for in a Product Opportunity

> The beginning is the most important part of the work.
>
> **Plato,** *The Republic*

The synthetic leather product Corfam was introduced by DuPont at the 1964 World's Fair following several decades of development work. The company believed that population growth would outstrip global leather production, creating a market for an expensive synthetic material. The enthusiastic consumer acceptance of nylon for women's stockings encouraged the company to pursue other apparel opportunities for polymers. The shoe industry was the largest consumer of quality leather, so it was the targeted application. On the positive side, their product was more durable than natural leather and retained a better surface finish under heavy-use conditions. However, it lacked the breathability and the conformability of the animal hide products. As a result, shoes made from Corfam were less comfortable than those made from traditional materials. The only reliable footwear application was in the military, where high gloss and abrasion resistance are prized.

The product was discontinued in 1971, resulting in a $100 million write-off for DuPont. At the high end of the market, Corfam's lack of comfort was a strong negative. Consumer acceptance fell short of projections. Leather prices failed to escalate as predicted by market research. At the low end, PVC footwear alternatives were considerably cheaper. DuPont's experience with nylon turned out to be irrelevant. Nylon had been considerably less expensive than silk and it was very functional in stocking applications. However, with Corfam, DuPont engineers were unable to significantly reduce costs as they scaled up a complex production process.

In 1979, Exxon announced the $1.24 billion purchase of Reliance Electric, a move that surprised the stock market. An engineering group within Exxon had invented an alternating current synthesizer (ACS), which it believed could drastically reduce the energy consumption of electric motors. However, it did not have the resources to develop the idea into a finished product. The company bought Reliance's production and distribution network to facilitate commercialization of the technology. Exxon believed that their discovery would be enormously profitable, so they did not consult with independent experts. After the merger, Reliance engineers discovered that the concept was not so useful. Exxon could not scale up the prototype design into a reliable and cost-effective commercial product. The company sold Reliance in 1986, after disappointing losses.

At both Exxon and DuPont, the R&D organizations developed solutions that the company had no experience in marketing or manufacturing. The top management acted on faith, based on optimistic projections. Both companies survived because they were profitable enough to absorb their failures.

There is no one unifying theme in brilliant new products. Sometimes an innovative concept falls flat on its face. At other times an old idea is repackaged and marketed well or it catches a new trend and goes on to become an iconic product. One visionary may gain acclaim for having foreseen a best-selling item when no one else believed in it. For every smash hit, there are scores of failures or even marginally successful launches that never pay back the initial investment. The latter cases are not well documented unless they crashed in a blaze of negative publicity.

RULE #22 NOTHING IS GUARANTEED

No one product is guaranteed commercial success when the payoff for research and capital expansion is years in the future. Modest errors in the estimates of the production costs and the prices of competitive alternatives can doom a new product, despite years of development work.

Prospecting for new applications:

- Evaluate end-user functions that require a multitude of laborious steps. If your material or systems approach can reduce time and labor costs then it may have value to the customer. Instant photography from Polaroid was a product innovation that provided a simplified solution.
    - Take advantage of competitors in your current market who are not focused on the end-user's needs because they lack FFU testing procedures.

- The price of your new solution can be much higher than that of the incumbent material if it creates value by adding significant functionality, saving time, or reducing costs elsewhere.
- Can value be added to a commodity via this approach? It is not necessary (or feasible) to convert the entire market. The convenience of canned spray paint carved out a niche within the paint industry, even though the solution was more expensive.
- Evaluate markets that are served by your core technology but in which you do not currently participate. Seek to understand and solve their problems by developing FFU tests.
- When a customer or an entire industry has an urgent problem that needs a solution, then they will be soliciting an expanded range of potential suppliers for help. Such opportunities arise after a sudden shift in the market conditions, which pushes people out of their comfort zone.
  - Each individual customer may grasp the first answer that appears viable to them. This can allow entry into a market where you have no prior presence.
  - As the situation stabilizes, the best solution or technology may consolidate its position across the industry. This creates an urgency for FFU testing and product improvement to position your solution as the optimal answer.
- Sometimes, a new technology comes along that offers the promise of improved efficiency or cost reduction to end users. However, if this application is not a pressing need, then they may not rush to implement the solution. This is especially true if an investment of attention or capital is needed on their part.
  - This is most evident when the application is not critical to the customer's core business plan and the incumbent situation is comfortable for them.
  - Customers with high growth rates and good margins may not be focused on cost control at this stage. Every company eventually becomes sensitive to these matters when its growth levels off and profits are squeezed.
  - Such situations may require patience and relationship building before the customers commit. Do not take initial indifference as a final answer if you have fundamental value.
- A medium to large market served by only one mass-produced product is a ripe opportunity. Some customers will be anxious for alternatives at either a lower price or having a wider range of options. Do not

be afraid to challenge the 800-pound gorilla. Talk to the customers and find out where they want alternatives. Try to service small market niches that have unique requirements.

- Recall General Motors becoming the larger company by offering alternatives to Ford's black Model T.
- Consider Dell, with their 1990s strategy of building customized personal computers to order as opposed to offering standard models.

• Discover opportunities to put your customers into a more auspicious environmental position versus your competitors.

- Paper wrappers for fast food replacing styrene foam clamshells— seen as more environmentally friendly in the 1990s as a renewable resource.
- Allowing customers to tout their product as requiring less CO_2 emissions.
- Warning—green products still need to be priced competitively.

• Use profiling techniques to describe untapped customer opportunities for the sales staff. These sales aids are useful when a product is highly successful with a few customers, but is not spreading quickly to other applications. Create a profile for successful (and unsuccessful) customers. Analyze the common factors that contributed to the success and those that hampered implementation.

- This technique is appropriate where a complex analysis is needed to evaluate potential solutions. These include situations that require customization, and significant engineering work. The salespeople may have a hard time determining promising opportunities.
- Consider profiling if all the initial customers were discovered and cultivated by special sales or marketing individuals. Later, when you release the product to the entire sales staff, they will need simple guidelines for what an ideal customer looks like.
- This is useful when the customer does not use this type of product in its core process. Consider the challenges of selling chemicals and equipment that are used to clean certain types of cooling systems. Many different companies have cooling systems, but some are a good fit for this product and others are not. The good applications are not all in one type of industry. Prepare a short list of questions that can be asked to identify potential customers.

RULE #33 TRADE-OFF SITUATIONS ARE LUCRATIVE OPPORTUNITIES

It is important to identify situations within an industry or a market segment where key performance trade-offs limit customer satisfaction with incumbent products. When new products resolve these annoying either/or situations, the market can move quickly to new solutions.

The helmets used by American football players have been criticized in recent years for not offering enough protection against concussion injuries. This deficiency is the result of a decision made decades earlier on a trade-off situation. The helmets were made with a very rigid shell, in order to prevent traumatic skull damage from the most violent blows. They also had to be durable enough to survive the hundreds of impacts that they receive in the course of their service life.

In order to better protect against the less violent impacts that cause concussions, the helmet would need to be softer and more compliant. However, this would offer less protection against the high-impact situations. Alternatively, the helmet could protect against both situations if it were allowed to fracture and deform in response to a violent collision, as is the case for bicycle headgear. However, this measure would drastically shorten the equipment's service life. The cost of this approach would not be affordable to the cash-strapped school programs where the majority of the injuries occur.

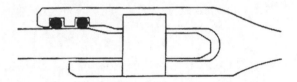

It can be embarrassing or even tragic when events reveal that your assumptions about the performance of a material are only valid in a specific operating range. Richard Feynman famously demonstrated such a trade-off during testimony before a committee of the U.S. Congress. The physicist compressed a segment of O-ring material in a C-clamp and chilled it in a glass of ice water. Upon releasing the clamp pressure, the elastomer did not immediately spring back to its original round profile.

The committee was investigating the loss of the space shuttle Challenger and its crew in a fiery crash. Joints between sections of the solid rocket boosters had leaked hot combustion gases, which then impinged on the shuttle's liquid fuel tank. Under the pressure of the launch, the joints tended to deflect slightly, creating a small gap, which was normally filled by the immediate expansion of the O-rings. Feynman demonstrated that the heat-resistant O-ring material lost its elasticity at cold conditions, such as those encountered on the fateful launch day. It is difficult to find materials that retain their integrity in the face of extreme heat while also remaining elastic when cold.

Examples of trade-off limitations:

- Automotive: Comfort of soft/luxury ride versus responsiveness and "road feel" of sports cars.
- Materials: Strength/stiffness/durability versus conformability/ softness/flexibility.
- Consumer goods: Light/small/easy to carry versus long battery life/ high performance.

Cold Calls by Outsiders

> Consultants have credibility because they are not dumb enough to work at your company.

> **Scott Adams**

United States Patent Office

Des. 219,428
Patented Dec. 15, 1970

219,428
HOLDER FOR BOTTLES OR THE LIKE
Bette C. Graham, 6055 Woodland Drive,
Dallas, Tex. 75225
Filed May 21, 1968, Ser. No. 12,028
Term of patent 14 years

In the late 1950s, IBM was the leading supplier of electric typewriters. They were approached by a bank secretary with a new product proposal. Bette Graham was correcting errors on bank documents by painting over them with a white liquid of her own invention. The pilot plant was a kitchen blender in her home. The solution was packaged in nail polish bottles.

IBM was not impressed by the humble origins of the idea and declined the opportunity to purchase the rights. They suggested that formulation refinement was necessary. The secretary continued to experiment with the idea and later opted to market the product herself. Within a decade, Graham's company was operating an automated plant to produce a million bottles of Liquid Paper per year.

Some companies actively seek to identify and license fresh product proposals ahead of their competitors. The difficulty is to sort the bad from the good. Time and resources are needed to determine the value and novelty of each idea. Reject proposals that fail to suit two of the following three items: your company's market focus, your production capabilities, and your distribution channel.

Guidelines for evaluating outside ideas:

- Big companies could be accused of stealing some fledgling inventor's idea by pursuing a similar concept in the future. Be careful of treating rejected outside ideas the same as internal ones. Protect the company either with a signed agreement governing each submission or review only those proposals that are already patented.

- Be leery of buying technology that is not yet patented. Never assume that a patent can be obtained, no matter what the inventors claim. Many procedural mistakes by an amateur inventor can undermine your ability to protect the intellectual property that you are buying.

- Ask for the records of the inventive efforts. Is the material appropriately documented?
  - Was any form of FFU testing performed? Were a range of alternative options evaluated before settling on an optimal selection?
  - Independently verify all claims made by the inventor before making commitments.
- Be careful not to undermine the creative initiative of your own employees, who may have advanced similar ideas in the past, but were denied resources to follow up on them.
- Avoid ideas that do not mesh with either your process technology or your market objectives.
- Avoid proposals that include only a raw idea, which have not been developed or refined.

Things get even harder if an intriguing concept is presented by an inventor who is secretive, unprofessional, or greedy. Any of these traits add great complexity to the negotiations and diminish the chances of a successful agreement. The worst situation involves a former employee or consultant who crafts patent claims around the boundaries of the company's process with hopes of extorting some royalties.

C. M. HALL.
PROCESS OF REDUCING ALUMINIUM FROM ITS FLUORIDE SALTS BY
ELECTROLYSIS.

No. 400,664.   Patented Apr. 2, 1889.

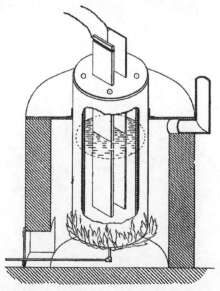

Make it easy for outsiders to know what technologies or opportunities you are interested in acquiring. Provide a means to contact the appropriate people in your organization to quickly decide if there is a fit and what additional work is needed. Cultivate a reputation among inventors as an open-minded company that is receptive to outside ideas. Rigid legal and administrative hurdles discourage independent thinkers.

In 1886, Charles Hall was 22 years old, but looked younger. He was trying to find support for his concept for smelting aluminium by electrolysis. This was difficult, because he had no credentials in the metal processing industry. Hall was such a poor salesman that he misspelled the name of the element in his promotional material.

Alfred Hunt, an MIT-trained metallurgist with the Pittsburgh Testing Laboratory, took an interest in the proposal. Hunt's credibility in the industry helped Hall to obtain financing and support to commercialize his idea. Hall's technology was better than his sales skills—their venture was extremely successful and the Hall–Héroult process remains the primary source of aluminum to this day. As a result of Hall's spelling error, the material's name in the United States (aluminum) is different from that in the rest of the world (aluminium).

Bette Graham and Charles Hall were blocked by the filters that established companies use to screen new opportunities. Graham was a low-status woman and Hall was young and inexperienced. However, both had valuable insights that would eventually become profitable operations.

Customer Requests

Everything starts as somebody's daydream.

Larry Niven

In the movie "Casino," Robert De Niro's character is a mobster who manages a Las Vegas gambling operation for a Mafia family. No detail is too insignificant to escape the attention of this workaholic. In one scene, he berates a kitchen employee because the muffins served at breakfast had variability in the number and distribution of blueberries. De Niro orders the casino bakers to put "exactly the same amount of blueberries in every muffin." The breakfast chef is stunned and perplexed by these demands. He cannot imagine a way to control the berry placement. Normally, the berries are mixed into a batter mixture, which is poured into the baking dishes. Being stuck in the frame of mind of the existing process, he is unable to consider alternatives, such as dispensing the solid additives separately from the liquid phase.

In a commercial environment, management must always reevaluate which quality attributes and product features are really important to its customers. Will uniformity of the blueberry placement in the breakfast pastry really result in more revenue and repeat visits to the casino? What are the important dynamics that influence people's value decision when they choose your product over the alternatives? Do noncustomers have a different set of priorities? Are the value gaps between your product and the competition narrow or large? How can you enlarge those gaps by improving the FFU performance? Will cosmetic improvements to muffin appearance improve your market position? Or, should you be looking for better muffin ingredients or baking technology?

When a customer with Robert De Niro's intensity demands something, you need to set your emotions aside and try to understand the big picture. Determine if this is an opportunity to exploit a hidden market need or just an irrational lunatic at the end of his rope. The two circumstances are often identical at first glance. Just because an end user is emotional and illogical does not mean that his or her point of view lacks merit.

In the eighteenth century, English craftsmen competed to design clocks that could keep accurate time aboard ships. Precise time keeping was essential for navigation purposes. Clocks powered by a traditional pendulum motion were not suitable because of the constant swaying motion at sea.

In 1761, John Harrison won the British Admiralty's £20,000 longitude prize that had been open since 1714. He had improved the coiled spring design used by earlier clockmakers to power his device.

In order to work properly, the coiled metal strips in these devices required precise thickness control and tempering. The technical requirements were far beyond the capabilities of most metal suppliers of the time. Instrument designers and metalworkers did not even share a common vocabulary. Clockmakers were compelled to devise their own processing techniques to achieve these tolerances. Incorporating experimental samples into their instruments was the only way to determine whether a material was good enough for the application.

Sometimes, customers contact a supplier after having experimented with their product in a new application. They always want some changes and

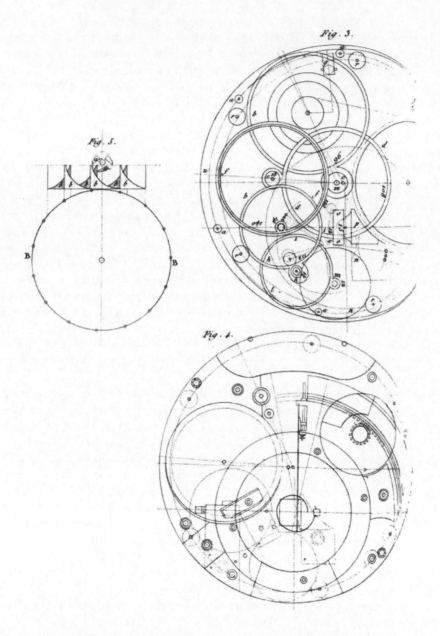

promise to have a huge volume application. These requests generate great excitement, but may be hard to implement.

Evaluating customer requests for product modifications:

- The customers may have only looked at one product batch. Provide them with additional samples for more complete testing. Ensure that they are satisfied with the full range of process variation.

- Never take the customers at their word that the product is ideally suited for their application. Evaluate FFU to understand which process parameters control performance.

- Determine your liability in the event of unforeseen accidents. The incidence of serious failure cannot easily be evaluated until the product is in routine use. Beware of secretive customers who may conceal details, risks, and promises made to their customers.

- Invariably, the customers ask for a few changes that require only minor process modifications. Upon receiving those samples, they request additional changes. Soon, simple adjustments have become a major development effort that saps resources from your strategic programs.

- Carefully evaluate customer requests for potential patentability. The best situations occur when outsiders bring their problems and your development staff finds solutions. The worst case is when the customers propose the solution, so they own the patent rights to your products.

Customer Modification of Your Product

Research serves to make building stones out of stumbling blocks.

Arthur D. Little

Oct. 23, 1956   H. W. COOVER, JR   2,768,109

ALCOHOL–CATALYZED *α*–CYANOACRYLATE ADHESIVE COMPOSITIONS

Filed June 2, 1954

Dr. Harry Coover experimented for years to find useful applications for cyanoacrylate while working for Kodak. In the 1940s, he tried to form it into the lens of an optical gunsight. However, the material was so sticky that it could not be processed. In the early 1950s, he attempted to make canopies for jet fighters from the resin. Again, the material was difficult to work with, due to its aggressive adhesive properties. A minuscule coating of cyanoacrylate caused two glass prisms to stick together, ruining an expensive reflectometer in the laboratory.

Coover was able to turn these failures to an advantage by formulating the material into a superstrong adhesive. Doing so took significant work, however. A new process was required to crack the rigid polymer down into a reactive monomer. This reactive agent had to be protected from water and metals that could prematurely catalyze polymerization. A special combination of free radical inhibitors, base scavengers, and special packaging was developed to stabilize the product for storage. Years after the initial discovery, the product was ready for its commercial launch as Eastman 910. It became famous as Super Glue.

The innovation initiative passed over to end users at that point. Customers discovered endless applications for the product, since its performance far exceeded that of incumbent adhesives. Eventually, it was approved for medical applications as a means to stop bleeding in a range of trauma situations. When deposited as a vapor, the product can reveal faint traces of fingerprints at crime scenes when conventional methods are not sensitive enough. Musicians and rock climbers have learned to coat their fingertips to create an artificial callus.

Chemist Robert Augustus Chesebrough spent 10 years developing Vaseline petroleum jelly before commercializing the product in 1870. He observed that the long-chain hydrocarbons in crude oil had very useful properties. This material was a by-product of petroleum distillation that remained after useful fractions had been boiled away. Chesebrough used vacuum distillation to remove the last of the volatile components. He then filtered it through bone black, a charcoal derived from bone material. The resulting product was a stable, jelly-like material that was free of the odor associated with animal fat.

Chesebrough erroneously promoted Vaseline as a miracle cure for skin injuries of all types. In fact, his product can keep cuts clean and moist, but the human body must still repair itself. The material was a huge commercial success because customers discovered countless applications that improved their lives. It is used as the base or carrier for many cosmetic and medical products. Vaseline acts as a lubricant, plasticizer, and protective coating in many household situations. Sportsmen use the product to keep dirt out of their eyes, prevent chaffing, and to make a pitched ball fly more unpredictably.

Some research suggests that true innovation mostly occurs to customers who modify existing materials to better suit their particular requirements. Lacking daily contact with the customer's application, the development engineers of a supplier are unlikely to make the same connections. Market-focused organizations watch what the customers are really doing with many different products. Customer innovation is difficult to manage and control to your commercial advantage. Responsive suppliers will manipulate process variables to create tailored materials for these novel applications.

What if your product is customized?

- You cannot patent your customer's ideas. You may be able to protect the most efficient process or optimal formula based on FFU test results. Or, you might purchase the rights to new technology, if you recognize its value early enough.

- Customer modifications reveal what end users really need. What they do is more compelling than what they say.

- Are some customers compelled to perform extra steps to prepare your material for use in their application? Consider further processing your product into a tailored solution to their needs.

- How many customers need the same solution? Can you service the diverse needs of each customer? Or will this create an opportunity for more nimble competitors?

- Can you offer flexible components that facilitate customization by the end users? A rigid supplier offers a limited number of flavors of a bottled energy drink in one serving size. An accommodating supplier would offer an unflavored drink mix, allowing users to formulate it to their own tastes and in their own quantity.

- Who is at risk if a customized product injures an end user?

Summary

- Understand the technology barriers that constrain market innovation and the trade-offs that hinder product performance. Removing them can trigger rapid market changes.

- Choose projects appropriate for the nature of your company. Market-driven companies will look for opportunity in different places than process-driven operations.

- When selecting projects, avoid simply copying your competitors or attempting to endlessly repeat your past successes.

- The comfort of the current way of doing things can blind you to opportunities to create truly revolutionary products.

- Carefully consider which options exist today because they really have value and which features are just artifacts of yesterday's technology. What new features or functionality are possible from the process manipulation within your control?

- New ideas can come from anywhere. Look beyond the usual suspects in R&D and marketing.

- Consider the maturity of the targeted market segment before deciding whether to develop a niche product or one that might dominate the market.

- Employ a consistent strategy for evaluating opportunities that are proposed by outsiders.

- Most innovation with your materials will be done by customers who are best positioned to solve the problem. Give them the tools to adopt your product to suit their needs.

- No one course of action is guaranteed to be successful, no matter how good the market research is. Cultivate a diverse range of projects rather than risking all on a single approach.

- Be cautious when attempting to enter a totally new market space because of the risk of a 100% loss of your investment. This is more likely when you do not understand the market needs.

5

Prioritizing Project Proposals

Which activities should be the R&D priority? How can we compare the potentials of dissimilar opportunities? What is the right balance of projects in the portfolio?

> The first principle is that you must not fool yourself and you are the easiest person to fool.
>
> **Richard P. Feynman**

Edwin Land developed two completely different processes for making polarized filters in the 1930s before he achieved fame with instant photography. The name Polaroid derived from the company's original technology. Land's first process employed a crystalline form of iodoquinine sulfate, which was known to act as a polarizer. Earlier researchers had tried and failed to grow crystals large enough for optical work. Instead, Land employed a multitude of tiny crystal needles. He mixed them with binders to form a thick colloidal dispersion. By forcing this mixture through a narrow slit, the crystals became uniformly oriented in one direction, creating the first polarized sheet.

Land had a very specific vision of the product that he wanted to create. He had extensive knowledge of the prior art and academic research into polarized light. Harvard University's resources were close at hand to provide advice and encouragement. With a well-equipped laboratory, he was able to control and understand all stages of the bench-scale development work to refine his ideas quickly. Land could also immediately evaluate the fitness of his samples. With this combination of tools, all the essential elements for rapid development were available to him.

Despite his technical achievements, Land's business model was not profitable. The company struggled for a decade, because automakers rejected Land's original vision for polarized headlights on cars. Ironically, Eastman Kodak kept the firm alive through the Great Depression, with orders for polarized camera filters. World War II made the company profitable with a tremendous demand for polarizing filters, glasses, and 3D imagers. Land's history argues for a diversified portfolio of projects, because even the best single-product idea does not guarantee commercial success in a predictable time frame.

Process-dependent products have an extra dimension of risk compared with others. Uncertainties include: the resource requirements for each project; whether the available process will be appropriate for the application; and

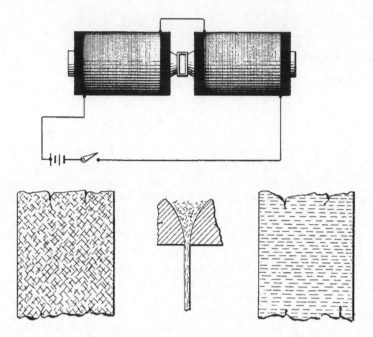

exactly how long commercialization will take. Because of these doubts, there is a strong temptation to concentrate on technology and market areas that have succeeded for you in the past. The choice between sticking to a proven formula and diversifying into new territory is never easy.

Prioritizing competing proposals:

- It is difficult or impossible to prioritize projects across all the diverse markets, divisions, and project types in a large company. Changing conditions will alter the relative priorities on a continuous basis. It is unwieldy to make these choices in a centralized technical function. The management of functional areas/regions should decide how best to allocate resources.

- Resources must be available to quickly pursue firefighting and short lead-time product modifications. Never commit all of your assets to a slate of long-term projects.

- Market-driven companies should pursue multiple projects to address their top new market needs. You cannot count on customers to embrace the one option that is optimal for you.

- Process-driven companies will pursue many different technologies in the hopes of finding one that succeeds in enhancing their capabilities and keeps them ahead of the competition.

- Development projects for process-dependent materials generally have higher risk and uncertainty than those for basic chemicals or assembled products. There are more steps to go wrong—fitting the product into the process, meeting cost targets, the functionality of the end-use application, and customer acceptance of your product.

- Periodically reassess the priorities. New findings during development may significantly alter the value proposition of a potential product. Market situations change constantly due to outside forces and competitive actions. Be prepared to demote activities as these changes are recognized.

See the two flowcharts at the end of the chapter for suggestions for evaluating proposals, depending on whether they are "pushed from within the organization" or "pulled toward an outside market opportunity." The two situations generally call for different evaluation procedures.

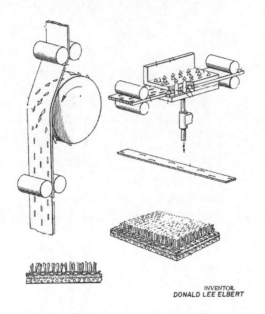

INVENTOR.
DONALD LEE ELBERT

Artificial turf for use on athletic fields was developed by a team at Monsanto in the 1960s. It is used at sports facilities around the world. Such a product is essential for indoor stadiums where sunlight is not available for natural grass surfaces.

Artificial surfaces like Astroturf are very attractive to schools and professional sports teams because they reduce the maintenance expense. Synthetic turf allows more frequent use of the fields than could be accommodated with natural grass, especially in wet conditions. This allows facilities to be employed for multiple sports, such as baseball and football. More frequent events and fewer cancellations due to the weather greatly improve profitability. However, athletes complain bitterly about the abrasive nature of the synthetic fibers. Sports purists often insist that the games are negatively affected by artificial surfaces. Such complaints have prompted a continuous improvement of the product's properties. However, some facilities have switched back to natural grass and single-use field designs to appeal to traditionalists. Maximizing economic efficiency does not please everyone.

RULE #55 NOT ALL CUSTOMERS WANT THE SAME THING

Listening to the end-user reaction to a product is often crucial to the development of new products. However, it is important to understand who actually makes the purchasing decisions and what motivates them. It is dangerous to extrapolate the preferences of a few customer voices onto the wider market.

Prioritization Styles

I don't do product research. I'll go and invent something and spring it on the world, and people will love it.

Edwin Land

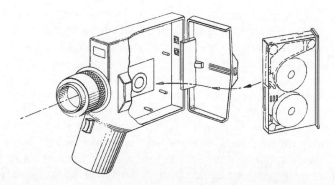

By the late 1970s, a mature Polaroid company had owned the market for instant photography for 30 years. They had progressed from the peelable black-and-white format to color systems where the print developed into a finished image while you watched. All these products represented a closed systems approach—a camera and a proprietary film pack. The formula was very lucrative for the company, one of the high-tech stars of the era.

In 1977, after years of development and at the insistence of its founder Edwin Land, the company rolled out Polavision instant movies. This system was a combination of a movie camera, a film cassette, and a tabletop viewer. Every element of the system was incompatible with the rest of the photographic industry. The film could not be viewed with the incumbent projection equipment, such as the widely available Super-8 format. The system did not offer sound recording capabilities and it had a marginal picture quality.

From a technical standpoint, the product was an impressive realization of a guiding vision. However, the unwieldy, expensive system was not what the public wanted. Its utility was substandard compared to video recorder systems that were then becoming available. It followed the company's history of commercializing the fruits of their founder's creative mind and relying on consumers to flock to his closed systems. In this case, the visionary was out of touch and customers had other options. Land was forced out of his management responsibilities. His habit of betting the future of the company on his intuition had finally failed him. Every company has its own procedures for rationalizing which projects to resource and which ones to starve. Its procedures need to evolve as the company grows and the project slate gets crowded.

The Bean-Counter Approach. Your goal is to generate revenue. The assorted project choices can be evaluated with a financial model to calculate a net present value for each project, as with any investment choice. Development projects can be funded to maximize the return on the research investment. The project expenses, capital requirements, and resulting revenue streams can be used to predict the profit potential of each proposal. That potential return should be discounted by the probability of the success of a proposal. This tool will provide an objective ranking of projects with a range of different time lines, market objectives, and risk levels.

Approved projects must be given the resources needed to achieve their results on schedule, to start delivering cash flow when expected. The best way to fail with this approach is to have too many projects, all of which are underfunded and chronically late. Process-dependent products do not fit easily into this methodology because the development time and the production costs are difficult to forecast. Realistically, this system can only work if there is abundant information generated by exploratory projects on which to base a financial analysis.

This approach might succeed in business environments that are relatively stable. In a rapidly changing market, the future profit of a product is impossible to estimate. Likewise, this technique will not work well for "blue sky" and early-stage research projects that seek to create enabling technology or explore new product concepts. They should be judged and funded separately from the rest.

Sharing the Pie. A decentralized company may divide its development resources between divisions, allowing each to use them as they see fit. The manufacturing group can be given a share for process improvements. The purchasing manager can have some engineers to evaluate and qualify cheaper materials. The marketing staff can explore improved products for existing markets and the market development group can evaluate projects ideas for future growth.

This approach is very democratic. It is much like investing in mutual funds for retirement, rather than trying to pick individual stocks. The technique will succeed by spreading the wealth around to the people who can best see what needs to be done in their individual areas. It is unlikely to achieve breakthrough ideas that change the direction of the entire organization. Companies in stable, slow-changing industries are suited to this means of allocating resources. Its most compelling advantage is that no one area is starved of resources because all the effort is focused on a flawed vision, such as instant movie film.

See the diagram on the "distribution of novelty" at the end of the chapter. Track how your technical spending is distributed across the 11 different project categories. Strive for a balance that is appropriate for your company's circumstances.

The Top-Down Approach. In a centralized organization, each functional group will submit its proposals for technical spending. These range from simple efficiency improvements to grand schemes for revolutionary new products. Periodically, the top management can view presentations and debate the merits of each. They choose between requests using one of many different techniques.

This method requires that the management group spend enough time to fully understand and compare all the options. They cannot be distracted by mergers, lawsuits, and disasters. A considerable amount of executive time must be invested. This method will not be flexible enough to accommodate projects that come up later in the year or those that undergo a significant change in scope. Extra money and manpower must be left to deal with unforeseen emergency situations. If not, then one or more of the projects that was funded will be robbed of its resources. In order to work, this approach requires a management team that is better positioned to see future needs than those who are closer to the action.

The Idea Exchange. Some companies harness the group consciousness to identify and prioritize projects. This approach assumes that polling the intuition of the entire organization will produce a superior result than polling the intuition of an individual or a small group. It also promotes buy-in among the workforce. The resulting strategy is the averaging of the workforce's collective suggestions.

This approach starts by clearly documenting each option from among the list of feasible projects. The risks, benefits, and requirements are stated in plain language. Every interested employee should be able to read and understand the concepts. Proposals should be vetted to ensure that they do not violate the laws of man, God, or nature. Balloting is conducted to select the top candidates. Management oversight can be maintained by allocating only a portion of the budget with this system.

Considering Risk

> Progress always involves risk; you can't steal second base and keep your foot on first.
>
> **Frederick Wilcox**

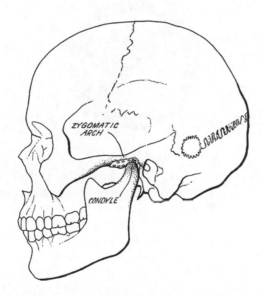

Vitek Inc. was a Texas-based medical start-up company founded by Dr. Charles Homsy. The company developed a process for molding plastic medical devices in the 1960s. In the early 1980s, they introduced an implant for the junction of the upper and lower jawbones. This device restored normal jaw function and mobility to patients suffering from temporomandibular joint (TMJ) disorders. Dr. Homsy used DuPont's Teflon resin, fiber, and film as components in this device.

Within a few years, complications were reported with some TMJ patients. In the problem cases, implants abraded after prolonged use. The resulting fragments caused tissue irritation and bone damage. The tragedy prompted lawyers to sue Vitek into bankruptcy by 1990. At that point the chemical giant discovered that the innovator's liability was now theirs.

Since DuPont was a player in medical research, it was easy for juries to conclude that they should bear some responsibility for the situation. The money that the company had originally made on the resin in each implant was not adequate to cover the legal costs, which reportedly cost the company tens of millions of dollars. This dictated a new corporate policy to limit which end-use applications the company's materials could be used in.

What are the downsides to unforeseen performance failures related to your materials? Will people die if the product fails? Could there be huge legal claims against the company in the worst case? Perhaps criminal liability charges against the development team? Or perhaps they will be among the fatalities discovered in the smoking remains of a spectacular process failure. Can the risks be anticipated and consistently controlled or will there occasionally be great harm from product outliers and process upsets?

In 1884, a French chemist, the Comte de Chardonnay, developed a process to spin nitrocellulose into silk-like fibers. His product was the first of many synthetic textile products that were introduced over the next 100 years. The fiber was woven into a beautiful fabric, which he marketed as "Chardonnay silk." With his scientific background, Chardonnay would have understood that the material was highly flammable. However, he must have rationalized that fashionable and expensive clothing would not be worn in close proximity to open flames. This fitness-for-use (FFU) assumption was soon proved horribly wrong by a series of fiery accidents, probably ignited by tobacco use. Popular opinion brought a quick end to Chardonnay's textile experiment.

It must always be understood that some of the products will be out of specification. This might represent a "one in a million" fluke or a statistically predictable event, depending on your outlook. At some early stage, project proposals need to be vetted to balance the risks and rewards. The downside factors must be identified at this point. Constant testing, verification, and review of these critical matters should take place at every step. Safe products have a wide gap between their process capability range and the onset of dangerous outcomes. Ultimately, it is the responsibility of the top management and the firm's owners to understand the risks inherent in any business opportunity.

Considering Process Factors When Prioritizing

The great tragedy of Science: the slaying of a beautiful hypothesis by an ugly fact.

T. H. Huxley

In 2002, an American fast-food chain rolled out two new sandwiches in the United States. These products employed a thicker meat patty than their existing products. The new formula was spiced and flavored to mimic the taste of a charcoal grilling process. The initial sales trials in selected locations were favorable, prompting a decision to launch the concept nationally.

However, the thick patty could only be cooked with equipment used in a fraction of the outlets across the country. To get around this obstacle, the original design concept was compromised to fit the available production scheme. The product was altered to accommodate two, normal thickness patties. Because this was similar to the existing products, an assumption was made that the grill cooks would not need special training or guidelines to achieve the desired product quality. However, this was not true, possibly because the spiced product did not have the same color as pure beef, making it difficult to judge the degree of doneness.

Without their normal techniques to evaluate processing, the cooks erred on the high side, to kill any bacterial contamination. The result was a considerable variation in the doneness of the meat from one location to the next, with many complaints that the meat was dry and overcooked. The concept was abandoned after 6 months, due to poor sales. It would seem obvious that important commercial decisions should be based on customer testing of the actual product made on the actual process. However, that exact product is rarely available when the decision is made. Instead, prototypes are tested, in the hopes that they will be a good simulation of the ultimate result. At that point, capital is allocated to build or modify the process that will eventually make the desired material.

This situation is common in a process-dependent environment. Expensive, full-scale trials on production equipment are often deferred until after small-scale experimentation has shown promise. Projects fail when the commercial process cannot reproduce promising early-stage results. The investment in commercial trials is often taken on faith that the development team will find suitable process conditions.

Evaluating the production needs:

- Will this product fit into your production wheel? How often must the product be produced and with what lead time? What impact will these have on the total product portfolio?

- What are the customer's quality expectations? Are we required to take back unsatisfactory products, even if they meet our specifications? Do customers understand the expected level of variation? Do we need to pay claims for their spoiled product if our material is out of specification?

- Can the FFU requirements be met by the process? Are the FFU needs the same for all customers?

- How will the changeover to and from this grade affect operations? Are more frequent changes required? What do they cost? What is the downside if errors are made?

- Do the anticipated batch sizes and lead times fit your operation?

- Will the new product run in a known process matrix or is it outside the comfort zone?

- How will the prototypes differ from the commercial product? Can you predict the likely impact of those differences? What factors could literally be killer variables?

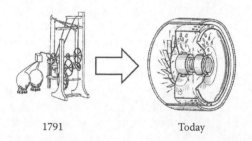

1791   Today

Gas turbine engines are widely used for aircraft propulsion and for the production of electricity. John Barber first conceived and patented the idea in 1791. His ideas proved to be impractical, because the available materials could not withstand the required service temperatures. Gas turbines were not successfully demonstrated until the early twentieth century. Barber was a century ahead of his time. Such situations are great for comparing yourself to Leonard da Vinci. However, if you want to generate cash flow, being ahead of your time is the kiss of death. If your product concept relies on process technology that exceeds current performance standards, then the chance of success drops significantly.

In the long run, the proposed product must be efficiently and consistently produced to sell it profitably. This might be done by an outsourced vendor or in your own world-class facility. Failing at this, all the other careful development work is in vain. It is best to reach this conclusion early in the project, rather than at the end.

Project Review

> Science fiction films are not about science. They are about disaster, which is one of the oldest subjects of art.
>
> **Susan Sontag**

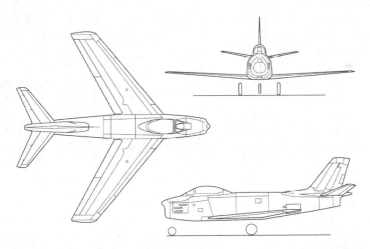

In the popular imagination, the outcome of a combat between two similar jet fighters is dependent on which pilot has "the right stuff"—a combination of extensive training, youthful reflexes, and natural aggressiveness. When the planes are different, we assume that speed and maneuverability will prevail. This changed in the 1960s when Major John Boyd of the U.S. Air Force and mathematician Thomas Christie conceived the energy-maneuverability theory to explain the relative fighter plane performance.

Boyd noticed that, in certain situations, high-performance aircraft could lose their duels with inferior models. The ability of an aircraft to gain and shed kinetic energy has always been vital for a dogfighting performance. Boyd and Christie developed graphs to quantify which velocities, altitudes, and approach angles were favorable or unfavorable during an encounter between one type of aircraft and another. Their work related the thrust, weight, drag, wing area, and other flight characteristics of an aircraft into a quantitative model. Based on these curves, they could demonstrate where one design had the most beneficial position over another. This was used to train pilots where to fight and when to run.

The management decisions needed for the development of process-dependent products can be equally complex. Just because one approach worked in the past, it does not mean that it will be the correct one for a completely different set of circumstances. Only by monitoring the effectiveness of different methods can you understand which one is best in any situation.

The organization needs to constantly evaluate why projects succeed and fail. These reviews must not be limited to the technical group's effectiveness. They need to focus on the bigger picture of whether the right projects are being resourced and the wrong projects are being culled. The reviews must also evaluate whether the organization is correctly implementing the new products and technology.

The success rate of management in choosing projects is rarely analyzed on a statistical basis over time. An unlucky executive may be discharged following the spectacular failure of a heavily advertised new product launch (witness New Coke). It is always easy to identify an especially poor decision in retrospect. However, a defective project selection system may be even more costly in the long run.

Effectiveness in absorbing new products is an important component of the business practice. This requires manufacturing to smoothly integrate technology into its operations. Sales must train its people on how to pitch the products to prospective customers and provide ongoing support. Logistical and customer service groups must adapt to novel products.

John Boyd's OODA loop—a system for reacting to events.

Observation: Collect data on the situation (incomplete and ever changing).

Orientation: Analyze the data to form a mental perspective.

Decision: Determine a course of action based on that mental perspective.

Action: Play out your decisions.

While this is underway, the situation will change. Loop back to the observation stage, building with each step, based on the feedback from the results of your actions. If you can loop faster than your competitor can, then you will have the advantage.

Strategies for evaluating the project's effectiveness:

- Monitor the history of projects both large and small. Classify them by type, sponsor, selection method, and participants. Categorize by novelty—see the chart at the end of the chapter. Track the resources expended versus the original plan. What was the return on investment on the development costs? When were milestones achieved versus the plan? Was the development methodology successful?

- Judge individuals and teams on their ability to meet the timetable and stay within the budget.

- Reward teams who discover that their mission is impossible early in the project. Penalize teams and managers who ignore the warning signs and continue to consume resources on a futile quest.

- Rank individual project sponsors by the rates of return on their initiatives.

- Did the project absorb a lot of extra man-days in the last stage to achieve its launch date? This could be a sign of inadequate project definition and planning in the early stages. In a panic, resources are thrown at the problem to meet the goals. The result is that other projects are starved of resources to accommodate this one project. Those other projects will invariably fall behind schedule.

- Compare the metrics of all projects. Look for root causes and take corrective action.
  - Seek systematic problems in the procedures for selecting, prioritizing, resourcing, executing, and implementing new products.
  - Do your development projects fail because insufficient or faulty groundwork was done in the earlier exploratory project?
  - Do not judge success by division or company profitability. Profitable operations will mask an inefficient and complacent development effort.

- Suppose that certain types of projects are always late and over budget, regardless of who is performing the work. This may indicate that incomplete information or unrealistic objectives are being proposed by the sponsors. It could indicate that the technical organization needs better training in project management. Often, it indicates that too many projects are being undertaken for the available resources. Are there management incentives to underestimate resource needs?

- Some projects may not be predicted to result in a measurable financial return, making it difficult to judge their success. They may be undertaken to support a product line or a large customer. In those cases, it may be better to track the sales growth versus the prediction of the sponsors.

- Suppose that projects are completed on time, but are slow to generate profit. This could be an indication that management is not focusing enough on the back-end implementation. Are the systems and culture conducive to new products? Are you underestimating the time needed for the market to adopt the new technology?

- Compare the projected probability of success with the actual success rate for various project types. Adjust your expectations to better agree with the actual performance.

Project prioritization is not just a matter of deciding which technology options to explore. It also extends to the choice of which promising exploratory results should be commercialized into products. A historically bad choice was made at Xerox in the 1970s. A team was formed to explore alternatives to the crude computer interfaces available at that time. This group had an open charter to explore a range of possibilities for the future of computing.

Within a short time, the company's PARC team integrated components into a set of innovative features that would seem familiar today. The group pioneered the graphical user interface, a multiple window format for displaying applications, a mouse and a local area network of connected computers. The only missing elements were computer viruses and spam e-mail. The laboratory gave tours and demonstrations of its technology, in an effort to generate commercial interest. The concepts were perceived to be too futuristic and did not directly support Xerox's core photocopier business. The company failed to pursue commercial applications, partially due to a bad experience in an earlier computer initiative. Apparently, Steve Jobs of Apple Computer was more interested than Xerox management. The revolutionary features of his Macintosh computer were inspired by the innovations at the Xerox laboratory.

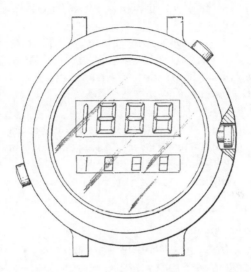

In the 1960s, the Radio Corporation of America (RCA) sponsored an internal research project to investigate liquid crystal technology for electronic displays. This material could predictably become opaque at certain temperatures, when subjected to electromagnetic fields. The team encapsulated promising liquid crystalline formulations between conductive glass sheets. This allowed them to identify promising materials and process paths. They demonstrated a prototype numeric digital display and a clear window that could become dark when current was applied.

The RCA management chose not to commercialize this research. The company continued to rely on cathode-ray tube display technology rather than proprietary new tools. Later, engineers at Sharp in Japan reverse engineered RCA's technology based on media reports. In the 1970s, Sharp's liquid crystal display (LCD) technology enabled the production of small calculators and digital watches with a much better battery life than the power-hungry light-emitting diode displays.

Many research organizations will demonstrate some interesting concepts that are not adopted. These misfit ideas seem to lack value or do not address the immediate needs. They may be ahead of their time, too dangerous, or deficient compared with the alternatives. In order to take advantage of them, a company must take a risk and plunge into the unknown. This can be a great success or a tremendous failure. These risks can be minimized by spreading your portfolio of projects across a map of project risk shown in the diagram at the end of the chapter. This approach diagrams the type of projects underway and their relative degree of risk.

Summary

- There are always more project proposals than a company can resource. It is essential to prioritize them based on their fit with a company's strategy.
- It is difficult or impossible to prioritize all the potential development projects across the diverse markets, divisions, and functions of a large company. Push these decisions down to the people who are close to the market and process the needs for each business unit, functional area, and geographical region.
- Consider in advance what risks you are willing to assume with each new product opportunity (risk of project failure or liability for damage caused by defects in your materials).
- Consider the implications for the manufacturing, sales, and support functions when selecting projects. Are you willing and able to expand or modify the process to accommodate the new products? Can you service the new business?
- Large organizations must constantly evaluate the success of their projects. Determine whether the right opportunities are being addressed and the weak proposals are being culled. Also evaluate the implementation of the new products or technology.
- Accept that there will be failed projects. A 100% success rate suggests that you are only pursuing low-risk options and proven technology.

Categories for new products:

1. New to the world—novel to both the market and to your company.
2. New product line—new product for your company that serves an existing product niche.
3. Extending a product line—modifying an existing process to serve a different market niche.
4. Improvement of an existing product for a similar end use.
5. Substitution—the use of an existing product in a new application, such as off-label drug uses.
6. Debottlenecking, cost reduction, and modest upgrade of existing products with proven technology.
7. End-user customization of your product for either a new use or an improvement for their needs.
8. Making an existing product with totally new technology.
9. Acquisition of a proven product and process from a current participant in a market.

10. Acquisition of a proven product, which you must adapt to your process.

11. Acquisition of a partially proven product concept and the development of a new process.

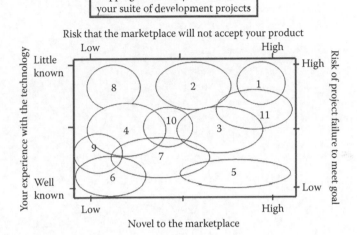

Categories for various examples:

Example Technology	Category	Comments
Polaroid movie system	1	High risk—new technology and new market
Cyanoacrylate adhesives	2 → 7	New product later enables end-user customization
Cracker Jack packaged snack	3	Existing product is modified to create a new market
Venetian cristallo glass	4	New glass formulation creates a premium product
Corning Pyrex	5	Existing technology is adapted to a new product
GE turbine refrigerator	6	Cost reduction falls flat due to inadequate FFU testing
Heat-treated rocks	8	New technology to make an existing product

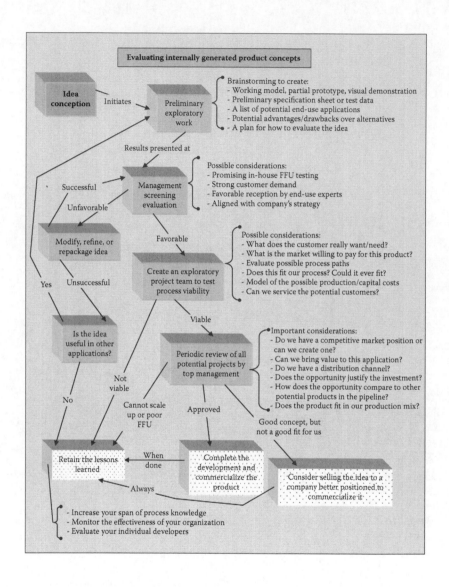

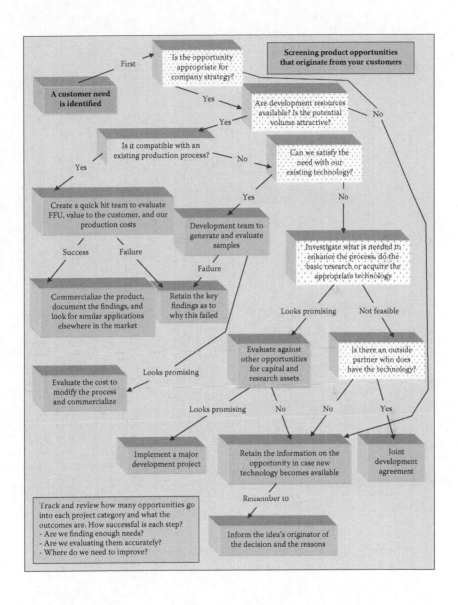

Screening product opportunities that originate from your customers

A customer need is identified

First → Is the opportunity appropriate for company strategy?

No

Yes → Are development resources available? Is the potential volume attractive?

No

Yes → Is it compatible with an existing production process?

No → Can we satisfy the need with our existing technology?

Yes → Create a quick hit team to evaluate FFU, value to the customer, and our production costs

Yes → Development team to generate and evaluate samples

No → Investigate what is needed to enhance the process, do the basic research or acquire the appropriate technology

Success / Failure

Failure

Looks promising / Not feasible

Commercialize the product, document the findings, and look for similar applications elsewhere in the market

Retain the key findings as to why this failed

Evaluate against other opportunities for capital and research assets

Is there an outside partner who does have the technology?

Looks promising → Evaluate the cost to modify the process and commercialize

Looks promising / No / No / Yes

Implement a major development project

Retain the information on the opportunity in case new technology becomes available

Joint development agreement

Track and review how many opportunities go into each project category and what the outcomes are. How successful is each step?
- Are we finding enough needs?
- Are we evaluating them accurately?
- Where do we need to improve?

Remember to → Inform the idea's originator of the decision and the reasons

6

Evaluating Critical Parameters in the Process

How do we determine which process knobs control the performance of our materials in the end-use application? Why are some process configurations more stable and successful than others?

> Believe nothing just because a so-called wise person said it. Believe nothing just because a belief is generally held. Believe nothing just because it is said in ancient books. Believe nothing just because it is said to be of divine origin. Believe nothing just because someone else believes it. Believe only what you yourself test and judge to be true.
>
> **Buddha**

This ancient wisdom anticipates the basis for modern scientific methods and engineering disciplines. Most failures of technology could be prevented by applying this philosophy. Only controlled testing will predict how to run a process to satisfy customer requirements.

Prior to the Industrial Revolution, high-quality swords were one of the most process-dependent items produced. Some producers were vastly superior to others. The best blades of the Middle Ages were those from Toledo, Spain, the Damascus region, and Japanese samurai swords. Sword performance was a life-or-death matter to the end users. Inferior blades shattered or bent under the impact of superior weapons.

The best metalworking techniques were trade secrets that could not be duplicated, despite the tremendous efforts to reverse engineer unsurpassed models. Some methods have been lost to history as a result of pervasive secrecy.

All three schools of metalwork relied on microlayering as the smiths folded the metal back on itself by repeated heating and hammering. Modern experts believe that ball-peening technology was employed in Toledo and was then rediscovered in the twentieth century. Carbon nanostructures have been identified in Damascus blades, along with traces of stainless steel additives. In Japan, the samurai sword makers learned to combine high- and low-carbon-content iron into a sophisticated composite using each material to its best advantage. In all cases, the routine was intricate and time consuming; each one requiring many years of labor.

Each region had its own legendary fitness test criteria. The best samurai katana were rated on how many victims could be killed with a single blow. A silk scarf was expected to be cut in half simply by floating down onto an upturned Damascus blade. The Toledo rapiers would snap back straight after having been bent 180°.

These production cultures did not comprehend the metallurgy behind their best practices. Generations of artisans experimented with different variations and passed the secrets to their apprentices. They could not always reproduce their best work after lucky accidents resulted in the perfect sword. The Damascus blades may have varied in quality with each shipload of Wootz steel imported from India. Samurai sword production was said to experience a 30% in-process breakage—often during quenching.

Today's market cannot wait for generations of craftsmen to slowly achieve perfection. We strive to understand the underlying science that controls performance. But, still we struggle to achieve consistency.

Most process-dependent companies run their production in the best way they can. The product mix is scheduled using the best compromise between

RULE #1 GROW YOUR SPAN OF KNOWLEDGE

#1

When developing dynamic products, one overriding rule applies. You must identify and quantify every process factor that influences product performance. Failure and performance variability will occur when you do not control every process knob and the secondary interactions between these primary factors. Fitness-for-use (FFU) deficiencies point to refinements.

customer urgency and operational efficiency. The production output is tested against a documented quality program. The plant discards, recycles, or downgrades any lot that exceeds the specification limits.

However, one or more customers are vocally dissatisfied with the results. Batches are periodically returned for failure to satisfy mysterious end-use requirements. The returned goods satisfied the same specifications as earlier shipments. They work well for other customers.

The manufacturing manager wishes that these whining customers would find another supplier. The salesperson constantly complains of lax quality control (QC). The situation follows the 80–20 rule in which 80% of the complaints are generated by 20% of the applications. Strangely, customers keep placing orders, hoping that the next batch will meet their needs. When it does not, they negotiate for credit or adjust to compensate for variability.

These are difficult situations. For some reason, only one product comes close to meeting the customers' needs. No one understands the key factors that discriminate between satisfaction and failure in these applications. The terms that the customers use to describe the performance issues are incomprehensible to the supplier.

There are two options to resolve the situation. The customer(s) can be fired, instantly eliminating 80% of the complaints for the product line. Or, a systematic study might uncover the value that your process brings to this application. Reasonable customers will cooperate to resolve the mystery. By undertaking a study, both you and the customer will learn a great deal about this application. They will either become your most loyal customer or discover how to satisfy their needs with another product. You must undertake this risk with open eyes.

The first step is to confess to your valued customers that you do not fully understand how to make your product. Like an alcoholic, the manufacturers of process-dependent products must admit the limits of their ability to satisfy customer needs. You must connect all the important details. This

RULE #2 CORRELATE THE CONNECTIONS

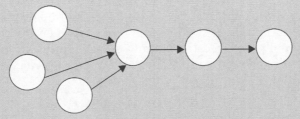

The process details only gain value when they are correlated together as a narrative model of product performance. By understanding how all the process variables affect the end-use performance, you have a chance to be successful.

is difficult because the essential information is very compartmentalized; the facts about the raw materials, the process, and the eventual use of your products are worthless in isolation.

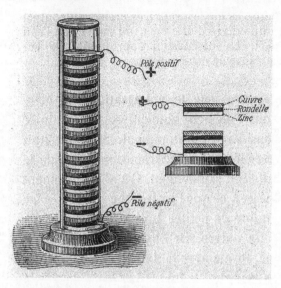

In 1786, Luigi Galvanti was dissecting a dead frog on a metal plate (or perhaps preparing lunch, depending on which version of the story you believe). When he touched the frog's leg with a knife, it twitched. This puzzling observation was reproduced and published in 1791. Galvanti speculated that electricity was somehow involved, but he incorrectly assumed that static electricity was stored in the frog's leg muscle.

Alessandro Volta studied this account, but disagreed with the conclusion. Volta carefully reverse engineered the experiment using the scientific method. The key was the interaction between two dissimilar metals (in the scalpel and the plate) in the presence of the frog's bodily fluids. The frog's leg was acting as an exquisitely sensitive measurement of an electric current.

Based on this insight, Volta constructed the first voltaic pile or battery. He stacked alternating zinc and copper disks between moistened paper. The resulting device proved the validity of his hypothesis and provided electricians with a reliable source of voltage, which was named for him.

Things We Want to Understand

Never assume the obvious is true.

William Safire

I shall assume that your silence gives consent.

Plato

Assumption is the mother of all screw-ups.

Wethern's Law of Suspended Judgment

Assume makes an ass out of u and me.

Anonymous

Therapeutic isomer   Birth-defect isomer

In 1958, West German doctors noticed an alarming spike in a particular class of birth defects. The victims were born with shortened and malformed limbs. Many people speculated that the radioactive fallout from nuclear weapons tests was the cause. Approximately 5000 German babies were affected by this epidemic. Globally, 10,000 infants were afflicted with similar problems in a variety of countries.

In 1961, Hamburg pediatrician Widukind Lenz determined the source of the epidemic. The deformities were correlated with the mother's use of a drug sold in West Germany under the trade name Contergan. The product was introduced in 1957 and quickly became the top-selling sedative in the German market. It was also effective as a pain reliever and cough suppressant. Doctors and patients were initially pleased with its performance. It was nonaddictive, nontoxic, and had few obvious side effects. Tests indicated that it was almost impossible to overdose on this compound. In short, modern science had discovered a miracle drug that improved people's lives.

Clinical trials were conducted on both men and women. Apparently, pregnant women were either not included in the studies or they were not tracked through to delivery. It was later realized that animal testing did not correspond well with human results in the matter of birth defects. Medical researchers did not understand that nontoxic materials could cause fetal deformities.

Contergan and other brands (generically known as thalidomide) were withdrawn from the market in late 1961. Research suggested that the production process created two mirror-image isomers of the thalidomide molecule. One version worked therapeutically as a sedative and the other version caused birth defects. This tragedy prompted many governments to strengthen the regulations on drug testing and approval. Cautious interest in thalidomide still exists to this day, although strict measures are taken to avoid exposing pregnant women.

RULE #53 QUESTION ALL ASSUMPTIONS. NEVER OVEREXTRAPOLATE GOOD NEWS

Here be dragons

Many people falsely assume that favorable test results in some aspects of an application predict safety and effectiveness across all situations. In fact, subset applications can react very differently than the mainstream. The worst outcomes may not become evident until long after the fact. Seek to understand all the mechanisms that control performance.

In the late 1990s, American cigarette companies were faced with yet another health concern. Unacceptable levels of nitrosamines (a family of carcinogens) were detected in tobacco purchased from U.S. farms. The raw material sourced outside the United States had much lower levels of this contaminant. The problem was traced to the flue-curing process employed by American farmers. Freshly harvested tobacco leaves are hung in heated barns for a week to dry and cure them into a suitable cigarette ingredient. Traditionally, the barns were indirectly heated with an assortment of different energy sources. The combustion gases were isolated from the product to prevent flavor transfer.

With rising energy costs in the 1970s, the farmers adopted a process where the exhaust of a propane burner was vented directly into the curing barn. This was far more energy efficient than using indirect heat sources. The propane combustion process was clean enough that it did not contaminate the leaves with a smoke flavor. However, it produced nitrogen oxides that reacted with the tobacco leaves to form nitrosamines.

The cigarette producers assumed that this new process would not change their raw material. When the nitrosamine problem was discovered, the farmers were compelled to retrofit their curing systems to prevent exposure to the oxides of nitrogen.

What combination of factors makes the product ineffective or unsafe?

- Correlate sample properties with performance:
  - What is the fundamental mechanism that governs material failure in each application?
  - *Strength of the effect*: How is product performance impacted by changes in concentration or duration? Example: if only a minute dosage of an antibiotic is needed to treat an infection, then it will have a strong effect.
  - *Breadth of the effect*: The effectiveness of the product across all applications versus the negative outcomes and cases of ineffectiveness. Example: a new antibiotic cures 97% of infections, but causes severe allergic reactions in 1% of patients, while 2% need alternative treatments.
- Characterize all the customers' applications:
  - Break down the applications into critical subgroups that require individual attention. Example: unborn children were the overlooked category for thalidomide testing.
  - Focus first on the subgroups that offer the most immediate commercial opportunity. Remember that the lack of negative results in one subgroup may mean nothing for the next subgroup.
- Determine how the process parameters (materials, equipment components, conditions, and procedures) affect the above items.
  - What range of process parameters and raw materials exhibit the best production stability? The higher the process variability, the more data required to prove safety.

- How best to start-up the process? How long does it take to reestablish stability after a transition? When a dynamic system is briefly unstable, you do not know what might happen.
- *The gain*: How does a change in the process conditions alter product performance? When a small change in the knob settings creates a big change in effectiveness or negative results, then you have a high gain.

- Evaluate the variability of application performance.
  - High variability means that two "identical" batches do not work the same in two consecutive end-use evaluations. For example, two different patients do not have the same response to a 10 mg dose of an experimental drug.
  - High variability can result when you are not controlling the correct process factors. Or this end-use application might be exquisitely sensitive to tiny factors, such as the twitching of the frog's leg. Exhaustive testing is required to ensure that the bell-curve extremes will not have dangerous reactions in the end use.
  - Even worse, the material may not work the same way from one time to the next. For example, a drug may work well initially, but decline in its effectiveness as the patient or the disease organism becomes resistant over time. It can be nearly impossible to sort out these complications from the process factors.

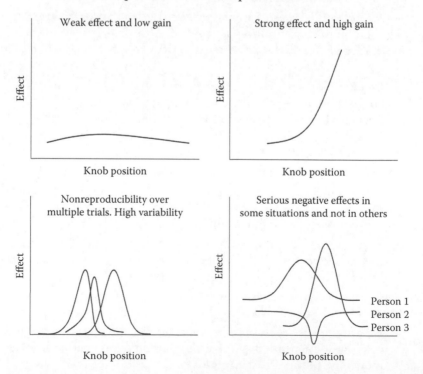

The ideal product should have a strong effect (a little goes a long way). It should have low gain (slight changes to the process do not cause a big change in product performance). It should work in a wide range of applications, with no negative effects (no allergic reactions or side effects). The process should be stable and easily adjusted from one product to another. What is the worst product? It would have a weak effect, with high gain, low stability, and many side effects.

High gain is often unfavorable because small variations in the process create a big effect on end users. A modest process fluctuation causes performance to change significantly. Undesirable side effects also vary in gain and strength with process variation.

Picture an opiate drug administered in an emergency situation to relieve the pain of an injury. A doctor must calculate the required dosage based on the condition, age, and mass of the patient. Either too little or too much narcotic will cause a problem. In this case, there are upper and lower limits on the acceptable effect. With high gain, the effect can easily jump out of the acceptable range. Thalidomide was mistakenly judged to be an ideal sedative because of the huge gap between an effective performance and dangerous results.

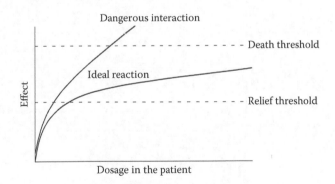

The development engineer team seeks out materials and a process where the effect strength is suitable for the application. The conditions and the procedures are then fine-tuned to lock in the proper effect level so that the process and material variations will not shift the effect outside of its desired range. FFU testing is a means to determine that the effects are within the proper range. This can be critical when adapting existing products to completely new applications.

Examples where negative effects are found in some cases and not in others:

- Exposure to trace amounts of epoxy and urethane monomers causes dangerous reactions in those people who have become "sensitized," but not in the larger population.

- Some agricultural chemicals can harm specific crop types if they are applied during certain stages of growth.
- Abrasive polishing compounds create scratches on a work surface if even minuscule amounts of large abrasive particles are present.

Proactive Record Keeping

Facts do not cease to exist because they are ignored.

Aldous Huxley

Facts speak louder than statistics.

Geoffrey Streatfield

Facts are stubborn things.

John Adams

Facts are stupid things.

Ronald Reagan

If the facts don't fit the theory, change the facts.

Albert Einstein

In 1878, the Procter & Gamble Company perfected a new bar soap formula called "White Soap." This product had a pure white color, because the manufacturing process removed impurities common in soaps of that period. The product was successful as an incremental improvement.

One day in 1879, customers came to the company raving about a batch of the product that excited them. These bars of soap floated, making them easier to find and use in the bath. According to legend, it was discovered that this batch of product had been left in the mixer for too long. A froth of tiny air bubbles had become entrained in the product and remained there while it solidified. The resulting bars had a lower bulk density, allowing them to float. This process deviation was turned into an iconic product because the company was able to tie the desirable characteristic to the specific production conditions.

There is no substitute for good record keeping in the process-dependent environment to understand the source of a problem or the benefit to an end user. Can the organization determine when an especially good or bad batch of material was made? Which line or machine produced that lot? How was the equipment configured at the time? Who was the operator? What were the process conditions during that shift? How did they change over time as the batch was processed? Were these conditions distinct from other batches that had a different performance? Which lot of raw materials was used? What were the QC measurements? How was the batch stored and transported afterwards? Is there more material from that same batch in the distribution channel?

By comparing the important details of particular batches, you can find the important connections. If not, then one of two things is going on. Either you have failed to collect the pertinent information or the issue originates with uncontrolled variables outside your control. Too often, we jump to the latter conclusion when we overestimate our depth of process knowledge.

Some companies can readily access all these details to analyze the root cause of performance variability. This is especially true if product defects endanger human life. Drugs and medical devices, aircraft components, sophisticated weapons systems, and certain safety-rated gears generally merit this attention. The cost of scrupulous QC, meticulous record keeping, and sampling systems drives up the price of such products.

There are relatively painless techniques to retain adequate records. For instance, maintenance craftsmen can maintain a log showing when critical parts are changed. The measurements of wear or the changes in specified tolerances can be measured and recorded at installation and removal. A daily report filled out by the operator could include raw material batch numbers, process conditions, and remarks. In combination with quality reports, a snapshot is retained of the process state.

On some control systems, the process conditions can be printed out at the touch of a button. However, in the author's experience, the printer is often broken or missing an ink cartridge. Allow the workforce to come up with practical solutions that work for them. Avoid the temptation to constantly burden operators with additional documentation, every time a new quality defect is reported. The information need not be called up instantly from a database. When there is a problem or an unexpected outcome, the hours of work needed to gather all the records are the least of your problems. The bigger problem comes when you have not retained the appropriate data or the information is incompatibly formatted.

The ultimate goal is not just to respond to major problems. Correlating customer satisfaction with operational aspects can provide vital guidance for improving a process and the product. However, there is no universal answer—what is useful to one operation is pointless busywork to another.

Gaining insight from process records is not new. In the early days of the Massachusetts Bay colony, the ultrareligious Pilgrim and Puritan colonists did not work on Sundays. This restriction extended to cooking meals. In order to have hot food on the Sabbath, women would prepare beans on Saturday night and leave the pot simmering over a low fire. However, long cooking times will soften beans to the point where the dish is an unpalatable slurry.

Early cooks recognized that this undesirable outcome did not occur every time. If the recipe included molasses or vinegar, the individual beans often survived intact. Food science tells us that acidity slows the legume-cooking process. The calcium in molasses reinforces the cell walls by cross-linking their pectin molecules. These factors preserve the integrity of the beans during long cooking cycles. This gave rise to the sweet Boston Baked Beans that we know today.

End-to-End Data Collection

Your most unhappy customers are your greatest source of learning.

Bill Gates

Unlike today's cautious and prolonged testing of new medical products, tests of the Salk polio vaccine in the 1950s were conducted on a crash basis. A huge trial was administered to 440,000 schoolchildren in 1954, to evaluate the vaccine's safety and effectiveness. The announcement of positive results electrified the American public on the morning of April 12, 1955. The U.S. government immediately issued licenses to six companies to manufacture the vaccine. The first shipment of the wonder drug was dispatched that same afternoon. Proposals were made to inoculate twenty to thirty million children over the next several months, prior to the expected summer peak of infections.

Two weeks later, the initial enthusiasm diminished when the vaccine made by Cutter Laboratories in Berkeley, California, was linked to the onset of polio in some of those vaccinated. Ten deaths and over a hundred cases of paralysis were eventually associated with an active virus in defective lots of Cutter's vaccine. Inoculations were halted across the country as an investigation evaluated the production process.

The polio virus was cultured in a solution made of pureed monkey kidneys to produce the vaccine. After a filtration step to remove the cultured cells, the virus was killed with formaldehyde. The investigators quickly focused on the plant's filtration equipment, which was less effective than the original laboratory-scale filters. The auditors found many deviations from Salk's recommendations when the six production companies struggled to produce commercial quantities of the vaccine. Generally, these improvised techniques accomplished the desired results.

However, at Cutter Laboratories, the process safety window was exceeded. The high-volume filters did not remove all the infected cells. Protected from the kill step inside clusters of cells, some of the viral material survived to contaminate two of eight batches of the product. The production process was hastily modified to provide better filtration, redundant kill steps, and additional tests for virus survival. By May 27, the government investigators certified that the problems were resolved and production resumed.

In the Salk vaccine case, all six manufacturing sites were apparently given the same guidelines for producing the product. However, each labored individually to implement the production of a temperamental new invention. The process window limits had not been fully explored and validated. No one was coordinating the critical procedures and best practices across all six locations. When visiting the various production operations, the investigators were surprised at how different the in-process batches looked.

Critical Questions: What process conditions/materials/procedures most urgently need to be correlated with product performance at the end-user stage? What am I going to do with all that data?

End-to-end data collection builds on the proactive record keeping described in the previous section. However, the earlier technique can sometimes take years to achieve results while companies wait for complaints or complements to be submitted by end users. The customer and the salesperson may feel that the response is too slow, especially if the supply chain is long or the feedback is fuzzy due to the time and the volume of data that are required.

End-to-end data collection requires accurate feedback on the performance of each batch of material. Getting uniform and detailed end-user performance results is always difficult. The customer and perhaps the field representatives need to be motivated to provide specific assessments of each shipment or batch. Rather than just hearing "good" or "bad," it is vital to get an objective ranking of the desired attributes.

In eighteenth-century England, Abraham Darby was a pioneering innovator in the production of cast iron. After learning the Dutch methods for casting brass with sand molds, he applied the technology to iron. This allowed him to cast parts with thinner walls than the traditional loam or clay mold materials. Darby overcame the shortage of charcoal for furnace fuel by implementing the use of coke. Iron production required large amounts of charcoal, causing rapid deforestation. The use of coke was feasible for Darby because of a fortunate coincidence. The coal in the vicinity of his Coalbrookdale facility was relatively free of sulfur, which has a negative effect on iron properties.

Darby's next problem was that different grades of iron ore each need their own special process accommodations. Darby and the French scientist René Antoine de Réaumur contributed to an empirical ore classification system, which clarified the matter. Ten grades distinguished between common impurities in European feedstocks. For instance, gray ore, containing graphite, was perfect for the cast-iron process. They could develop this information by experimenting with the various ores and noting their effect on the processability and finished-product properties. With this system, iron makers could choose the optimal raw materials for their process or select alternative end uses, based on the available feedstocks.

Performance feedback guidelines:

- The customer has to be cognizant of when one batch stops and another one starts. Someone needs to provide objective commentary on what is really going on at the customer's end. Perhaps the customer had issues and was able to compensate for the problem. Needing production, they ran the batch, rather than returning it.

- Objectivity is hard to find in an adversarial vendor–customer relationship. Purchasing agents hate when their own production managers admit, "that one was our fault."

- Detailed information should be retained on every lot. This must include anything pertinent to everything—the raw materials used, process conditions employed, equipment configuration, operator names, QC measurements, weather conditions, and even what products were run prior to this batch.

- It may also be necessary to perform additional tests beyond the normal QC measurements. Save and label samples for later evaluation such as FFU tests or specialized analytical methods. Having uniform information in one database facilitates the drawing of accurate conclusions.

- Obtain the defect samples, especially when differences in terminology exist. A detailed postmortem examination can reveal failure mechanisms and suggest root causes.

- Do everything possible to ensure consistency in the performance reporting from the customers. The frequency of the reports may drop over time when customers get tired of complaining or see no feedback for their effort. In others, the volume level of the complaints will rise when end users get angry that the problem has not yet been solved. Neither situation helps you to understand and solve the problem.

Auditing Product Performance

It is unwise to be too sure of one's own wisdom. It is healthy to be reminded that the strongest might weaken and the wisest might err.

Mahatma Gandhi

Cardiac stents are one of the most intensely audited engineered products ever produced. They are used to treat blockages that reduce the flow of blood to the heart muscles. These tubular meshes are guided into the clogged blood vessels and are expanded by pressuring an internal bladder. Afterwards, the bladder is withdrawn and the stent clings to the internal wall of the artery. This procedure can improve the blood flow and heart function. In the early 1990s, the mesh was bare stainless steel. Within 10 years, more complex composites were designed to slowly emit a steady dose of drugs to prevent a reoccurrence of the blockage.

These coated stents were later blamed for dangerous blood clots, months or years after their implantation. The causes and risks have been debated in the media, the courts, and medical journals. This is a situation where safety and performance can only be discovered by implanting the product into a large population of sick patients and then auditing the outcomes over a period of years. Laboratory animals may be useful for initial testing of the stents, but only full-scale human use will provide valid long-term results. In the course of their treatment, extensive medical records are maintained for each patient. These provide a vast amount of data. On the positive side, medical researchers can document the outcome of the procedures for the duration of the person's life. On the negative side, these records are cited in court cases if the patient dies.

In the world of normal products, much less data are available. Most complaints fail to communicate enough information to understand and resolve the situation. This is not the fault of the customers; they have better things to do than documenting the issues caused by your product. Their vocabulary, outlook, and motivation are out of sync with yours. They may correlate the problem with some product feature that is immaterial to the problem—"The drums with the green labels give us more problems." Perhaps the green label stock is only used when the plant runs out of the blue label stock.

Auditing implies a systematic and aggressive procedure for investigating and experimenting on product performance at customer locations. Audits can take a variety of forms. Send your application experts to watch a random selection of your product in use at the customer's location to observe problems in use. Or, observe a batch that was known to have issues at this customer's facility. Better yet, audit the performance of a batch of material that was closely monitored in production to understand the complete span of knowledge. Perform audits on samples from your designed experiments. Pare down the experimental variables to a manageable number through internal FFU testing.

> *Engineers can gain tremendous perspective on the performance of a product by visiting the customer and watching the product run for hours or days at a time. Visitors can inspect the "failed" products and compare them with the "good" ones. Tallies can be made of the number of items run versus those that are rejected. This will give an objective comparison of one lot versus another.*

Auditors will observe people using or abusing the product in ways that were never expected. The proportion of the product that is damaged in shipment will be clearly apparent. The effects of aging, settling, and leakage from containers will be obvious to people who have seen the making of the product. By using the product in the same order that it came off the production machine, then patterns in defects may become obvious, as never before. The audit is a relatively safe way to test beta versions of a developmental product. Two or three different versions can be compared with one another in a carefully supervised test.

Some vendors use the occasion of an audit to identify customer problem areas. The experts who perform the work may have been in any number of similar plants during their careers. Based on their observations and experience, the experts can make valuable suggestions to the customers on ways to improve their operation.

Active Experimentation

> There is something fascinating about science. One gets such wholesale returns of conjecture out of such a trifling investment of fact.
>
> **Mark Twain**

Prior to the dawn of recorded history, mankind learned to construct permanent shelters with mud bricks. The process was old when it was documented in the Bible's book of Exodus. A mixture of clay and straw fibers is shaped in a mold and dried in the sun. In arid climates, structures made with these bricks can last for centuries. However, they are suboptimal in wet conditions and in large structures with high dynamic loads.

Process innovators discovered that heating the bricks in a kiln fused the clay into a new state, creating a stronger and more water-resistant product. However, the original firing process was very fuel intensive, making this solution expensive. Initially, fired bricks were used only where necessary, such as exterior surfaces and points of heavy loading. In Mesopotamia, the fired bricks were 30 times more expensive than mud bricks around 2000 BC. By 600 BC, the Mesopotamian producers employed more energy-efficient kilns resulting in a cost differential of three to one. During imperial Roman times (300 AD), the ratio was two to one. This continual decrease in

the relative cost of the firing technology suggests process efficiency improvements from active experimentation.

Active experimentation is faster than passive data collection. However, it is costlier in sample generation and material that is given away for customer evaluation. In this method, you must steer the process to the extremes, making samples that will give the most information. The effects can be measured across a wide span of process conditions to determine which factors most strongly influence the performance in the end-use application.

The challenge is to understand which process parameters to adjust and how far to go from the normal aim point. Gather a group of people who have insight on those factors. This group should represent a wide range of operators, QC inspectors, process engineers, application specialists, and other knowledgeable individuals. Identify all the parameters that can vary from one batch to another. The list should be exhaustive and must not be limited by any self-imposed constraints.

Look for all types of variations within each sample population. Use internal FFU tests, customer performance, QC tests, and material properties as measurement tools. Big variations within a batch might indicate instability in the customer's operation or undiscovered factors in your process. If some of the best samples have low variability, then the group has picked the important factors and that the experiment was largely free of "noise" confuses the results. Positive results must be replicated with longer runs to demonstrate that you can control the process and have really discovered all the uncontrolled parameters.

In a perfect world, some samples will be very well received by the customers and others will be rejected as being the worst that they have ever received. More typically, the results will suggest how to manipulate the properties needed for individual applications.

Products sold to different applications might not be sensitive to the same parameters. The changes put in place to satisfy one customer may actually hurt the performance of other customers. Avoid changing the conditions on those other products until the requirements are determined.

Jan. 10, 1961   W. L. HAWKINS ET AL   **2,967,845**
ALPHA OLEFIN HYDROCARBONS STABILIZED WITH CARBON
BLACK AND A CARBOCYCLIC THIOETHER
Filed Nov. 29, 1956

Contrary to popular opinion, plastics do not survive forever in the environment. Polyethylene is susceptible to oxidation, which results in a decline in product strength. In the 1950s, researchers at Bell Laboratories were anxious to find ways to extend the life of plastic-coated telephone wires. Traditionally, telephone lines were jacketed with lead to insulate and protect them. However, lead coatings were heavy and expensive, and created environmental concerns.

W. Lincoln Hawkins and Vincent Lanza were tasked with solving the polyethylene problem. This coating process had originally been developed for military applications in the 1940s. In telephone installations, the wires were expected to last for decades. Instead, they tended to deteriorate in less than a year, especially when exposed to direct sunlight.

Hawkins and Lanza correctly diagnosed the root cause of the problem as oxidative degradation. The typical antioxidants of that period became ineffective when blended with the carbon black that was commonly used to prevent deterioration due to the ultraviolet radiation in sunlight. The team evaluated potential candidates by exposing various polymer blends to pure oxygen at elevated temperatures. By monitoring the oxygen uptake, they could quickly screen the relative performance of their trial samples.

Additives such as benzyl phenyl sulfide and dibenzyl sulfide were identified as the most effective solutions for telephone cable insulation. This was a situation where accelerated aging tests were the only feasible means to compare potential solutions.

See the "concept map of product development communication" at the end of the chapter, which graphically details the ideal project communications situation versus a dysfunctional chain of communication. The developers need an intimate knowledge of the end-use performance to successfully customize a product for each application. This requires that they must see the impact of every decision.

Customizing Products for Niche Markets

The more alternatives, the more difficult the choice.

Abbe' D'Allanival

In the late 1870s, George Eastman was a well-paid banker in Rochester, New York. As an amateur photographer, he made an investment in time and money to learn the intricate methods for preparing, exposing, and developing photographs with the wet-plate techniques of that time. Dissatisfaction with the incumbent techniques led him to investigate improvements to image-processing technology.

He first studied the literature and came across a British publication that described a photographic plate with a dry emulsion coating. While keeping his day job, Eastman experimented in his mother's kitchen to duplicate and improve upon this technology. Over the course of several years, he identified the most appropriate chemical coatings and invented a process to apply very uniform and consistent coat weights. Using his own camera and developing equipment, he was able to test experimental samples to measure his progress.

By 1881, he had been granted U.S. and U.K. patents and had set up his first production facility. He also acquired a partner who financed his business. Ironically, this financial backer was the second-largest producer of buggy whips in the country. Only then did Eastman quit banking and devote his full attention to the company that would become Eastman Kodak. His product was successful because photographers no longer had to prepare the plates, expose them, and then develop the prints in quick succession. By customizing photographic products to the needs of the amateur, Kodak exploited a specialty niche, which had not been served until then. This application grew into a mass market.

Customers will patronize companies that have the ability and willingness to tailor products to meet their requirements. The challenge for management is to decide which applications are right for this approach, without gambling their existence on every decision. Eastman patiently protected his primary income source until he had the right technology to service the market.

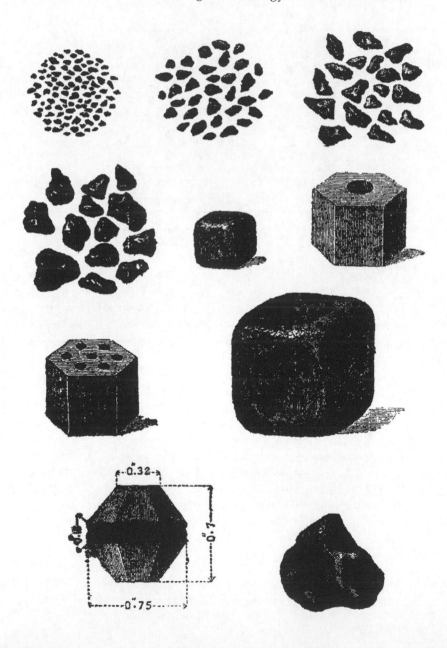

Gunpowder was introduced into Europe in the 1200s. At first, the traditional Chinese method of using a finely ground powder was copied. By the 1400s, European powder mills had discovered improvements to propellant technology. They mixed the powder with various liquids to fuse the individual components, and then ground the chunks into corns of various sizes, depending on the end-use application.

Small guns such as pistols needed a fine-grained product. Artillery called for larger pellets, with a slower burn rate. Tailoring the particle sizes to the gun chamber allowed for the most efficient burn. This minimized unburned powder from being blown out of the barrel before contributing its energy to the projectile.

The corning process kept the powder mixture from segregating during shipment, improving its shelf life. Eliminating fine particles also improved its processability and reduced the risk of accidents.

Critical Questions: What customer situations will prompt me to offer customized products and in what situations will I stick with the one-size-fits-all approach that optimizes my production process?

Requirements for customizing a product:

- The company needs to decide if it should customize products to end-use needs or be a commodity supplier. Meeting special needs requires a service-driven mindset operation and a flexible process. A customization strategy must be the shared goal of the entire organization.

- Your existing product line and service must be adequate enough that customers trust you to solve their problems with a customized solution.

- Your process needs to be efficient across a wide process range to make the required product variables. It is almost always possible to create prototypes of an awesome product. But, with an inadequate production scheme, your solution will either be too expensive or too variable.

- The process must be nimble enough to accommodate frequent changes, because customized solutions often require short runs. More process and application specialists are needed toward this end than operations that are geared toward a mass market.

- Your business should be structured to deliver and support each market segment. A production process alone does not represent a customer solution. The process must be fronted with a sales and marketing system that can identify unmet needs and service the solutions.

- The organization must constantly evaluate if potential applications are suitable to its capabilities. Production will struggle with an unmanageable collection of incompatible products if you get this wrong.

- You must have an outlet for changeover scrap from frequent grade changes on a dynamic process.

Tracing the Process Paths

We have a habit in writing articles published in scientific journals to make the work as finished as possible, to cover up all the tracks, to not worry about the blind alleys or describe how you had the wrong idea at first, and so on.

Richard Feynman

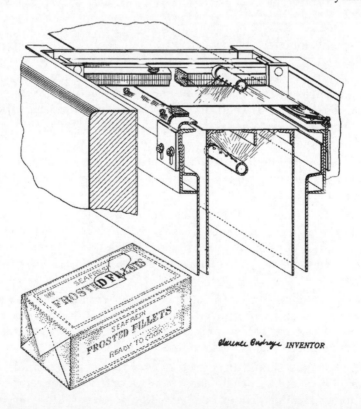

Clarence Birdseye INVENTOR

In the early 1920s, Clarence Birdseye understood the critical process parameters for freezing fish for long-term storage. Traditional freezing techniques produced large ice crystals, resulting in poor taste and texture. Birdseye understood that by freezing the product rapidly, the crystals remained small. To accomplish this, he packed the fish in uniform rectangular shapes and drew away heat by conduction through cold steel plates. This isolated the product from the toxic refrigerant chemicals while rapidly removing heat.

Initially, the business was not successful. The process was laborious, resulting in a high unit cost. Also, the public avoided frozen fish due to their experience with the incumbent technology. Birdseye automated the process, which greatly reduced the packaging costs. When the limiting technical factors were eliminated, public perception could be addressed with marketing.

The company was later successful in preserving fruit juice concentrate after further process innovation. But, market penetration was limited because the Birdseye process was not appropriate for vegetable products. Within a few years, Donald Tressler and others had discovered the required process steps. A brief blanching step prior to freezing curtailed an enzyme reaction that degraded product quality, even at subfreezing temperatures.

Commercial value requires both favorable process economics and good FFU. By constantly experimenting with all the possible options of an existing process, it is often possible to optimize both cost and performance.

Suppose that you need to create a piece of paper to handwrite a note. Because of the flexibility of human handwriting and our innate ability to discern shapes on paper, the FFU range of this paper product is very wide. The task can be accomplished with almost any variety of plant fiber. Every fiber process, from a handheld screen to a huge paper mill, can do the job. The resulting products might not be beautiful, but they will work in any dry environment.

Critical Questions: What is the magic combination of low cost and high performance for each customer application? What is the most efficient way to optimize that solution during the product development project?

In the real world, process choices are complex because so many factors influence the decision. In the final analysis, only a few options can run on the process and compete in the marketplace. A development engineer can cost the company a huge amount of money over the life of the product line by overlooking one combination of factors that is more economical or functional than those that are chosen. A good development team will evaluate and eliminate whole groups of process paths to work down to a manageable number of viable choices.

Plotting out the possible process paths can be a powerful project management tool. These charts can reveal development options that are being overlooked. They can also provide the team with a common vocabulary for discussing run strategies. A team leader might see that an engineer is focusing too narrowly, rather than exploring the whole field. Early in the project, a wide range of potential paths should be investigated. Toward the end of the project, these should be narrowed down to a few good ones.

Every situation calls for a different format to display the process path options. Your project will be defined by the important decision of which possibilities to include and which to omit. The process path shown below is hypothetical. It represents an early decision as to which type of process to use. As the choices are narrowed, some columns will be deleted and others will be added.

Predicting the viability of a process path from first principles is often impossible. Past experience can be deceptive because no two situations are

Example of a process path selection					
Process type	Process conditions	Materials	Feeding procedures	Reactor conditions	Finishing conditions
Sequential batch process	High temperature/ low pressure	Formula #1	Proposal A preheat the materials to 50°C	Introduce catalyst before low-intensity mixing	Cool to 35°C prior to depressurizing with no mixing
Large batch process	High temperature/ high pressure	Formula # 2	Proposal B no temperature control on feedstock	Introduce catalyst after low-intensity mixing	Cool to 35°C prior to depressurizing with low-intensity mixing
Medium batch process	Low temperature/ low pressure	Formula #3	Proposal C cool the feedstock to 10°C	Introduce catalyst before high-intensity mixing	Flush reactor immediately into separator
	Low temperature/ high pressure	Formula #3A		Introduce catalyst with high-intensity mixing	

identical. Usually, only empirical evidence and experimentation will prove or disprove the utility of a path. In a commercial situation, many potential process paths are nonviable, especially in a market that is constrained by competition. Team thinking is often biased toward a particular process line and the available feedstocks.

In the next example of process paths, a 2×2 designed experiment is being run to determine the effects of two process extremes. The team wanted to explore what would happen if they had no temperature control on the feedstock. By looking at the minimum and maximum possible temperatures, they can establish if this variable needs to be controlled.

They also wanted to investigate whether postreaction mixing was required during the cooling step. There were conflicting opinions on whether there were benefits or drawbacks to the mixing requirement. The feedstock temperature was known to be a factor for the advantages that were claimed in the two reports. This 2×2 designed experiment will separate these two factors and allow any interactions to be determined.

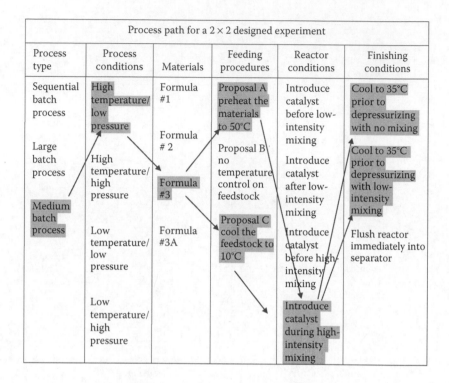

Process path for a 2 × 2 designed experiment					
Process type	Process conditions	Materials	Feeding procedures	Reactor conditions	Finishing conditions
Sequential batch process	High temperature/ low pressure	Formula #1	Proposal A preheat the materials to 50°C	Introduce catalyst before low-intensity mixing	Cool to 35°C prior to depressurizing with no mixing
Large batch process	High temperature/ high pressure	Formula #2 / Formula #3	Proposal B no temperature control on feedstock	Introduce catalyst after low-intensity mixing	Cool to 35°C prior to depressurizing with low-intensity mixing
Medium batch process	Low temperature/ low pressure	Formula #3A	Proposal C cool the feedstock to 10°C	Introduce catalyst before high-intensity mixing	Flush reactor immediately into separator
	Low temperature/ high pressure			Introduce catalyst during high-intensity mixing	

Summary

- One overriding rule applies when developing products on a dynamic process. You must identify every process factor that can influence product performance. Failures and performance variability will occur if you fail to grasp the effect of every parameter in the process and the secondary interactions between these primary factors.

- When evaluating the process parameters, consider the effect strength, the gain or loss of effect with knob position, the performance variability from one end user to another, and the likelihood of negative effects.

- Good record keeping in the process-dependent environment is a key to understanding the source of the problems or the benefits that are discovered with individual batches of product at end-user locations. Knowing which information to record and retain is often determined by trial and error.

- On-site FFU testing and audits can help to correlate the information in your records with the performance of each batch.
- Active experimentation with the process is often the best way to determine which process variables control the end-use performance, but this requires significant effort.
- Carefully consider when and if to create customized products for individual customers.
- You should map out the possible process path options to ensure that all options are understood and evaluated.

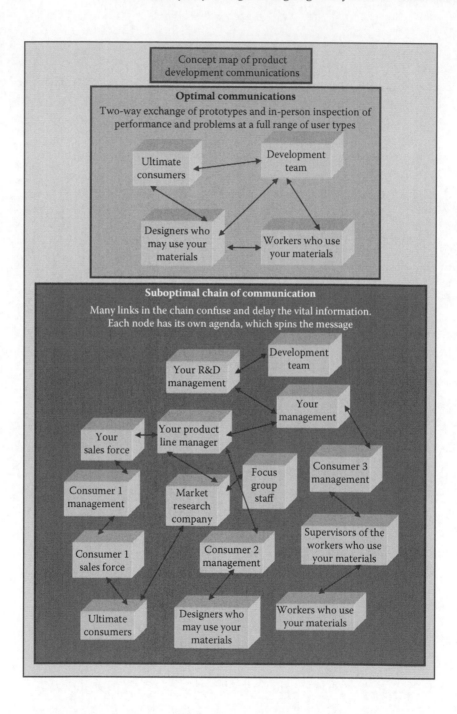

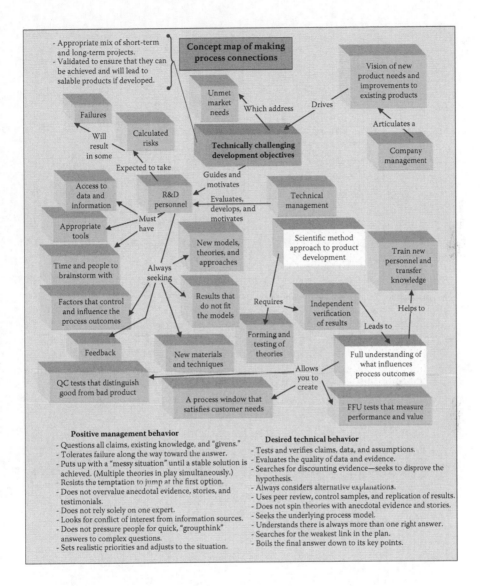

- Appropriate mix of short-term and long-term projects.
- Validated to ensure that they can be achieved and will lead to salable products if developed.

Concept map of making process connections

Vision of new product needs and improvements to existing products

Unmet market needs — Which address — Drives

Failures

Will result in some

Calculated risks

Technically challenging development objectives

Articulates a

Company management

Expected to take

Access to data and information

R&D personnel

Guides and motivates

Evaluates, develops, and motivates

Technical management

Appropriate tools

Must have

Scientific method approach to product development

Train new personnel and transfer knowledge

Time and people to brainstorm with

Always seeking

New models, theories, and approaches

Factors that control and influence the process outcomes

Results that do not fit the models

Requires

Independent verification of results

Helps to

Feedback

New materials and techniques

Forming and testing of theories

Leads to

Allows you to create

Full understanding of what influences process outcomes

QC tests that distinguish good from bad product

A process window that satisfies customer needs

FFU tests that measure performance and value

Positive management behavior
- Questions all claims, existing knowledge, and "givens."
- Tolerates failure along the way toward the answer.
- Puts up with a "messy situation" until a stable solution is achieved. (Multiple theories in play simultaneously.)
- Resists the temptation to jump at the first option.
- Does not overvalue anecdotal evidence, stories, and testimonials.
- Does not rely solely on one expert.
- Looks for conflict of interest from information sources.
- Does not pressure people for quick, "groupthink" answers to complex questions.
- Sets realistic priorities and adjusts to the situation.

Desired technical behavior
- Tests and verifies claims, data, and assumptions.
- Evaluates the quality of data and evidence.
- Searches for discounting evidence—seeks to disprove the hypothesis.
- Always considers alternative explanations.
- Uses peer review, control samples, and replication of results.
- Does not spin theories with anecdotal evidence and stories.
- Seeks the underlying process model.
- Understands there is always more than one right answer.
- Searches for the weakest link in the plan.
- Boils the final answer down to its key points.

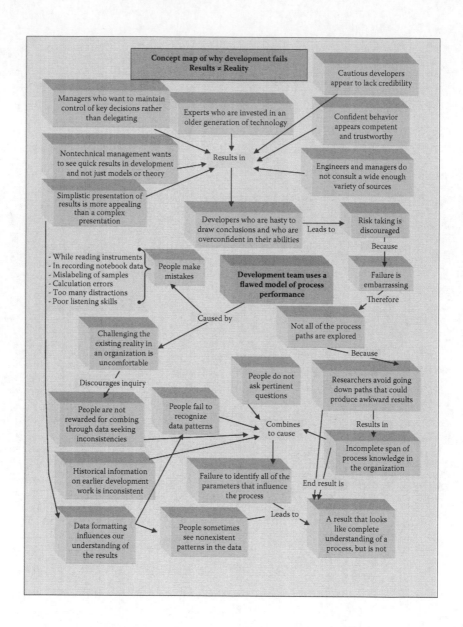

Concept map of why development fails
Results ≠ Reality

Cautious developers appear to lack credibility

Managers who want to maintain control of key decisions rather than delegating

Experts who are invested in an older generation of technology

Confident behavior appears competent and trustworthy

Nontechnical management wants to see quick results in development and not just models or theory

Results in

Engineers and managers do not consult a wide enough variety of sources

Simplistic presentation of results is more appealing than a complex presentation

Developers who are hasty to draw conclusions and who are overconfident in their abilities

Leads to

Risk taking is discouraged

Because

- While reading instruments
- In recording notebook data
- Mislabeling of samples
- Calculation errors
- Too many distractions
- Poor listening skills

People make mistakes

Development team uses a flawed model of process performance

Failure is embarrassing

Therefore

Caused by

Not all of the process paths are explored

Challenging the existing reality in an organization is uncomfortable

Because

Discourages inquiry

People do not ask pertinent questions

Researchers avoid going down paths that could produce awkward results

People are not rewarded for combing through data seeking inconsistencies

People fail to recognize data patterns

Combines to cause

Results in

Incomplete span of process knowledge in the organization

Historical information on earlier development work is inconsistent

Failure to identify all of the parameters that influence the process

End result is

Data formatting influences our understanding of the results

People sometimes see nonexistent patterns in the data

Leads to

A result that looks like complete understanding of a process, but is not

7

Organizing Development Projects

How do you organize, plan, and evaluate development projects? How do you define project deliverables, specifications, quality levels, and cost targets?

> If the finished parts are going to work together, they must be developed by groups that share a common picture of what each part must accomplish. Engineers in different disciplines are forced to communicate; the challenge of management and team-building is to make that communication happen.
>
> **K. Eric Drexler**

Thomas Edison is not appreciated as a production process innovator. In fact, the novelty of his inventions required him to develop new materials that optimized device performance. He was not the first to demonstrate an incandescent light bulb. The earlier models used inappropriate filament materials and only operated for minutes before burning out.

Edison famously screened 6000 materials to select the optimal filament. However, popular accounts neglect the process that he developed to condition the filaments prior to use. Most substances have tiny voids or occlusions in their internal structure. When current is applied, the minute traces of gas inside these defects can expand. This will damage the fragile elements and contaminate the vacuum inside the bulb. Most of Edison's competitors did not realize that trapped gas pockets could drastically shorten the life of a bulb until they read his patents.

One of Edison's teams concentrated on extending the bulb life by preconditioning the elements under vacuum. The entrapped gases were coaxed out of the filament by slowly heating them with increasing levels of applied voltage as the off-gassing was pumped out. Only with this procedure could the team evaluate the relative merits of each candidate material and steadily increase its service life.

Other teams were developing the other essential elements of lighting technology and power distribution in parallel. Edison coordinated their efforts toward his overriding vision to create a total systems solution to replace gas lighting.

In the 1930s, a different situation confronted the early developers of the polyethylene production process at Imperial Chemicals Industries (ICI) in the United Kingdom. Michael Perrin, John Paton, Edmond Williams, and Dermot Manning were working on a pilot plant reactor to scale up a polymerization reaction based on earlier ICI bench-scale research. The process

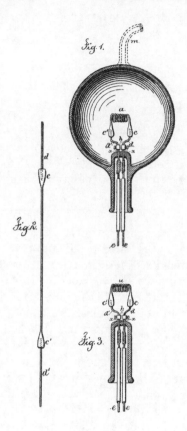

T. A. EDISON.
Electric-Lamp.

No. 223,898.   Patented Jan. 27, 1880.

employed an autoclave, where ethylene gas was subjected to very high temperatures and pressures.

$$\left(\begin{array}{cc} \text{H} & \text{H} \\ | & | \\ -\text{C} - \text{C} - \\ | & | \\ \text{H} & \text{H} \end{array}\right)_n$$

Regardless of how the development team held the conditions constant, the reaction dynamics were highly variable from one trial to the next. Eventually, they discovered that trace amounts of air sometimes leaked into the ethylene gas feedstock. A threshold concentration of oxygen was needed to trigger the polymerization process. However, a slightly higher oxygen level was found to poison the reaction. The polymerization kinetics were highly nonlinear with respect to trace contaminants. For both the light bulb filament and the polyethylene reaction, controlling tiny amounts of residual air was the key to successful processing.

Until the team discovered oxygen's key role in the process, their work over a period of months was a frustrating series of failures and nonreproducible results. However, they knew from earlier experiments that a solution existed and they believed that there was a lucrative market for the product. For this reason, the team persisted until they discovered the important process knobs. The next problem was to control the reaction, once it was triggered. The autoclaves would occasionally overheat and explode. It was necessary to efficiently remove the excess heat from the exothermic reaction to prevent catastrophic runaway reactions.

The storied American chemist Carl S. Marvel at the University of Illinois was able to produce and identify polyethylene by a different synthesis technique several years ahead of ICI. Rather than commercializing it, he dropped this line of inquiry believing there were no applications for polyethylene.

INVENTOR.
EARL S. TUPPER

A decade later, most polyethylene production by the ICI process was being used in military applications, such as wire insulation for sophisticated radar systems. After the war, the American entrepreneur Earl Tupper used polyethylene slag, a waste product of petroleum production, to mold a variety of cheap consumer products. He had developed a process to refine this greasy scrap material into a polymer that was clean enough for the public.

Tupper produced a wide range of products that could be made with his chosen raw material and process technology. He pioneered methods for molding hinged closures and snap-fit lids. Initially, the company supplied useful, but inexpensive molded shapes that were given away as promotional items by consumer brands. Eventually, one of his product ideas became more successful than his other ideas, generating impressive sales volumes. The public was excited about his nonbreakable food storage containers. These products facilitated the easy storage of leftover food with less mess and spoilage. By the 1950s, the company had become purely market focused on this application. Tupperware gained a dedicated following, with a large sales force that hosted parties in consumer's homes to pitch the concept.

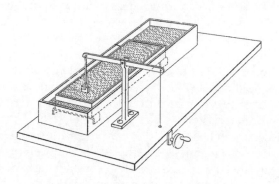

Nov. 5, 1940.   K. B. BLODGETT   2,220,860
FILM STRUCTURE AND METHOD OF PREPARATION
Original Filed June 11, 1937

During the 1930s, Dr. Katharine Burr Blodgett conducted research for General Electric (GE). She had the good fortune to be mentored by the Nobel Prize-winning chemist Irving Langmuir, a pioneer in surface chemistry. Blodgett was allowed to explore basic technology, without regard to short-term commercial needs and customer problems. She pioneered new processes for applying monomolecular coatings onto surfaces. With this new technology, GE developed impressive new products, such as the first nonreflective glass in 1938.

These developments were useful in advanced optics applications such as telescopes and microscopes, where glare distorts the image. Blodgett's innovations created a new product category—industrial coatings and antiglare technology. To accomplish this, she developed process and analytical

RULE #36 DETERMINE WHICH COMES FIRST, THE CHICKEN OR THE EGG

A development project starts in one of two ways. There may be a picture of the desired end result, such as electric lighting, in Edison's case. In this market-pull situation, the goal is to identify the best process to service that application. Or there may be a promising new material or process path with uncertain end uses, as was the case with polyethylene, plastic molding, and monomolecular coatings. These technology-push projects seek customer problems that can be solved with the new technology.

techniques to apply and measure exceedingly thin coating layers. When it works, fundamental research enables revolutionary new products. However, the failures of this approach are never as well publicized as the success stories. Even successful programs can take decades from initiation to commercialization.

The developmental route for either approach is never certain. A systematic exploration of all the process paths can lead to the optimal solution, if one exists. Often, the commercial outcome is different from the original goal, because of development findings and end-user inventiveness. Initial commercial assumptions will be proved false and unanticipated options will be revealed. The organization must be flexible to pursue solutions that are workable, economical, and appropriate for its situation. Suboptimal paths and potential applications must be explored, documented, and rejected in a timely manner.

The development strategy is different for each product type. Approaches that work well for mechanical designs, electronic devices, or software projects will not be appropriate for process-dependent products. Even within a company's suite of projects, the development procedures must suit the circumstances. A project to improve an existing solution will be executed one way. An exploratory effort to develop a totally new process for a new market follows a different methodology.

Early on, the team must establish whether the project is viable or not. Its work can be stopped or redirected if the findings or changing market conditions dictate. The organization should have a mechanism to periodically monitor the project's status, spending, and prospects for success. The management must understand the trade-offs and alternatives that the team uncovers. And in return, the team should be kept informed about changing priorities and market requirements.

A poorly managed team could flounder and fail in a hundred ways, even while it appears to be working hard and is making impressive presentations. Dysfunctional teams will deemphasize the important considerations while immersing themselves in minutia. In the worst case, they will realize the futility of their task, while the management continues to push the project.

The team and the company leaders need two-way communications regarding the important elements and opportunities around the project. The relationship depends on the nature of the company, which, in turn, dictates the character of the R&D organization.

RULE #29 LIST AND REFINE PROJECT REQUIREMENTS

A well-managed project team should be focused on discovering and constantly refining a list of requirements for the desired product, a scheme for its production, and a model for the costs. It must seek to understand whether the systems available to it are capable of producing a result that is aligned with the company's goals.

The "concept map of R&D mission statements" at the end of the chapter describes how you should choose one guiding principle and stick with it. The following two graphics detail the requirements for project deliverables and constraints. In all three cases, you need to focus on a few critical items.

Defining the Product

> An undefined problem has an infinite number of solutions.

Robert A. Humphrey

> ... constraints help us reframe the problem and discover new opportunities in the process.

Roger Martin

What is not constrained is not creative.

Philip Johnson-Laird

Reportedly, the Boeing Company had a very terse set of design specifications for its 727 airliner project in the early 1960s. The plane was required to fly 131 passengers nonstop from Miami to New York City's LaGuardia runway No. 4-22. The story is simplistic, considering the stringent requirements for the licensing of new airframe designs. However, it is a good model to follow—*clearly articulate the critical requirements so that the designers have a clear vision to craft an elegant solution.*

Before undertaking any development project, a comprehensive set of product guidelines must be drafted and prioritized to articulate a guiding vision. Management must invest time and effort in setting appropriate (or even inspirational) goals if it expects to achieve favorable results. It must ensure that the deliverables match or exceed customer expectations and that the

RULE #50 CAREFULLY CONSIDER "DON'T WANTS" WHEN COMMISSIONING PROJECTS

In addition to the deliverables, consider what to avoid—the constraints that are put on the development team. Mature organizations are very good at creating constraints, even while being visionary with the deliverables. Every constraint should be documented and justified at the start.

product can be produced efficiently. Deciding what activities you do and do not want in the scope of a project is crucial.

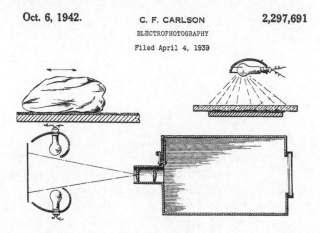

Oct. 6, 1942.   C. F. CARLSON   2,297,691
ELECTROPHOTOGRAPHY
Filed April 4, 1939

In the 1930s, Chester Carlson was working in the patent department at the electronics firm P. R. Mallory & Co. He was frustrated by never having enough copies of key documents. This prompted him to seek a novel technology to create duplicates on demand. From a patent search, he concluded that photographic film technology had already been extensively researched and patented, so he avoided the photography approach.

Following up on the basic research of Hungarian physicist Paul Selenyi, Carlson explored the change in electrical conductivity that occurs when light strikes a photoconductive material. He envisioned projecting the likeness of the document onto a surface made of such materials. In a series of experiments, he demonstrated that pigment particles could be preferentially attracted to the charged areas created by the projected image. In subsequent tests, the particles were transferred onto paper. Later still, they were permanently fixed to the paper. Eventually, Carlson and physicist Otto Kornei perfected every aspect of the process, one unit operation at a time.

With his background in patent law, Carlson documented every step of the discovery in secret. However, his concern for security was unfounded. No one else was interested in the concept, including the 20 companies that he approached with the idea. After years of effort, he teamed up with Battelle Memorial Institute and Haloid, a small photopaper company. Haloid later renamed itself Xerox and revolutionized the office environment with its hugely successful 914 plain paper copier in 1960.

Important questions:

- What are the cost or price targets?
  - What are the basic assumptions?—freight on board, location, volume, and seasonality of demand.

- Can the cost be higher if performance goals are exceeded or if superior properties can be achieved?
- Will any subset of the market accept a less functional offering at a lower price? Would such a product be a disruptive innovation? If so, would it help or hurt our market position?
- What is the scarcest resource that constrains the project?
  - Time to market?
  - Project expense?
  - One of the key human resources?
  - A particularly critical piece of equipment?
- What critical trade-offs might the project face? Which side will be favored if they occur?
- What are the "kill factors" that will end the project if they come about?
- Who are the targeted customers?
  - Do we know what is really important to them?
  - What factors/situations do the customers want to avoid at all costs? What constraints do they place on this product? Do all customers have the same constraints?
  - Who are not the customers for our products today and what features or price point would it take to win them over? Should the project team try to address their need?
  - Which customers will let us see their operation and test our prototypes in their facility?
- What key problem does the proposed product solve and what alternatives do the customers have?
  - What are the cost/benefit compromises of the alternate solutions versus our product?
  - What things constrain our competitors? What do they do well and what are their weaknesses?
- What properties are required?
  - Example: bulk density not to exceed 1200 kg/cm^3 at ...
  - Example: must have a water activity factor of X and a specific gravity between ...
  - Example: coefficient of friction against Teflon at 50°C must be between 0.2 and 0.25.
  - Example: must resist saltwater corrosion under internal test method X for 14 days.

- What are the end-use requirements?
  - Example: must be able to process on customer ABC's XYZ machines at a rate of 190 cycles/minute with no more than 1% misfeeds.
  - Example: must have sufficient strength to support 100 t/m load as specified by test method ASTM # ...
  - Example: When mounted on the following vehicles, it must be able to withstand the 100,000 km road test at ...
- Timing requirements
  - Alpha prototype to be accepted by the market manager by June 1.
  - Field test production must commence no sooner than May 12, but before the customer's summer shutdown.
  - Full release to sales by ...
- What are the ancillary requirements?
  - Example: must be packaged safely in the following container sizes ...
  - Example: must be able to withstand air shipment/mountain pass atmospheric pressure conditions to an altitude of 3000 m without ...
  - Example: must have a 25°C, 40% RH storage life of ...
  - Example: must be available in the following textures, surface finishes, and colors ...
- Production needs/constraints
  - Must be manufacturable with no more than a 30-minute change-over cycle on the following production trains ...
  - Must not require the introduction of new hazardous materials into the plant inventory.
  - Raw materials must be delivered from silos D and F ...
  - Must run within the available utilities in plant ...
  - Capital upgrade costs may not exceed ...
- What intellectual property position do you want/need to have?
  - Under an umbrella of existing patents—often fast and low risk, but not innovative.
  - Using known technology—hard to get competitive advantage.
  - Novel, proprietary position—time consuming and lower probability of success.
  - License and customize existing patents—less risky, if you can find and afford them.

- Service requirements/constraints
  - "Product will be shipped in the following transport modes ..."
  - "Must be compatible with the following distribution and retailer support systems ..."
  - "User's manual and data sheets must be approved by the following deadline ..."
- Financial concerns: Should we build a high-speed production train that is efficient for long runs at high output rates? Or, do we design for a series of small, nimble lines that can switch from one product to another very quickly? In the first case, the unit production cost is low, but there is a lot of capital tied up in the inventory because the plant cannot make products to order.
- Profitability concerns: Do we serve the entire market by offering a wide variety of options? Or, do we concentrate on the highly profitable 20% of the end-user applications that generate 80% of the cash flow? Are we market focused or process focused?

Many projects are based on assumptions about the properties, attributes, and constraints required to satisfy the customer's needs. If these assumptions are valid, then there is a chance that the development team can reach an acceptable conclusion. This is often the case for projects that attempt to make modest changes to an existing product. More ambitious leaps in performance are less likely to meet the targets within a reasonable time frame.

Every project team should be tasked with the goal of proving that the proposed goals are flawed. If a team succeeds in doing so, then the company will have saved a significant amount of time and money. In practice, this is rarely ever done unless the errors are too obvious to ignore. Corporate culture frequently prevents teams from questioning the wisdom of their charter.

The development team should be empowered to ask for clarification or additional guidance on items that are not clear to the team. It should also have the flexibility to answer the questions itself, if the sponsor is not able to do so. By necessity, the project time line and budget cannot be finalized until all of this basic information has been established.

In some situations, the project team may be expected to develop these guidelines based on its own research into customer needs. Ideally, this is done by a small market research team beforehand.

Conflicts will arise when two important properties are found to be negatively correlated. For instance, achieving a particular end-use performance goal might require a very expensive raw material or a very poor process yield that will cause the product cost to exceed the target. Decisions are needed at an early stage as to whether one should be optimized at the expense of the other or if a compromise is appropriate. If the team is polarized into two camps over the conflict, then no progress will be made.

Conflicts in Creating Specifications

Oh no! You did it exactly the way that I told you!

Unknown

The external fuel tank on NASA's space shuttle is an extraordinary collection of materials, every one of which is mission critical. The huge container provides fuel for the first 2 minutes of the mission, before it is dropped into the ocean. The design requirement is to contain the liquid propellant, with the minimum amount of packaging weight. The distinctive orange color is the natural result of ultraviolet damage to the polyurethane foam insulation skin, rather than paint. Without the insulation, water vapor would freeze to the skin, due to the cryogenic temperatures of the liquid fuel inside. Ice would add weight and break off during take-off, damaging the orbiter.

Serious problems also occur when fragments of foam come loose at the peak velocity prior to separation. After decades of use, it was determined that no amount of process adjustment could guarantee that every speck of insulation would stay in place. Invariably, urethane foam will develop voids, cracks, and localized adhesion failures in the months between its fabrication in Louisiana and the launch from Florida. The process of mixing the urethane precursors, dispensing the rising foam and shaping it into place is subject to a number of uncontrolled variables. Factors such as a pressure drop across the nozzle, mixing uniformity, shear stress during dispensing, and the temperature fluctuations of the two components are integral to the formation of foam defects.

Eventually, NASA concluded that the urethane approach had insufficient strength and redundancy to withstand the supersonic airflow conditions. The cohesive failure of the foam could and would occasionally cause serious damage to the orbiter's heat-resistant tiles. This insurmountable performance shortcoming was a major reason for retiring the shuttle fleet, rather than extending its operational life. This failure can probably be traced back to a specification decision early in the shuttle program. It took decades for the outcome of this safety versus weight trade-off to become evident. Poorly chosen product requirements, deliverables, and constraints inevitably lead to performance problems.

Sept. 25, 1951   W. SHOCKLEY   **2,569,347**

CIRCUIT ELEMENT UTILIZING SEMICONDUCTIVE MATERIAL

Filed June 26, 1948   3 Sheets-Sheet 1

At the end of World War II, Bell Laboratories commissioned the most influential civilian research project of the twentieth century. A team headed by William Shockley sought semiconductor materials that could act as solid-state analogs of vacuum tubes. Within 2 years, it had succeeded in creating a crude working transistor.

This first discovery, built on technology in an obscure 1924 Canadian patent by physicist Julius Lilienfeld, was the field-effect transistor (FET). While the properties of this approach were encouraging, the performance had serious limitations. Exploratory work continued to look for a better solution. By 1951, the team had perfected a doped germanium matrix that was contaminated with gallium and antimony. This p-n junction transistor was the basis of commercially successful products, rather than the team's original 1947 discovery, which is often cited as the birth of the silicon revolution.

The time-consuming element of the team's research involved a process for growing perfect single crystals with a sandwich structure of germanium containing different impurities. The crystal formation step was followed by an etching process to expose the innermost layer in order to attach a contact wire. The final challenge was to achieve precise control of the center layer thickness. This parameter was the key to accommodating rapid signal changes in an electrical circuit.

Despite winning the 1956 Nobel Prize, the team did not meet the original goal of mimicking the vacuum tube performance. Semiconductors responded differently and lacked the power capacity of the incumbent technology. Instead, device developers took advantage of the material properties to create totally new products. Electrical designs had to change to accommodate this powerful tool, rather than just replacing vacuum tubes with transistors. The transistor ultimately revolutionized electronics because of those differences.

Conflicts in project goals:

- Do project sponsors carefully think through all the ramifications of their requests? Does anyone test and verify their assumptions? Is the refinement and clarification of goals left to the project team?
  - Defining the customer needs can add significantly to the project duration. The longer the project is, the less appropriate the original charter becomes.
  - As the team evaluates requirements, "scope creep" sets in. The participants find additional features and applications that could be accommodated. Projects become endless.
- Individual items on the requirement list are not prioritized. Some people treat them as suggestions and others feel that they are must-do priorities. The longer the list is, the more confusing this becomes.
  - Requests for clarification from the project team sound like criticism to the sponsor.
  - Items high on the list could be regarded as being more important than those that are at the end of the list. The person who compiles the list from assorted sources might not have that intention.
  - Not all the decision makers have the same understanding of what the requirements mean or how they control product performance for the end user.
  - There is a natural bias toward including more items, rather than less.
- Items on the list are poorly thought out.
  - Detailing one way to achieve the objective, rather than describing the desired result.
  - Individuals from diverse backgrounds may interpret the words in different ways.
  - Lists from previous projects are pasted together, without adequate integration.
  - Boilerplate items are included, regardless of their value to the customer.
  - Failure to discover what customers really need or which new attribute would delight them.
  - Customer needs from yesterday are listed rather than tomorrow's reality.

- The needs of all stakeholders are not taken into account.
  - The full scope of customer requirements is not clearly defined at the outset.
  - Product attributes are listed, but service needs are not defined.
  - Complex product concepts may not be fully understood by some participants until they can actually see the nearly finished product in action. Only then will the opportunities and drawbacks occur to them.
  - The more revolutionary the concept is, the more it will evolve during the development cycle. The finished product might have little relationship to the original project proposal.

Streamlining project deliverables:

- Allow all groups to comment on project requirements before final approval. Include all affected functions. Resolve problems with ambiguity, contradictions, and fuzzy logic before the work begins. Features that add significantly to resource requirements or time to market should be defended by the sponsor before a management review.
  - Manufacturing should outline the needs for any new product that it will produce.
  - The sales group should specify its requirements for samples, brochures, advertising photographs, and other collateral material. This is vital if the sales cycle revolves around key deadlines dictated by trade expositions, catalog deadlines, or seasonal offerings.
  - Regulatory and safety compliance officials within the organization should review the product's implications for their areas of responsibility at an early stage. They should be kept abreast of changes in the schedule and scope.
- Create discrete subprojects to define customer needs for later development efforts. The deliverables could be a background knowledge of the applications, fitness-for-use (FFU) test methods, or an understanding of specific end-user requirements. This preproject activity delineates clear project goals for product development. The more ambitious the project proposal is, the greater the need for a definition effort.
- Allow free-market principles to guide the deliverable list. If the sponsor loads up the list, then the project duration and the development costs will increase appropriately. In doing so, the projected return on investment will decrease and the relative justification for the product will erode.

- Deliverables should describe the project outcome from the customer's perspective.
  - Detail user wants and needs in order of priority. The list should start with the "must have" and end with the "it would be nice to have" features.
  - Never confuse the "wants" of your organization with the "needs" of real customers.
  - Express what is required in the product, not how to achieve the end result.
  - Deliverables should be measurable and verifiable goals for the project team to work toward. Whether they were met should not be a matter of debate.
  - "Other goals will be added as we think of them," is not an acceptable option.
  - Avoid ambiguous requirements, such as "the customer must be happy with the results."

Four critical project parameters:

- *Unit Cost*: What is the maximum acceptable production cost per unit of product?
  - The higher the sales volume is, the more you can reduce costs with an economy of scale.
  - What is the customer's maximum acceptable price point? The customer never tells you this directly. You must estimate the costs of their alternatives.
  - How much could the competition lower their price in response to your new product?
  - Process automation and tooling reduce production cost at the expense of the following three items.
- *Time to Market*: When must the product start satisfying customer needs?
  - What key deadlines dictate product availability requirements?
  - Can the development time line be met, considering the following:
    - The amount and quality of the technical resources employed.
    - The availability of small-scale or pilot plant equipment.
    - The number of features or the technical complexity requested in the deliverables.
    - Accurate and comprehensive project objectives.
    - The amount and quality of the existing process and material knowledge.

- – The availability of the appropriate equipment and validated procedures for development activities and FFU testing.
- *Project Cost*: What resources must be spent to complete the project? (money and man-days)
  - Raw materials, supplies, plant labor, etc.
  - Include up-front costs such as production tooling and process reconfiguration.
- *Functionality*: How good should the product be?
  - The perceived value to the customer will increase the sales potential and price points.
  - Functionality is FFU, customer value, and low variability.
  - Increasing functionality generally takes more time and raises production costs.

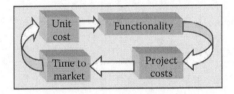

Obviously, these four key items cannot be evaluated in isolation from one another. They are linked and must be addressed as a group. This linkage sometimes escapes the notice of those who are too close to the choices or too focused on their own specialty. The single-minded optimization of one parameter at the expense of others is usually foolish. The importance of each item depends on whether the company is market focused or process focused.

Critical questions for project prioritization:

- What products and process improvements are most urgently needed to satisfy our growth strategy or protect profitable business? What are the rate-limiting factors for their development? Do we have the basic technology that creates a market advantage over competition? Are resources being dedicated to accomplishing these goals at the expense of all others?
- Which material properties are vitally needed in these solutions to address the most important customer problems? Which features can be deferred to later releases to gain incremental growth in the future, without delaying the core product's introduction?

- What is the market's volume versus its cost sensitivity? How much time and money should be spent to minimize the unit cost in order to maximize the return on investment?
- What will be the return on investment from the allocation of additional development resources to complete these projects sooner? How elastic are the project needs? Are there inherent development rate limitations that cannot be overcome with more people working on them?

Costing Concerns for Dynamic Products

Prediction is difficult, especially the future.

Niels Bohr and others

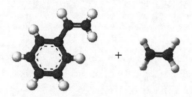

In the 1990s, a global petrochemical producer introduced a family of polymers produced with a proprietary metallocene catalyst technology. These materials were copolymers of ethylene and styrene monomers. The performance characteristics were not revolutionary in the sense of a Teflon or Kevlar. However, the product offered a combination of an efficient polyethylene production process with specialty properties.

The supplier negotiated with a number of plastics processors to develop applications for this new material prior to the commercial launch. The plan was that a market base could be established quickly to justify a large reactor dedicated to this product line. These partners used the supplier's pilot plant samples to develop products, basing their commercial plans on the promised resin price.

Several years into this cooperative development, the company discovered that it could not produce the polymer for the price that was originally forecast. The process was still so inefficient after optimization that costs were several times higher than anticipated. Some customers failed to find a good fit. Others were not interested in the much higher price and the entire product suite was scrapped with huge losses.

RULE #13 ALWAYS KEEP PRODUCT COST IN MIND DURING PRODUCT DEVELOPMENT

Correctly forecasting and achieving the production cost targets for new products is a vital development goal. This is especially true when significant time is required before the product runs on a full-scale process. Development teams must constantly reevaluate the cost model assumptions at every step of the project. An early resolution of costing uncertainty issues must be a top priority.

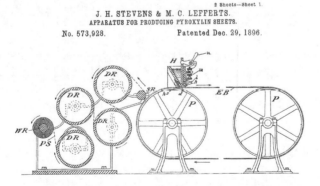

2 Sheets—Sheet 1.

J. H. STEVENS & M. C. LEFFERTS.
APPARATUS FOR PRODUCING PYROXYLIN SHEETS.

No. 573,928.   Patented Deo. 29, 1896.

The Englishman Alexander Parkes gave birth to the modern plastics industry with the commercialization of celluloid, the first synthetic polymer. In 1862, he won a bronze medal at the Great Exhibition in London, which generated considerable interest in his revolutionary material. When softened with heat and a solvent, Parkes could roll and form the polymer into useful shapes. However, his company struggled to advance the concept from a novelty to a viable commercial reality. The process was inefficient and expensive. As the pioneer in a new technology, equipment and procedures had to be invented for every step. A series of cost-cutting efforts resulted in performance problems and drove away customers. Bankruptcy soon followed.

In the wake of Parkes' failure, John Wesley Hyatt and Daniel Spill worked separately to make celluloid, with Spill also declaring bankruptcy at one point. Hyatt famously succeeded in using celluloid as an alternative to elephant ivory in billiard balls and piano keys. Garment applications such as collar stiffeners, cuffs, and false shirtfronts contributed significantly to volume growth. Hyatt mastered the critical techniques for using camphor as a plasticizer and then driving it out of the material during processing to produce a stable finished product. He invested in efficient production technology to hold costs down.

The more innovative the process is, the more difficult it will be to estimate the product's mature production cost. Entrepreneurial organizations forge ahead on faith, in hopes that the costs will be reasonable or that they will find a compelling application where the material can command a high price. This is a fundamental disconnect that can kill a radical new product late in the development cycle. The sales and marketing people think they can see a lucrative opportunity for the new technology. However, they cannot discern the market's price sensitivity until they have a product for customers to try. The technical people can envision a production route to make the desired material. But, because of process dependency, they cannot calculate the cost to satisfy all the end-user requirements prior to scale-up.

The management must obtain accurate answers to reduce the uncertainty. How much money must be sunk into its development before we can determine if the proposed product will be viable? What is the expected production cost to achieve good performance? These questions are fiendishly complex when you lack an understanding of both the process technology and the product's value proposition.

This great uncertainty magnifies the importance of selecting the right process for the job. It is madness to build a totally new process before you can estimate what its output will cost and what price the product can be sold for. If a pilot plant or an existing manufacturing process can be modified to provide limited quantities of products, then these products can be test marketed to establish end-use utility and thus the target sales price.

Planning for a Quality Outcome

> Don't undertake a project unless it is manifestly important and nearly impossible.
>
> **Edwin Land**

$$\left\{ O - \underset{\underset{R^2}{|}}{\overset{\overset{R^1}{|}}{Si}} \right\}_n$$

A building supply company developed an enhanced roofing membrane that was innovative and functional. Initial samples of the invention were refined and improved to make a product that could be produced at a price appropriate for the application. The optimal material was a combination of silicone-coated polyethylene and a tar compound. In the production process, it was necessary to cross-link the silicone, transforming it from a free-flowing liquid into a tightly bonded matrix. To accomplish this transformation, a platinum catalyst was added at parts per million levels.

After the company began commercial production, problems started to appear. The customers complained of variability from one delivery to the next. Some batches of silicone did not seem to cross-link properly. When this occurred, the tar's hydrocarbon components permeated through the polyethylene, fusing the rolls together. Extra testing was required to determine which batches of product were suffering from the issue. No amount of tinkering with the silicon formula could resolve the situation.

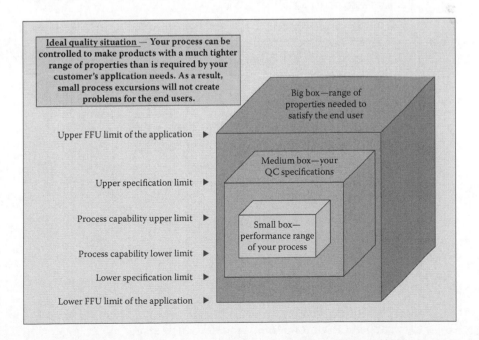

Ideal quality situation — Your process can be controlled to make products with a much tighter range of properties than is required by your customer's application needs. As a result, small process excursions will not create problems for the end users.

Big box—range of properties needed to satisfy the end user

Medium box—your QC specifications

Small box—performance range of your process

Upper FFU limit of the application ▶

Upper specification limit ▶

Process capability upper limit ▶

Process capability lower limit ▶

Lower specification limit ▶

Lower FFU limit of the application ▶

Eventually, the root cause of the problem was traced to the polyethylene, rather than the silicone. Minute amounts of specific additives in the plastic resin (antioxidants, amines, and certain pigments) tended to poison the silicone cross-linking reaction. Two identical rolls of film might have different additive packages, depending on the supply-chain factors. Few users of plastic film were concerned about such arcane formula fluctuations, because their applications were insensitive to these factors. After that expensive education, the roofing company had to carefully screen and scrutinize the additive choices of its film suppliers.

The manufacturer learned that quality in a dynamic process is not just a matter of maintaining machine conditions in a tight range and testing product properties. The factors that control product performance must be identified during the development phase. Understand what is required to delight the customer before making promises. It is not a quick or easy quest.

Developing quality:

- Learn all the product attributes that are crucial for performance in the application.

    - Product attributes are fitness requirements, not quality control (QC) specifications, which come later.

    - Note the phrase "all the product attributes." It is easy to identify the primary aspects of performance. Uncontrolled secondary factors also cause quality issues.

    - Mock up sample products early in the project, and use FFU tests to determine which attributes are important to performance. Concentrate on finding process options that minimize the variability of those critical properties.

    - Evaluate a range of feedstock alternatives to understand how trace contaminants will influence the product during its life span. Spike the materials with controlled levels of potential contaminants to determine the problem thresholds.

    - Look for performance characteristics that allow the product to set future customer expectations, not just meet their requirements today.

    - Run trials in a range of different weather conditions to anticipate the environmental effects.

- Map out the high and low limits of the attributes that define acceptable performance (FFU range).

    - Find the boundaries of the material properties around the application requirements—these are customer needs, not the capability range of the production process, which comes later.

- Determine if the product attributes are independent or if they interact with one another.
- Develop objective ways to measure performance in each application. Document and validate these test methods.
- Understand if all customer applications have the same requirements.
- Build the process, select the materials, and outline the operational procedures so that the product properties are always kept well within the application limits.
  - Decide carefully on retrofitting an existing process or purchasing a new one.
  - Do not seriously compromise the design to make it compatible with existing equipment.
  - Engage the plant personnel to obtain insight and to give them ownership. Never craft a process in a vacuum and assume it will work the same way in a plant environment.
- Set the production specifications tighter than the FFU limits and wider than the process capability range.
  - In a market-driven environment, the specifications derive from the application needs, not the capability of an existing process. Upgrade the process equipment to meet those specifications.
  - Your specifications must totally define the product requirements. Nothing can be "understood" or "unwritten."
  - Everything about the application must be documented. Be holistic and include considerations such as packaging, labeling, and shipping requirements. These concerns must be factored into the process design from the start, not as an afterthought.
  - Designate absolute requirements separately from features that are merely desirable.
  - Use a separate specification and product designation for each different application.
- Continuously refine the process to narrow property variation.
  - Continuous quality improvement prevents your competitors from having a static target.
  - Customers who are accustomed to a consistent performance will not easily change to another material with a higher variability.
  - Focus on improving attributes with the best return—either by providing performance benefits or by improving the capability to meet the most stringent product attributes.

- Make sure that the people doing the improvements are in tune with the strategic product line needs. Do not spend money on improvements that are not valued by the customers.
- Quality is not limited to the attributes that are known and controlled today. Keep as much flexibility as possible in the process to accommodate future needs.
  - Flexibility does not mean giving the operator more freedom to disrupt the process by adjusting the wrong knob. Lock-out control settings that should not be changed.
  - Flexibility can be the option to reconfigure the line in a matter of hours, not the capability to do so in the middle of a production run.
  - Use modular components. These can be removed, replaced, or modified to deal with different requirements.
  - Never skimp on corrosion and wear resistance for long-life production equipment.
  - The same can be said for control systems. Leave a few extra slots in the cabinet and some spare input/output ports that can handle extra control loops and interfaces. Oversize the utilities.
- Publicize and document the process and performance problems that are discovered in development. Prevent their reoccurrence during commercial production and the sale of the product.
  - People naturally tend to conceal their failures and embarrassments.
  - Unexpected product failures are often predicted by issues that were seen during experimental testing, but which were ignored or forgotten.
  - Murphy's laws teach valuable lessons. Learn from them.
- Keep the quality assurance organization separate from the production management. QC personnel should be isolated from demands to meet ship dates and optimize production metrics. It is far easier to pressure a QC inspector to release a defective lot than it is to fix the systematic problems.

Summary

- Product development can start with the goal of a desired product and the search for a viable process. Or it can begin with interesting technology and the search for a profitable application.

- Flexibility is important, because the optimal process route cannot be predicted in advance.

- A well-managed project team will focus on discovering and constantly refining a list of product attributes, a scheme for production, and a cost model.

- Before undertaking the development work, clearly delineate the project deliverables to match the sponsor's expectations and the ultimate customer's needs.

- Document constraints that should limit the team's freedom of action. Restrict the number of constraints that can be imposed by the internal customers.

- Developers must carefully balance four key parameters: project cost, project duration, product cost, and product functionality.

- Consistency and FFU should be designed into the process early in the development phase.

- Celebrate your development stage failures and mistakes. The problems that are discovered during the experimentation need to be documented and retained because they always tend to reoccur during commercial production and use of the product.

- Obtain an understanding of the product costs as early as possible in the project. Do so before committing significant capital or making pricing promises to customers.

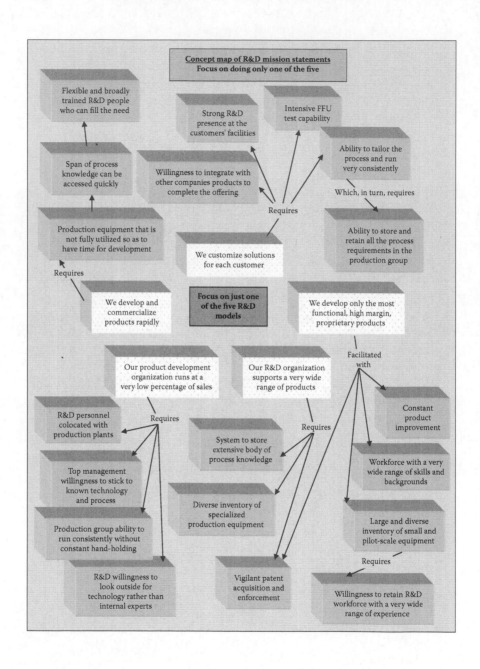

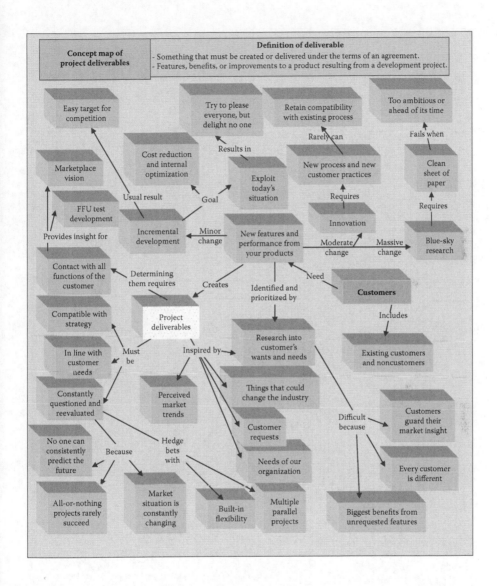

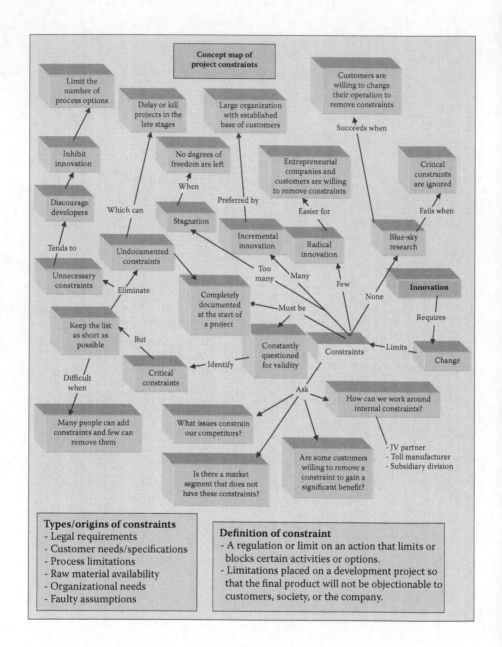

Concept map of project constraints

Types/origins of constraints
- Legal requirements
- Customer needs/specifications
- Process limitations
- Raw material availability
- Organizational needs
- Faulty assumptions

Definition of constraint
- A regulation or limit on an action that limits or blocks certain activities or options.
- Limitations placed on a development project so that the final product will not be objectionable to customers, society, or the company.

8

Project Execution and Oversight

How should we run dynamic product development projects? How do we select members and organize teams? What are the basic steps to carry out a development project?

> An imbecile can manage his own affairs better than a wise man the affairs of other people.
>
> **Arabian Proverb**

During the battle of Waterloo, Napoleon Bonaparte was concentrating on the task of breaking the line of British and Allied infantry regiments, which blocked his route to Brussels. In the midst of the engagement, an aide drew Napoleon's attention to a column of troops that were approaching their right flank from some distance away. After a quick look through his telescope, the emperor made a snap judgment that these were a detachment under Marshal Grouchy, rushing to link up with him. No action was taken as a result.

Decades of campaigning and assorted health problems had put Napoleon into autopilot decision-making mode. A messenger had reported Grouchy's position to Napoleon a few hours earlier. If Napoleon had consulted a map and calculated their likely rate of march, he would have realized that his wayward regiments were too far away to reach the battle so quickly. Only later, after precious time had passed did his cavalry scouts report that these were actually his Prussian foes. Some 30,000 fresh troops were rushing to support the British, turning the tide of the battle against the French. Later that same evening, Napoleon fled from the Prussians ahead of his defeated army.

Like the French emperor, many business leaders make reflexive project decisions based on what has worked or failed in the past. Or, perhaps, they rely on the many books, articles, and software tools dedicated to project management. While many of these techniques are applicable to construction projects, they have limited value when doing development in a process-dependent environment. A capital project consists of a great many predictable tasks arranged both sequentially and in parallel. Planning charts work in construction projects because the resource needs and the duration of each task can be accurately estimated.

Material development projects can have activity loops, which are repeated until a successful result is found. Competing options may be under consideration at the same time. It is rarely possible to predict the number of iterations

needed or if any one line of inquiry will succeed. Each nested development loop represents uncertainty for project duration and cost.

Projects with a triple-nested loop, shown below, are notoriously difficult to budget and forecast. In general, the more looping nodes there are in a project plan, the less certain the outcome will be. Not every series of development trials has a practical solution. Failing to find one, the team has to either set off in a new direction or be reassigned to a more promising new project.

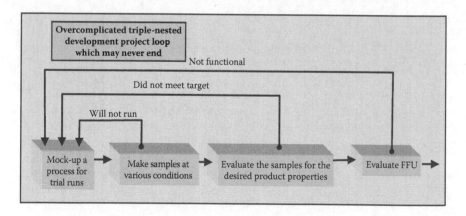

Project Problems

Never underestimate the power of human stupidity.

Robert A. Heinlein

NASA learned many costly lessons during the development and testing of the Apollo command module in the 1960s. The program was primarily driven by NASA's determination to beat its Soviet competition to the moon. To do so, NASA had to develop and test new technology, while simultaneously building the equipment, training the crew, and inventing procedures for things that had never been done before. In short, the project resembled the triple loop shown above.

In Project Apollo, the team culture neglected the prime rule of space flight—the crew must survive the mission. January 27, 1967, was not an exciting or glamorous day to be an astronaut. The Apollo 1 crew was bored and frustrated as it lay sealed in its capsule during a prolonged series of systems and communications tests. That changed in a flash when an electrical short circuit sent fire roaring through the command module. The three men died before they could be rescued.

An investigation revealed that a host of errors had been committed in the rush to meet project deadlines. For the sake of simplicity, the crew compartment atmosphere was pure oxygen. As a prototype for the new series, the capsule's wiring was spliced and frayed from countless changes, tests, and corrections. And finally, the astronauts were confined to a sealed chamber that was crammed with flammable materials, including their clothing. In retrospect, tragedy was inevitable in an environment with so many hazards. Project problems are often obvious in retrospect, but are not acted upon.

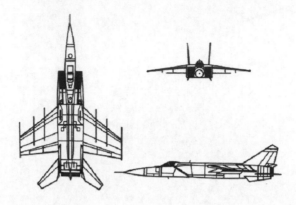

In the 1960s, the Soviet Union designed, built, and deployed the MIG25 fighter jet. This interceptor was intended to defend against the high-flying American B70 Valkyrie supersonic bombers and the U2 reconnaissance planes. The MIG25 was superbly suited for the job, but by the time the MIG25 became operational, the B70 program had been cancelled and the U2 planes were no longer flying over Soviet airspace.

In order to meet design specifications and to achieve high speeds, the MIG25 was primarily built of stainless steel, limiting its range and maneuverability. NATO was alarmed when it discovered how fast the plane could travel. It incorrectly assumed that the plane would also be light, nimble, and capable of long-duration flights. The F15 fighter plane was initially designed in light of these false assumptions. Given the blanket of secrecy, the long time from conception to deployment, and the high stakes of failure, it is not surprising that military projects often have inappropriate goals. They overcome this liability by spending billions of dollars on multiple versions of each platform and by pursuing continuous modernization projects to correct the flaws in earlier designs.

Warning signs for development projects:

- The Mega Project. Beware of any project plan that has fundamental research at the front end, the construction of a new process plant and simultaneous product development, and the launch of a major new product platform at the back end. This is way too much stuff to cram into a single project. The Apollo mission worked this way. Break your projects into manageable pieces with decision points between them so as not to be caught up in endless loops.

- The Project for Project's Sake. A dysfunctional company will enthusiastically complete a major project and reward the participants for achieving all the original goals. However, the project deliverables may be useless because of changes in the market. These situations are generally apparent to the participants, who are powerless to stop the train. The sponsors are too fixated on important milestones to realize that they have lost their way, as with the MIG25.

- The All-or-Nothing Project. Because of process dependency, the success or failure of a project may not be apparent until the end. Only when the finished product is in front of the customer can you determine if it brings value to the market. In such cases, it is crucial to have a reality check before every stage, so that value can be proven before too much money is spent. Ideally, the most risky steps should be near the beginning, so that the project can be killed before it absorbs too many resources.

- The A-Team. Some companies have superstar engineers or teams. Every executive demands these people for their project. This is often lazy, wishful thinking. No one can be an expert on every type of project or technology. The perfect team for one project may fail at a different type of project. Try to share best practices as widely as possible. Provide incentives to the top performers to train, mentor, and season the inexperienced participants. Recognize how one type of project differs from another and use the appropriate people and techniques for each category.

- The Wandering for 40 Years in the Desert Project. Many teams reach a dead end. There is a strong incentive to cobble up a new plan and send the project off in a new direction. Roadblocks are an ideal point at which to reevaluate the strategy and value of a project.

- The Waterloo Project. A dominating project leader or product manager can run the project by the force of personality. This looks good until the leader makes a bad choice and takes the team in the wrong direction. In other situations, the team sits idle while waiting for the designated person to make an assessment. There is no point in calling them a team if one individual makes all the decisions and crushes all dissent.

- The Pushed Project. An R&D organization can sometimes undertake a research project based on its vision of how a new technology will find a market niche. The sales/marketing organization does not sponsor the effort. However, once the project is complete, the sales staff ignores it. Customers will not buy the product if it fails to suit their needs. Failure is normal for research projects. But, it is clearly wasteful to develop and commercialize products on faith alone.

Project Organization

> ... automation applied to an inefficient operation will magnify the inefficiency.
>
> **Bill Gates**

In the 1850s, an ambitious project was launched to lay a transatlantic telegraph cable from Ireland to Newfoundland. A serious error was made early in the enterprise. Wildman Whitehouse was named as the chief electrician, responsible for specifying the materials, operating procedures, and equipment. Whitehouse had two serious flaws. He had little practical knowledge in the field and he would not accept advice or criticism from those who were knowledgeable.

The team included the noted scientist William Thomson, the future Lord Kelvin. Thomson accurately predicted the failure of Whitehouse's plan, but was too polite to take his objections to the top management.

Whitehouse chose a cable diameter that was too small for the requirements. He allowed the manufacturers to use low-purity copper, which increased the electrical resistance. He failed to specify which direction to spiral the individual wires inside the cable bundle. One vendor twisted the cable to the right and the other twisted it to the left. This created serious splicing issues when the cables were joined. Finally, Whitehouse responded to a weak signal strength by increasing the line voltage. When it reached an estimated 2000 V, the cable shorted out and went dead in 1858, after only a few weeks of operation. Another effort finally succeeded in 1866, without Whitehouse.

Software billionaire (and space tourist) Charles Simonyi invented the concept of metaprogramming in the 1970s. In his vision, a sole software architect would envision a finished application and set all of the coding rules and requirements. The metaprogrammer would review and direct the work of a subordinate team of code writers. The team members would individually shape components of the application by following the leader's guidelines to the letter. The idea was to cut out unnecessary communications and inappropriate compromises, allowing the leader to make all the key decisions.

Simonyi's system required that the team leader be brilliant and totally knowledgeable about all aspects of the application. By Simonyi's estimate, the metaprogrammer needed to be correct on 85% of all technical decisions in order for the method to work better than the traditional software techniques. The leader needed to be significantly smarter than the rest of the team, or he or she risked being second-guessed.

In a process-dependent environment, the odds are stacked against the project leader to a much greater degree. Projects should be guided by performance testing conducted by the team members. It is not possible for a centralized leader, who is getting the results on a secondhand basis, to hold all the information.

Since no leader can participate in every activity, a more collaborative approach is required for process-dependent situations. The members must share their information, experiences, and ideas freely among themselves to reach a consensus solution based on their shared learning. But, even their collective observations are not sufficient to make the final decision in an industrial organization. Representatives from functional areas must be consulted to ensure that the product can be manufactured, distributed, and sold by the organization.

Projects of importance should have multiple levels of oversight. This is not a sign of a bloated bureaucracy or an overstaffed middle management.

Rather, it is a division of labor that prevents the errors in judgment by one individual from crippling a key program.

It is possible for one person to perform all these functions, especially in a small business. But, in a large organization, it is difficult for a single individual to track all the key metrics. The exact duties of each person will overlap and vary from one organization to another.

The three overseers can be called by many names. Here, they will be termed the *project leader*, the *management mentor*, and the *product line manager*. *Project leaders/program managers* are often senior engineers or technical managers who lead the day-to-day activities. Most of their time will be spent working on the project in a technical function. In addition to their own specialty, they monitor, assist, and coordinate all the work and act as the main point of contact.

The project leader should have the breadth of experience to understand the functions performed by all the team members. Good interpersonal and organizational skills are essential to keep the group working together and progressing toward the goal.

Management mentors/project champions are generally technical middle managers who have been given the responsibility to oversee one or more project teams. They will often have direct reporters serving on the teams, but they may not have administrative responsibility for all the team members. The oversight of project teams will only represent a faction of their total workload.

This individual will periodically meet with the entire team. These meetings monitor the progress toward goals within a work plan and budget. Operational goals always change during major projects because of what the team discovers, and the shifting realities of the marketplace and company priorities.

The management oversight should ensure that the team has the appropriate resources and that it is making progress toward a strategic goal. This individual is the team's advocate to upper management and keeps the team updated with the information that it needs to guide its work. On their own, mentors should not make significant changes in the direction or scope of the project.

Product line/marketing managers are responsible for a brand, a product line, or a specific product. They ensure that the project suits the market needs. In a market-driven organization, the new product must solve the customer's problem and fit the company's business model. In a process-driven organization, the critical need is for the product to fit into an existing production scheme. Ideally, the product line manager will ensure that a bottom-line benefit will result from the resources expended. This oversight is needed most at the beginning of a project to understand the market requirements. When the project is near completion, close attention is needed to ensure a successful handoff to sales and manufacturing.

Some members work part time on a given project. These people will be called on to perform their particular function or expertise when needed. The rest of the time, they will be working on other assignments, projects, or staff functions. This works well for people whose narrow specialty is peripheral to the team's activities. Projects rarely have a uniform workload for every specialty. However, conflicts occur when the team needs that specialist at the same time that someone else has pressing priorities for him or her. Organizations need a mechanism to resolve such conflicts smoothly.

Teams that are entirely composed of part-time members are best suited to projects that are low in priority and have a long duration. The activities may consist of short periods of intensive activity on a plant trial or a field test. These may be followed by long periods of inactivity while waiting for management, customers, or vendors to evaluate the results and make decisions.

In some situations, it is appropriate to dedicate people to a project and totally immerse them in the mission. These individuals will work on a wider range of activities than their normal functional expertise. They must fully understand all aspects of the assignment. In such dedicated situations, it may be appropriate to move the team into a common work space so that they can interact closely together. The relocation of personnel works best when the team members normally reside in different buildings and rarely see one another.

A full immersion/colocation strategy has its benefits and drawbacks. Participating in all aspects of the project facilitates communication and ownership. The outcome of poor decisions on trade-offs becomes immediately obvious. The work of some specialists can suffer when they are cut off from their coworkers, laboratory space, and their department support structure. Some sales and marketing people react badly to being relocated to a workshop or production environment. Others thrive on the change of perspective and keep the team focused on the true needs of the customer.

Dedicated personnel assignments work best when the focused efforts of a cohesive team are needed to achieve quick results. This succeeds when the resources and the work pace are controlled by the team. If delays are imposed on the team by the organization, the customers, or the nature of the work, then it is a less useful approach. According to Wernher von Braun, "Crash programs fail because they are based on theory that, with nine women pregnant, you can get a baby in a month."

In a hybrid situation, there will be a small core of team members almost totally dedicated to the job. They will get assistance from part-time members or specialists from the larger organization. As the needs change, people can be added or subtracted from the core team.

Howard Hughes was one of history's least flexible project managers as the head of Hughes Aircraft. The reclusive billionaire insisted on micromanaging many aspects of aircraft design while maintaining obsessive secrecy and privacy. Often, his most trusted managers could not contact him for urgent decisions during emergency situations.

In 1944, the company was developing two different aircraft under contract with the military—the XF11 reconnaissance plane, pictured below, and the HK-1 transport plane. On May 7, Hughes crashed his car, resulting in the

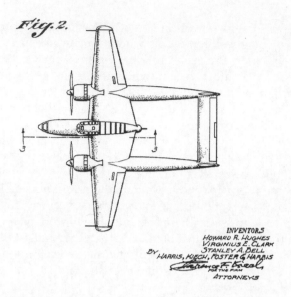

Fig. 2.

INVENTORS
HOWARD R. HUGHES
VIRGINIUS E. CLARK
STANLEY A. BELL
BY HARRIS, KIECH, FOSTER & HARRIS

ATTORNEYS

eighth major head injury of his adult life. A doctor advised him to take a vacation from the intense pressure of running his business empire. Hughes did so, without advising his subordinates where he was going or when he would return. He resurfaced a year later, after the end of the war. The development projects suffered, ending Hughes' dream of becoming a major player in the aircraft industry.

On a positive note, Hughes ignored his aircraft business for the remainder of his life, concentrating on his airline and movie projects instead. Free from his micromanagement, the company became a successful defense contractor, focusing on the less glamorous task of developing aircraft controls, satellite circuitry, and missile components.

Flexibility and timely decision making are critical for effective, independent teams. It is unreasonable to expect that every project in a company should be staffed the same way. A mixture of dedicated teams, part-time participants, and hybrids may be required to address the full range of needs.

Personnel guidelines for teams:

- Team members must communicate and cooperate freely and honestly. Colocation may help, but it will not force incompatible egos to cooperate.
  - Populate the team with people who have a history of working together. Consider letting the team leader recruit people into the group. People who volunteer to work together out of respect and a shared history are less likely to require "team-building" exercises and management intervention.
  - Look closely at the people who are never chosen by their peers to be team members. Are they shy, quiet individuals who work

better on their own? Or are they selfish, disruptive, or antagonistic people who no one wants as a collaborator?

- Team participation should not become a route for weak employees to evade professional development and evaluation.
- Use a mix of experienced (this must include the project leader) and newer employees. Combine the technical depth with the fresh perspective of the newcomers. Hopefully, this will not merge staleness and cynicism with inexperience.
- Teams should be chartered with a very specific goal to accomplish. They must be given the resources to swiftly accomplish a specific goal. Additional tasks should not be given out piecemeal as the old ones are finished or are found to be dead ends. The lines of communication and authority get scrambled when a multitude of long-standing teams compete with the existing department structure for organizational attention.
- Management should resist the temptation to create small (one- or two-person) "teams" to address issues. This tactic is seductive if it can bypass the normal oversight and budgeting systems. Mini-teams can absorb resources without delivering benefits, if they are not coordinated with other areas.
- Beware of department agendas or internal politics in teams pulled together from feuding organizations. Problems arise when members are split between the positions taken by the team and the demands of the bosses who pay, evaluate, and promote them.
- Take the time up-front to have the team fully talk through, discuss, and debate the best course of action. The project sponsor, the management overseer, and the product manager should be present to explain the background. The advantage of having a team is lost if its plans were created by outsiders or a minority faction. The team discussion and goal setting may expose serious personality conflicts that need to be dealt with at the outset.

See the "concept map of development teams" at the end of the chapter for a graphical overview of the roles, goals, and relationships in a process-dependent environment. A review structure is needed to monitor effectiveness and make systematic improvements.

Project Execution Guidelines

There ain't no rules around here! We're trying to accomplish something!

Thomas Alva Edison

Mr. Edison was explaining that there is no one formula or system for turning clever ideas into products. He did not follow a cookbook recipe to achieve innovation. He was a restless, innovative genius lacking the patience to put up with theoretical nonsense. Working at the dawn of the electrical age, he was happily ignorant of environmental impact statements, workplace safety laws, heavy metals regulations, or guidelines for electromagnetic interference shielding. Edison's approach was right for him and his team, but it leads to disaster when applied inappropriately.

In April 2010, British Petroleum (BP) was rushing to complete drilling operations on its Macondo oil well project in the Gulf of Mexico. Having reached a suitable oil and gas reservoir, the company was anxious to cap the well and move the Deepwater Horizons platform to another job. After the drill rig was moved out, a production platform could be brought in to begin oil and gas production. The process of sealing the well with a concrete plug was the critical final step in an expensive project. However, this procedure was poorly executed, resulting in the loss of the rig, 11 lives, and an environmental disaster.

A presidential commission subsequently identified a series of project management failures committed by BP and its subcontractors. Specifically, the investigation determined that,

> ... changes to drilling procedures in the weeks and days before implementation are typically not subject to peer-review ... such decisions appear to have been made by the BP Macondo team in ad hoc fashion without any formal risk analysis or internal expert review. This appears to have been a key causal factor of the blowout.*

---

* *Final Report of the National Commission on the BP Deepwater Horizon Oil Spill and Offshore Drilling*, p. 122.

The commission went on to blame the failure to carry out appropriate testing on the cement plug, which was the only barrier to the eventual blow-out. The critical "negative pressure test" was performed based on a terse e-mail, which contained only 24 words.

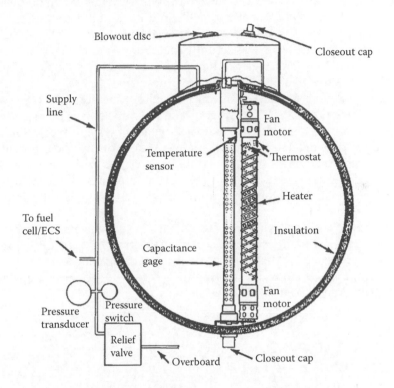

Several years after the Apollo 1 fire, NASA was preparing the Apollo 13 moon mission for launch. Here again, important decisions were made without risk analysis or expert advice. Following a series of ground tests, liquid oxygen was being removed from two tanks in the service module. The first tank was decanted using standard procedures. However, defects in the second tank prevented it from draining normally. Even the application of 5 atm of gas pressure would not force out the cryogenic liquid.

This particular tank had previously been considered suspect and was removed from the Apollo 10 mission. A later analysis suggested that the drain problem was caused by an internal tube that did not fit properly and jarred loose during handling.

Rather than swapping out the defective equipment, a decision was made to accommodate the situation. Engineers concluded that the fill tube problem would not affect flight operations. An ad hoc process was developed to empty the tank. Electrical power was applied to an internal heater to boil off the liquid oxygen over an 8-hour period. (*Note:* this is a suspiciously arbitrary time period, possibly corresponding to one shift on the work schedule.) This procedure was performed on two occasions.

Unknown to anyone at the time, the improvised process damaged the tank's internal components, due to a design error. After the oxygen boiled away, the system overheated and melted the electrical insulation on the adjacent wiring. It was later estimated that the interior circuitry exceeded 1000°F during this process. The engineers relied on the heater to turn itself off, but they did not verify that this was actually occurring. An external temperature indicator could not display any values higher than 80°F, making it difficult for technicians to detect excessive temperatures.

After completing the tank-draining procedure, the equipment passed a routine battery of tests. This evaluation did not anticipate the possibility of damage from high temperatures inside a cryogenic system. Weeks later, the spaceship and its defective oxygen tank were fired toward the moon with a crew of three. At 56 hours into the flight, the damaged wiring was routinely energized. This generated a spark in the presence of a highly concentrated mixture of fuel and oxygen. The resulting explosion crippled the ship and nearly killed the crew.

A number of design flaws combined to cause the drainage and overheating problems. The disaster occurred because the output of a dynamic process was not appropriate for the end use. The engineering team treated the oxygen system as a black box and trusted that it was functional without doing appropriate fitness-for-use (FFU) tests.

RULE #45 THERE IS NO ONE UNIVERSAL DEVELOPMENT TOOL

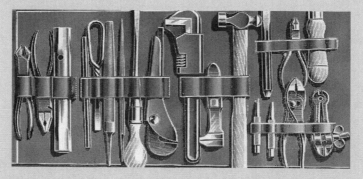

No single development format, procedure, or formula will fit every situation. Each project is unique and must be evaluated and executed appropriately. That is why this book is laid out as a series of suggestions for things to think about, rather than the step-by-step cookbook approach.

Before starting a project:

- Establish the mode and frequency of communication/documentation within the team. Will decisions made in verbal conversations be recorded or just remembered?
- The team should agree on known parameters—the things that all members of the team accept to be true about the product, the process, and the application. There should be a list of what is not known and the things that are in dispute.
- Review the archival knowledge—check notes, previous project records, literature, industry journals, and interview researchers who have done previous work. Look for errors and contradictions.
- Where is additional research needed? Do some areas have conflicting information? This could suggest that primary mechanisms are confounded by secondary interactions.
- Create an action plan to locate the missing pieces. Consider how to calibrate the process control knobs and verify/falsify your basic assumptions.

During a development project:

- Use screening trials across the work space to understand where the process runs well and where it can be controlled. What changes could expand this window?
- The information gained from early trials and FFU tests will change your view of the product. Seek a balance between premature specification lock-in and open-ended projects that never make a commitment.
- Schedule meetings and reports around the development cycles (run a batch of samples, evaluate FFU, decide what to do next) instead of arbitrary calendar dates. Look for ways to accelerate these loops by planning and preparing for the next round of tests and evaluations. Identify bottlenecks that slow each loop. Minimize handoff delays.
- After each loop, evaluate progress relative to expectations.
- The entire team should review the trial plans to ensure that the end result will teach them something, prove/disprove a theory, or provide materials for testing.
- Tests of material properties and/or performance results must give direction for the next round of process trials. If they do not provide direction, then you are running the wrong trials, doing the wrong tests, or laboring under false assumptions.
- Overcome the desire of functional experts to make decisions in isolation. Write down the proposal, let everyone have input, and be happy with the plan. The process engineer may prefer a certain

sequence of samples to run a trial with optimal efficiency. However, if this plan fails to suit the needs of the FFU expert, then conflicts will arise.

Market-focused organizations should strive to scale-up prototypes into batches that can be field tested at the customer's location. If these tests are encouraging, then sales is usually anxious to lock-in a finished version and push the product toward commercialization. This zeal creates a conflict between the market drivers and the technical group. The team must deliver both a production-friendly process and a workable product. Additional optimization is often needed to resolve the problems of stability, cost, and compatibility.

No. 875,881.   PATENTED JAN. 7, 1908.

E. G. ACHESON.

METHOD OF DISINTEGRATING GRAPHITE.

APPLICATION FILED DEC. 17, 1906.

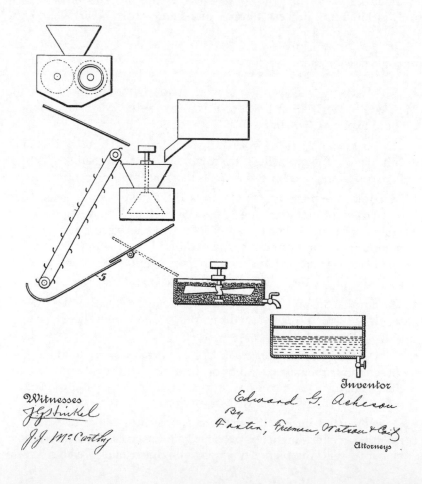

Inventor

Edward G. Acheson

By

Foster, Freeman, Watson & Coit

Attorneys

Witnesses

J.G. Stinkel

J.J. McCarthy

A similar dynamic occurs when a team is working to optimize an existing production process. This creates a need to test all of the products to determine that the changes have not impacted on their ongoing end-use performance. The requirement to do all of this testing and to not compromise process efficiency can be very frustrating for those in production management.

A third conflict comes about when a team is focused on process improvement, but discovers some unique and interesting product properties. Invariably, this curiosity has either no immediate commercial application or it will require a considerable development effort to evaluate its utility. The team usually needs to press forward with its original assignment, but document its unexpected results for future investigation. Accidental discoveries have shown that the opposite approach has merit.

In the 1890s, Edward Goodrich Acheson was attempting to produce diamonds by heating powdered carbon to high temperatures with an electric arc. He failed in this objective, because high pressures are also needed to generate synthetic diamonds. However, when Acheson added clay and salt to the mix, he created blue-black crystals of silicon carbide. The artificial diamond project was dropped and the new material was pursued instead.

Acheson found that his discovery was an excellent abrasive, which he promptly patented. He went on to develop an effective production process and founded a company to sell the material under the name carborandum. Continuing his experiments, Acheson discovered that the silicon component would vaporize at 4150°C, leaving behind graphite. In 1899, the Acheson Graphite Company was created to sell this product as a lubricant.

Critical Questions: What important process or application information was learned by the team in the course of its project? Was this information retained in a format that will be available to assist future developers? What would we do differently?

Basic Steps

It is not enough to do your best; you must know what to do, and then do your best.

W. Edwards Deming

During the first decade of the twentieth century, Thomas Edison was looking for something to occupy his time. He had just closed down a 10-year

effort to exploit low-grade iron ore deposits in western New Jersey. Technical difficulties and the discovery of rich ore fields in Minnesota had ended his hopes of becoming a mining baron. His systematic product development focus turned toward transportation technology.

Edison was convinced that electric cars were superior to the gasoline engine approach. To realize that vision, he had to overcome the obvious deficiencies of the lead–acid battery. Its battery life was short, the device was heavy, and the electrolyte solution was corrosive.

The inventor gathered a large staff in West Orange, New Jersey, including university-trained chemists and physicists from the United States and Europe. He set them to work investigating every possible combination of materials and processing techniques. Eventually, they narrowed the choice to a single best option. The preferred product featured a nickel oxide cathode, an iron anode, and a 20% potassium hydroxide electrolyte solution. Endless problems were encountered and overcome, before production could be initiated. The batteries were manufactured on a process that was custom built to Edison's specifications.

Commercial delivery vans in New York city were the initial application. Retailers were ready to give up horse-drawn delivery wagons and their messy waste products. Customers were initially receptive, but end-use problems quickly became evident. As complaints grew, defective batteries were taken back and examined to determine their failure mode. This analysis prompted a production stoppage of several years, while Edison and his staff struggled to retool the process to eliminate the faults.

Eventually, they resolved the issues and production resumed. Edison's battery proved reliable and formed the basis for a profitable company. The venture serviced a range of markets until 1975. However, these applications did not include the passenger car market and the delivery van market that were the goals of the project. Edison's friend, Henry Ford prevailed in the marketplace with his gasoline-powered approach.

Edison's organizational skills were meticulous and followed a master plan. No one conducted industrial research any better. All the right steps were taken to optimize and produce the best possible product. However, like Edwin Land, the legendary Thomas Edison found that the evolving marketplace can leave even the best developers behind, if they do not act quickly enough.

The key goals of project teams (the following are not in order of importance or sequence):

1. Gathering information about the performance of various sample products in customer applications.

2. Evaluating sample product properties and correlating them with customer satisfaction.

3. Determining how the process parameters (materials, conditions, equipment, and procedures) affect the above two items.

4. Characterizing the variability of product performance and seeking process parameters that minimize the variation.

To fully achieve these goals, the compilation of a huge amount of data is required. It could take years to accumulate all of this knowledge in a new technology area. With complete understanding, the team can easily define the product, the process, and all the inputs. A market-driven company is likely to start with #1, then look for #3 to optimize #1. Later, it will work on #2 and #4. A process-focused company will start with #3 and seek to minimize #4 and optimize #2. It may not work on #1 at all.

The flowchart at the end of the chapter details the course of a typical project (as if there is such a thing!) from start to finish. An evaluation is required at every stage of the project. Sometimes, an aggressive project team is so focused on overcoming obstacles that it does not realize that the ultimate goal cannot be achieved. A management review will judge whether this particular idea is still a better one than other ideas that are waiting to be resourced.

The cancellation of a project should not be looked on as a failure for the team members. No one can predict a commercial outcome months or years down the road. The work on a project is an effort to answer that question. If the answer is not pleasant, then stop spending money on the project.

Each project should have its own work plan, based on the needs of the company, the technology, and the customers. This must include projected dates to achieve certain goals and expectations of how much time and money will be spent at each stage. These estimates should be updated periodically and reviewed by management. Initially, this is little more than guesswork. As the team learns more about how to make the product, the work plan becomes more accurate.

Intellectual Property

> An organization's ability to learn, and translate that learning into action rapidly, is the ultimate competitive advantage.
>
> **Jack Welch**

At the start of the Iron Age, our distant ancestors struggled to process iron ore into useful implements. Most of the process techniques for copper, bronze, and gold were not appropriate for iron, due to its superior hardness and elevated melting point. The early smiths probably experimented for generations before they mastered the skillset needed to consistently make iron tools and weapons.

In particular, iron needs a flux, such as limestone, when heating the ore, to remove nonmetallic contaminants. Advanced furnace designs and bellows were required to reach the appropriate temperatures. The lump that emerged from the reducing furnace needed to be repeatedly heated and hammered to drive out cinders and slag. A long, slow carbonization process could boost the carbon content and uniformity, which greatly improved the iron properties. Finally, quenching and tempering were discovered as ways of achieving its superior strength and ductility.

Each of these techniques was probably discovered and improved at divergent locations across the Old World. The best practices were handed down from master to apprentice in secrecy across hundreds of generations. Breakthrough techniques were jealously guarded to protect market share from competition. Technical organizations struggle today with the same issues of discovering and protecting process secrets as the ancient smiths did.

A NON-RUSTING STEEL.

Sheffield Invention Especially Good for Table Cutlery.

According to Consul John M. Savage, who is stationed at Sheffield, England, a firm in that city has introduced a stainless steel, which is claimed to be non-rusting, unstainable, and untarnishable. This steel is said to be especially adaptable for table cutlery, as the original polish is maintained after use, even when brought in contact with the most acid foods, and it requires only ordinary washing to cleanse.

Thousands of years later, steel replaced iron and was produced on a massive scale. Developing process techniques to control the carbon content was critical to this transition. Early in the twentieth century, a number of experimenters in the United States, the United Kingdom, France, and Germany independently developed improved steel alloys, which are now known as stainless steel. This material is much more resistant to corrosion than normal steel grades are, creating a wide range of end-use applications.

With so many people publishing results and introducing products at the same time, it was very difficult to convince the patent authorities that any one invention had the novelty and nonobvious criteria required. In England, Harry Brearley tested different alloys in search of the ideal material to make rust-resistant rifles for the British Army. He documented an extensive range of additive levels that would deliver improved performance and still be compatible with the standard unit operations of the industry. The army turned down the discovery, but others were anxious for steel that would not tarnish.

Brearley's employer, Brown Firth Research Laboratories, did not allow him to apply for a British patent on the research findings. However, Brearley was able to obtain a broadly worded U.S. patent. This success was probably due to a combination of good timing and a wealth of experimental data demonstrating how the various alloy components influenced material properties. In particular, he found that the carbon content of the finished product needed to be very low and that the chromium levels needed to be in a very specific range. Then, as now, extensive experimental data provide credibility for patent applications.

In 1860, Frederick Walton received a British patent for a process to polymerize linseed oil into a rubbery compound called linoxyn. Walton had first noticed this material as a skin on the surface of oil-based paint when the cans were left open. Oxidation will slowly convert liquid linseed oil into a polymer, by linking the ends of the chains together.

Walton accelerated the reaction by adding lead acetate and zinc sulfate. His process involved spraying the mixture onto sheets of cloth that he hung vertically in a hot chamber. A skin of linoxyn built up over a period of months. This rubbery residue could be scraped off the fabric and dissolved in boiling benzene. Afterwards, the pure linoxyn was compounded with powdered fillers and coated onto canvas to create a patented floor covering called linoleum. With patents on both the process and the product, Walton's market position appeared to be very strong.

Linoleum became trendy and very profitable for Walton. Inventors researched process options to work around his patents. In 1871, William Parnacott patented his own linoxyn production process. He bubbled hot air through a vat of linseed oil to oxidize the mixture. Afterwards, a sticky mass of linoxyn formed as it cooled in trays. This process produced a cheaper product that was less functional than the original. Within a few years, Walton was under serious pressure from multiple competitors. Some floor coverings were made without the canvas backing to both avoid Walton's product patent and provide a cheaper alternative.

Walton failed to obtain sufficiently broad patent claims on either his polymerization process or on the finished floor-covering product. His initial patents only covered the first options that worked well. Alternate process paths and product formats were not fully explored or patented. Walton suffered a further blow to his intellectual property position when an American court ruled that linoleum could not be treated as a brand name. The proper application paperwork had not been filed in Washington. Also, the linoleum name was already in popular use for a range of floor-covering materials from assorted suppliers.

Expending resources to fully explore process alternatives conflicts with the goal of bringing new products to market quickly. Instead of creating improved versions of the product, money is being spent on expanding the knowledge base and defending an existing market position. This dilemma is crucial to setting organizational strategy. Companies with a dominant position in a large and profitable market try to protect their technology with a broad patent base. However, they are vulnerable to competitors with alternate solutions that employ novel technology.

RULE #37 CONSIDER PATENT ISSUES EARLY IN THE DEVELOPMENT PROJECT

It is easier to generate the different samples and data needed to fully document patent applications while the development work is underway. Investigate alternate process paths and product configurations. In the case of integrated systems of products that work together, it is better to develop them with patentable features, rather than to graft proprietary features on after the fact.

Intellectual property techniques for process-dependent environments:

- Create a database of all patents and patent publications that are applicable to your products, processes, and the related industry. This can be as simple as a binder full of papers or as complex as a large searchable database. Whatever form it takes, the most important documents should be easily accessible and well known to the development professionals. The more extensive the collection is, the harder it is to locate the useful pieces.

- Some commentary should be attached to each important patent in the collection. A lawyer might blanch at this suggestion. However, it is a daunting task for a new engineer to read a stack of legalese to glean information. It helps immensely if experienced people have screened the prior art and interrupted it. The comments might be very simple, for example, "This one only applies to water based coatings on blow molded substrates," "The claims here appear to be written around our XYZ product!," "We don't think they use this process for products that compete in our market," or "We need to mention this patent number on all of our 'eco-blend' products." Reviewers should not express legal opinions or speculations.

- While conducting development work, the team members should be aware of prior art. They should look for product performance and effects that exceed or contradict those published claims. Such new knowledge represents an opportunity to acquire additional intellectual property rights.

- Consider that a competitor's patent is often written around its process considerations, and not yours. Identify which aspects of the patent are relevant to your situation and which are not. Keep this in mind especially when licensing technology.

- Employees in frequent contact with customers should be especially well briefed on which information they can share, which is proprietary, and what type of intelligence they should gather.

- The same is true for customer information that the field contacts send back to the organization. The customers' confidential information must be prevented from becoming public.

- Any information considered to be a trade secret should be documented centrally. Engineers should be fully aware of what information they can and cannot share with vendors, customers, and nontechnical coworkers.

- Top management should set priorities for which products and processes merit protective patents to protect its commercial position.

Development Resources

Any problem can be solved using the materials in the room.

Edwin Land

To invent, you need a good imagination and a pile of junk.

Thomas Alva Edison

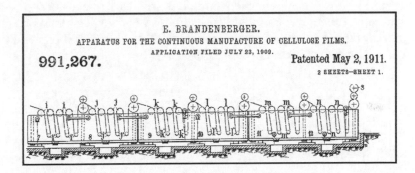

Cellophane, the original polymeric film, was developed by the Swiss chemist Jacques Brandenberger during the first decade of the twentieth century. The development cycle took 10 years, because the production process had to be constructed, tested, and refined before the finished product was available to customers. This was truly an industrial development, rather than a bench-top chemistry experiment. The cellulose was transformed in a continuous process, rather than a sequence of discrete batches. A viscous pulp solution was forced through a slot die into a bath of sulfuric acid, where it congealed into a thin sheet. This material was pulled through a chain of tanks, where it underwent a sequence of reactions to emerge as a continuous sheet of thin film.

The system was unique in that the processing time for each step was controlled by the overall line speed and the path length of the tanks. If one stage needed twice as much exposure time as another, then more rollers were added to extend the path length and the residence time. All of this equipment was custom made for Brandenberger's development trials and it had to resist the corrosive process conditions.

Without customized equipment, it would have been impossible to create uniform and consistent samples to find end-use applications for the new technology. Over time, cellophane production lines grew wider and faster than Brandenberger's original equipment. Totally new methods and equipment were required to measure the properties of the finished product and to convert large mill rolls of film into useful product formats for customers.

On the downside, the cellophane process employed a toxic cocktail of chemicals to perform the magic transformation of raw plant fibers into a clear plastic film. The disposal of the spent chemicals generated a significant water pollution problem. This factor damaged cellophane's market position in comparison to that of polyolefin polymers as stricter environmental regulations were implemented in the middle of the century.

There are often trade-offs and compromises when running development trials on small-scale equipment or large production lines. Even if the production equipment is available, it may consume a great deal of material, time, money, and effort to generate each sample. In the early stages of a project, it is generally advantageous to screen as many variations as possible. This requires configuring the prototype process with flexibility and efficiency to create a wide range of different samples.

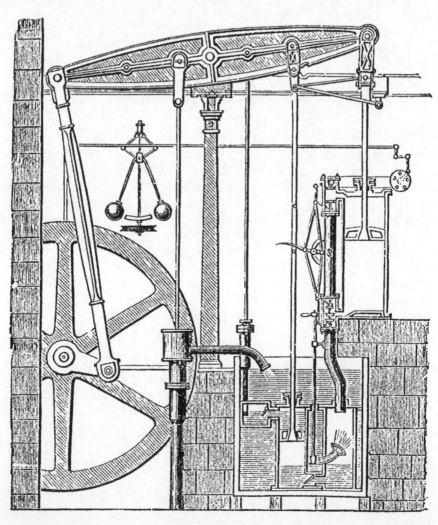

The "R&D structural balance diagram" at the end of the chapter shows a spectrum of R&D organizations ranging from process focused to market focused. The "concept map of R&D organizational trade-offs" shows how resources can be distributed among the various functions.

James Watt is widely known as the father of the steam engine. But, there were several types of commercial engines in operation, pumping water out of coal mines before Watt became involved. Between 1763 and 1765, he had access to a working half-scale model of a Newcomen engine at the University of Glasgow. This exposure allowed him to realize that the design was inefficient, wasting energy by heating and cooling a large iron cylinder with each stroke of the piston. In quantifying the efficiency, he devised the unit of measurement for power that bears his name. Watt envisioned an improved version that condensed steam in a separate chamber, greatly increasing the work that he could get out of a ton of coal.

However, converting clever ideas into working equipment was extremely costly. Watt needed to partner with John Roebuck and, later, Matthew Boulton, in order to obtain capital and access to their iron works. With that help, he spent years struggling to build working prototypes. In particular, it was very difficult for Watt to get a good seal between the piston and the cylinder wall. John Wilkinson's precision-boring process provided the solution. Wilkinson had invented a system to cut cannon barrels to exacting specifications. This new technology was the key to making pistons that fit precisely inside the cylinder. Without the contributions of Roebuck, Boulton, and Wilkinson, Watt could not have initiated the steam revolution.

During development, the appropriate equipment is needed to evaluate and compare a wide variety of raw options. The widest possible range of process conditions should be investigated. When process functionality is readily available, then the development team can make rapid progress.

Even the best process simulator requires people to run the machines, record the conditions, collect the samples, make observations, and label the resulting output. Automation seems like a good way to churn out lots of samples. However, research on new materials and processes is rarely suited to automated equipment. Process exploration is best performed by well-trained laboratory technicians and operators. They should be able to set up and operate many different equipment configurations and to supervise operators in running production trials as the process is scaled up. This is especially important when development runs stretch around the clock.

These technicians also test material properties and perform FFU testing. Their degree of specialization and the breadth of their responsibility will vary from one company to the next. Experienced, flexible technicians can be kept busy all the time in a development organization.

Development laboratories should be separate and distinct from production quality control (QC) laboratories, even in small organizations. QC operations focus on doing standard tests in a very repetitive way on a continual stream of samples that come from the production floor. The interests of space and efficiency require that they maintain a strict discipline of clearing out samples after they have been tested. Parts, tools, and standards for the tests should be carefully shielded from anything that might change them from the approved standard conditions.

Development laboratory operations are completely different. The equipment is constantly being reconfigured in the search for a better way of doing things. Samples need to be labeled and retained. A given sample might be constantly retested in different ways to understand what makes it unique from other samples. Parts, tools, and components may be frequently adjusted, modified, or recalibrated to new settings. These two groups should be very concerned when they share facilities and equipment.

The same clash of cultures could also occur between the production operations and the development team. Production is about doing things the same way every time in the single-minded pursuit of consistency. They should not want nonstandard components, materials, tools, and supplies in their work space. A tired operator or material handler will eventually grab a familiar-looking box or component and spoil a production batch. Development is about trying different things to see what might happen—making knowledge, not products. A lack of distinction between the two functional areas is a bad sign.

Summary

- Product development in a process-dependent environment cannot be scheduled and planned with the same tools and techniques used for a construction project.

- Development teams should be composed of people who work well together. They should be given clear goals that are aligned with company strategy.

- One team member should be responsible for managing the team's daily work. A member of management should ensure that the group has resources and that it is progressing toward the desired goal. The product line manager ensures that the project stays focused on the needs of the market.

- The key goals of development projects are:
  - Gathering information about the performance of prototypes in real or simulated applications.
  - Evaluating material properties and correlating them with customer satisfaction.
  - Determining how the process parameters affect the above two items.
  - Minimizing the variability of product performance and properties.

- Provide the development teams with the widest possible range of raw material choices, process options, data-gathering methods, modeling tools, and FFU tests.
- Anticipate conflicts between the goals of satisfying the customer and optimizing the process. The development team is often caught in this paradox.
- Evaluate as many potential process paths as possible. Beware of premature optimization.
- Consider patent issues early in the development project.

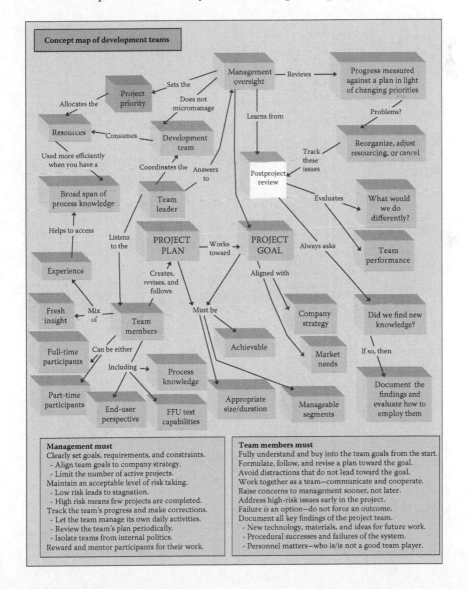

Concept map of development teams

Management must
Clearly set goals, requirements, and constraints.
  - Align team goals to company strategy.
  - Limit the number of active projects.
Maintain an acceptable level of risk taking.
  - Low risk leads to stagnation.
  - High risk means few projects are completed.
Track the team's progress and make corrections.
  - Let the team manage its own daily activities.
  - Review the team's plan periodically.
  - Isolate teams from internal politics.
Reward and mentor participants for their work.

Team members must
Fully understand and buy into the team goals from the start.
Formulate, follow, and revise a plan toward the goal.
Avoid distractions that do not lead toward the goal.
Work together as a team—communicate and cooperate.
Raise concerns to management sooner, not later.
Address high-risk issues early in the project.
Failure *is* an option—do not force an outcome.
Document all key findings of the project team.
  - New technology, materials, and ideas for future work.
  - Procedural successes and failures of the system.
  - Personnel matters—who is/is not a good team player.

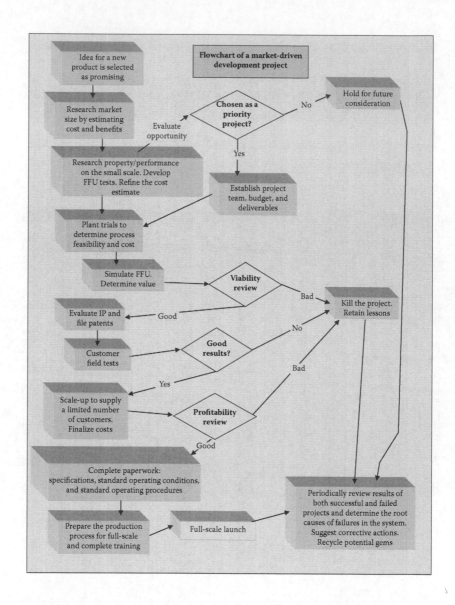

Flowchart of a market-driven development project

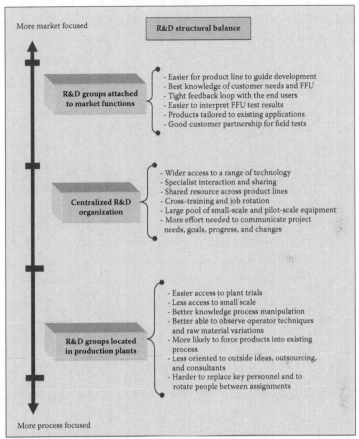

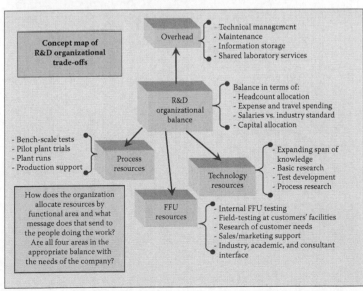

9

Small-Scale Trials

*How do we use small-scale equipment to evaluate process options and create proto-
type samples? How do we identify the best process alternatives to commercialize our
new product ideas?*

> ... the seeds for failure are sown very early in the life of a particular
> system—during the concept development and design phases.
>
> **Professor Robert Bea**

In the early 1830s, there was a great surge of interest in creating products
from latex rubber in both the United States and Europe. Rubber was a miracle polymer from South America that outperformed leather, oil-impregnated
fabrics, waxes, and tars in many flexible waterproofing, gasket, and sealing
applications. New companies were formed and astonishing products were
launched in an effort to be the first to market. The rubber boom of this period
was very similar to the Internet bubble of the late 1990s, including the resulting crash.

Consumers soon discovered that uncured rubber goods had serious flaws.
In summer, they became sticky as the latex first softened and then deteri-
orated into a foul-smelling liquid. Cold weather robbed them of flexibility
and caused cracking. The market for rubber products collapsed as customers became disillusioned. In 1834, the Roxbury India Rubber Company gave
credit for $20,000 worth of defective materials. As one of the largest and best-
capitalized companies, it was barely able to stay in business, unlike most of
its competitors.

In the wake of the crash, an impoverished inventor, Charles Goodyear, was
able obtain raw rubber very cheaply for his experiments. He toiled for years
to overcome rubber's drawbacks, hoping to develop functional consumer
goods. Goodyear tried to improve latex by blending it with every conceiv-
able additive and evaluating every feasible process path. On several occasions during this decade, he attempted to commercialize a promising recipe,
but met with dismal failure as the old problems resurfaced with a change of
weather.

In early 1839, Goodyear famously cooked a blend of latex, lead oxide, and
sulfur on a stove top and found that, rather than melting, it cured into an
ugly solid plaque. But, curiously, the resulting sample still retained its elastic
properties, even when exposed to extremes of heat and cold. This discovery was what the industry had been seeking for years. The process became

known as vulcanization and remains the fundamental technology for curing both natural and synthetic rubber products.

With his breakthrough, Goodyear hoped to quickly achieve fame and fortune, after years of poverty. However, he was constantly disappointed by the extremely process-dependent nature of vulcanization. Turning the concept into useful products was very frustrating for him and even more so for his financial backers and partners. Today, we know that sulfur atoms cross-link the isoprene molecules in latex, forming a very stable thermoset elastomer. To do so, the appropriate amount of sulfur must be uniformly dispersed in the latex matrix, while the temperature is maintained in a narrow range for a certain period of time. End-use performance is dependent upon each of these variables.

Given the process equipment available at the time, this process window was difficult to achieve and consistently repeat. The products that Goodyear heated in his oven would slump and fuse before they were fully cured. One product in a batch might be blistered and charred while others had soft, tacky spots that were not suitably cross-linked. Techniques were developed to maintain a uniform heat history from one piece to the next. Numerous tests were done to find just the right combination of additives and the appropriate compounding techniques. Goodyear struggled to master the process details with small-scale experimentation before he finally applied for a patent in 1844.

Once the basic parameters were demonstrated, commercialization followed quickly. Goodyear lived to bask in the glory of an admiring public. However, he was plagued to his death by lawsuits that sprang from sloppy control of the intellectual property rights. While seeking financial support during the long struggle, he had provided samples to his competitors, who were trying to carve out their own patent positions.

Truly novel products often spring from small-scale experimentation, which identifies essential interactions between raw materials and process conditions. It is cheaper and safer to do such work on a laboratory bench or in a pilot plant, rather than on manufacturing lines. Suitable production equipment was nonexistent in Goodyear's case. Ground-breaking material development usually requires that existing tools are modified to suit the process needs of new situations.

The first priority of a material development project is to generate a large volume of ideas through literature searches and creative speculation. This list should be comprehensive to the point of being intimidating. These options must be narrowed with experimentation.

RULE #5 PRUNE THE OPTIONS CLUTTER

The initial experimentation stage should be geared to exclude potential raw materials and process paths that are not optimal. When working with a clean sheet of paper, there are many options to explore—often too many. You must prune the weak options with a defined program of experimentation. Afterwards, the work can be focused on the most successful routes that remain. An exhaustive list of things that work and those that do not work is crucial to writing good patent applications.

Imagine the job of product development at a microbrewery. As a beer lover, you are responsible for creating distinctive seasonal beverages that appeal to like-minded customers. Even in a small operation, specialty beers and ales are fermented in large vats. New product development constantly updates the selection of funky beers. The plant's capacity is far smaller than those of the mass-market beer brands. However, even these small batches are too large to make experimental samples for taste testing and recipe refinement.

The brewmaster experiments with small batches to perfect new offerings. It is important to ensure that these test runs are made under conditions similar to those in the subsequent production runs. The small volume of test lots allows the developers to screen different recipe and process variables before choosing their next seasonal offering. This determines a preferred set of ingredients and conditions for scale-up to full production. Even with laboratory-scale brew tests, the product is not approved until it has been brewed successfully in the production vat.

During a typical laboratory trial phase, a multitude of samples will be generated and tested for their suitability. This includes their processability, stability, safety, and fitness-for-use (FFU). The development team should choose materials or families of materials after evaluating the widest possible range of choices. Ideas for potential process paths will be identified by laboratory testing, literature searches, and contacts with outside experts. These options should be driven by economics, safety/environmental factors, the patent landscape, and product performance. Prototypes are then compared to evaluate FFU performance.

Small-scale development work sets the tone and the path for the entire project. The kernel of an original idea will be fleshed out to the point where the value (or lack of it) should become obvious. Small-scale process experiments reveal fatal drawbacks in ideas that looked promising initially. Not every concept can become a commercial product. For any 10 options, 5 or more should be terminated at this stage.

Tools for Small-Scale Development

It doesn't matter how beautiful your theory is, it doesn't matter how smart you are. If it doesn't agree with experiment, it's wrong.

Richard Feynman

In the early 1950s, the German chemist Karl Ziegler pioneered a new catalyst system to improve the polymerization of polyolefins over the earlier Imperial Chemicals Industry (ICI) process. His transition metal catalyst could function at lower temperatures and pressures than the incumbent techniques. These discoveries were crucial to improving production economics in the coming years. As an academic, he (like Carl Marvel before him) was not interested in commercial applications. However, in Ferrara, Italy, Giulio Natta seized Ziegler's findings and hurried to apply them. Natta recognized that Ziegler's small-scale results could be the basis for a profitable business. He worked in the catalyst research laboratory of Montecatini, which is a large Italian chemical company. This facility had the equipment, the expertise, and the inclination to pursue implementation of the technology.

In 1954, Natta's team polymerized polypropylene by building on Ziegler's exploratory work. Montecatini patented the process and commercialized it within 3 years. It was soon selling licenses to companies around the world, triggering the rapid growth of plastics in our daily lives. Italy was an early leader in this revolution, based on the country's expertise in making extrusion equipment for pasta production. Natta continued to refine the product properties by adjusting the catalyst to tailor the arrangement and randomness of the methyl side-group placement along the polymer backbone. The Ziegler–Natta catalyst technology was further improved by laboratories around the world and remained a key route to making olefin polymers for the rest of the century. Both Ziegler and Natta shared the 1963 Nobel Prize in Chemistry.

RULE #34 KEEP THE APPROPRIATE DEVELOPMENT TOOLS CLOSE AT HAND

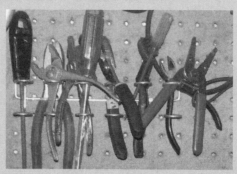

The early stages of development are all about experimentation. Samples need to be created under a wide range of process conditions and then evaluated. Flexible tools must be available before the need arises, or else significant time will be lost in acquiring them. Suboptimal equipment generally degrades the quality and pace of small-scale experimental results.

Equipment considerations for small-scale development:

- Experimental equipment is only useful if it can simulate a portion of a current or future production process. It is pointless to create a wonderful sample that cannot be duplicated on a commercial scale. In fact, customers become demoralized when they see a wonderful prototype product but cannot buy it, because commercial production is not feasible. One answer is to buy or design production equipment that exactly duplicates the size and important features of the laboratory equipment. However, this is rarely feasible.
- Ideally, the pilot plant should be more flexible than the current production plant.
    - Laboratory equipment should be able to generate numerous small samples with a range of different raw materials and/or process conditions. This may require the facility to easily clean out the system and start another run.
    - The equipment should be reconfigurable to run with an assortment of interchangeable components.

- The small-scale equipment should have better instrumentation than a production line. The idea is to capture every aspect of the process as it generates a given sample.
  - Think outside the box—consider unusual add-ons like high-speed video camera mounts or the ability to precisely monitor the timing of each event in a sequence.
  - Can movable or multiple probes be used to measure conditions across the width of a pipe or a reaction vessel? This can map gradients, allowing process modeling for scale-up.
  - Rotary equipment should be able to be returned to a home or zero position. Timing data should be monitored and recorded based on the angular position of the process.
  - Freeze or capture samples from discrete segments of a dynamic process for analysis.
  - A chromatograph or spectrometer may be positioned to monitor the composition of a flow stream in real time. This may allow you to cycle through different process states while looking for the optimal levels of specific intermediates or by-products.
- Some pilot or laboratory equipment is intended solely to generate data.
  - Design test fixtures to simulate degradation, color change, by-product formation, or the breakdown of materials inside a process. This will allow the screening of raw materials or new formulations before they foul the production trains.
  - Specialized components can simulate one aspect of a process. For instance, molten polymer can be injected between two closely spaced plates to predict how it will flow into thin-walled molds. Samples can be evaluated over a range of temperatures and shear rates.
  - Evaluate the materials used to make contact surfaces in the production process. A very small chamber containing metal plaques and corrosive raw materials can be held at the most extreme process conditions to determine negative interactions.
- Prototyping equipment converts your experimental samples into copies of your customer's products. This is a powerful means to extend your span of knowledge.
  - Making a finished product allows you to visualize the job from new perspectives.
  - These tools validate your theories and demonstrate the product to end users.

- Modify or purchase laboratory versions of your customer's equipment.
- Totally new materials, process paths, and by-products can be dangerous, because there is no operational experience. Laboratories need a thorough safety analysis and training program.

In the late 1940s, the Naval Ordnance Test Station at China Lake, California, was tasked with developing small rockets for military use. These were powered with sensitive solid propellants. The consistency of these materials was critical, because even small variations could significantly affect the ballistic performance.

The development efforts were hampered by the fact that the propellant was produced on the other side of the country. Turning development plans into experimental samples often took months. In 1948, the China Lake facility was authorized to upgrade its propellant laboratory. The new capabilities allowed it to mix small batches, blend the resulting paste, and extrude it into cylinders of the appropriate size for development projects. Afterwards, the team could perform physical, chemical, and combustion tests on the samples. The best candidates could then be packed into rocket bodies and fired on the range soon after. This relatively modest capability improvement drastically reduced the time needed to develop and modify weapons systems.

Procedures for Small-Scale Development

If we all worked on the assumption that what is accepted as true is really true, there would be little hope of advance.

Orville Wright

The English industrialist Henry Bessemer made his first fortune with a pigment called "false gold." This product was a finely powdered bronze that gave color to gold paint. It was in short supply in Victorian England, despite its growing popularity for decorative uses. The only source of supply was a guild of secretive German craftsmen.

Bessemer researched the incumbent production methods, so that he could set up domestic manufacturing. His investigation started in a library. Five hundred years earlier, the process techniques were described in an encyclopedia compiled by the German monk Theophilus. The traditional technique required bronze ingots to be pounded into a thin sheet. These sheets were then mixed with honey and ground in a mortar. The honey prevented the bronze particles from fusing together during the grinding process. However, a laborious washing step was needed to purify the final product. This process had changed little in half a millennium.

Bessemer then examined the German product under a microscope. In doing so, he established the appropriate particle size and shape that was functional in the paint applications. Only then did he experiment with production techniques that could make an equivalent product at a lower cost. A steel-roller assembly was employed to squeeze the metal into a thin sheet or leaf. A low-intensity tumbler reduced the sheets into powder. This approach required trial-and-error experimentation with process variables to create particles of the correct size and shape. However, since Bessemer knew what success looked like, he had a clear goal to work toward. Small-scale experimentation rarely follows the same path twice, so Bessemer's experience is instructive, but it is not a template for all such work.

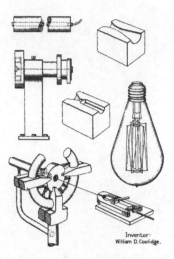

Early in the twentieth century, General Electric was working to improve the filaments in incandescent light bulbs. Tungsten was widely recognized as an ideal material because of its high melting point. The only tungsten wire then available was made by sintering powdered metal. It was very brittle and would fail when the bulbs were shaken. Clearly, a better wire-forming process was needed.

William Coolidge researched this problem and was inspired during a visit to the dentist. He was impressed by the processability of the mercury amalgams used to make fillings. Coolidge subsequently found that a cadmium amalgam was a good processing aid for the formation of low-temperature tungsten. This mixture was extruded into strands. Subsequently, the cadmium and mercury were extracted by heating under a vacuum. The resulting tungsten blanks could be drawn into strong, thin wires.

Despite theoretical predictions, the resulting pure tungsten wire tended to sag during FFU testing. Continued development found that doping tungsten with other materials at parts per million levels solved the sagging problem and created a solution that has endured for over a hundred years. These bulbs consumed one-third less energy than the carbon filament version and had a longer operating life. In addition, the tungsten technology enabled the development of better x-ray tubes for medical imaging applications. It was also crucial to the vacuum tubes used in a wide variety of electrical equipment.

Techniques for development screening:

- It is not always necessary to mock-up or simulate the finished product in the early stages. Look for ways to create discrete aspects of a product and test those for useful attributes. This approach can zero in on the optimal solution by establishing boundaries.

- The whole team needs to be on board with this approach. Team members who specialize in testing the finished product can become discouraged with nothing to experiment on.
- This phase should not continue for a long time, especially if there is some question about the validity of your model. Create complete prototypes and make sure that the attribute testing agrees with the FFU targets.

- Look for ways to mathematically simulate the industrial process to predict whether the experimental results will scale up to a commercially viable size.
  - Simulations only give a partial answer while failing to model other process aspects that are commercially important, such as human touch and feel attributes. Treat this technique as one tool, not the only source of information.
  - Modeling may not predict dynamic instability and transition points, unless the database contains huge amounts of experimental data.
- Screen many materials, conditions, and equipment configurations during the initial evaluation to get the widest possible view of the landscape.
  - Do this before zeroing in on promising routes.
  - Document what worked and what was eliminated and why.
- Establish which material properties are important for the end-product performance. Set targets and limits on these properties.
- It is possible to tinker with the process configuration, the raw materials, and the process variables. But, it is tedious to adjust all three at the same time. Try to lock in one of the first two, to reduce the complexity. Create multidimensional charts of the experimental space.
- Run a consistent control sample in all the trials. Chart the control's performance from one trial to the next. If the results drift over time, then the experimental apparatus or conditions are changing in some subtle way.
- When a promising candidate is identified, be sure to evaluate samples made at the specification limits. Determine if there are any sharp changes in properties or performance that represent high-gain situations.
- Any significant finding from an experiment should be replicated and confirmed. Repeat the experiment, varying the conditions to ensure that the benefit is not limited to some transient set of conditions.
- Understand the experimental space and track which parts of it have been evaluated.

- Industrial experimenters can easily overlook prime real estate in the available landscape, especially once they find promising results in one subset of it.

- Being locked into one process configuration by the limitations of the existing equipment deters an evaluation of alternate process paths.

- Researchers in different organizations will define the experimental space differently, based on their process experience, limitations, and commercial focus.

- The space is often multidimensional, making it hard to visualize and document.

- The more profitable the product is, the more options for competitors to duplicate it on less efficient processes. A thorough process evaluation leads to broad patents.

- Constantly seek the optimal place to run—the point at the center of the process window that gives the greatest operating latitude and avoids inflection points. Narrow the choices steadily.

- Pay as much attention to each unsuccessful sample as you do to the promising ones. Try to understand why it failed. What inherent property or attribute caused it to fall short? Can you make an even bigger failure by maximizing this factor? Is there another parameter that multiplies or reduces this effect? With a full understanding of all the keys to failure, the experimenter will know what to avoid.

- Understand the fundamental mechanisms that control performance.
  - Evaluate where the process runs well and makes a good product and then try to understand why.
  - Test the ability of your proposed model to predict both its success and its failure.

In 1856, William Perkin was conducting bench-scale research to create a synthetic version of the antimalaria drug quinine. Most of his trials were failures, but one left a tar-like coating in one of his valuable flasks. While cleaning out the residue with alcohol, he noticed that the solvent took on a purple color. His experiment had created the first synthetic dye.

Perkin seized on this curious phenomenon and patented it. The following year, he opened a factory to produce mauveine dye. The willingness to switch directions and explore interesting observations made Perkin a wealthy man and led to a series of new, artificial colors for fabrics and other applications.

Barriers to Effective Projects

You want it Good, Fast and Cheap. Pick any two of the three.

Unknown

Alfred Nobel had serious barriers to his development efforts. His explosive products periodically destroyed his process equipment; in one case, killing his brother. His customers were liable to cancel orders and bankrupt his family business when they unexpectedly ended their wars. Governments could and did abruptly ban his materials and evict his factories, even during a century with notoriously weak product regulations and safety requirements.

Nobel's breakthrough was a process to stabilize nitroglycerine into a form that was safe to transport and store. He absorbed three parts of the liquid explosive into one part diatomaceous earth, a form of natural silica, to create dynamite. This stabilizer allowed nitroglycerine to retain its explosive power, while greatly decreasing accidents. By adjusting the blend ratios, the explosive yield could be customized to end-use requirements. By forming the product into a cylindrical shape, it was well suited to fit into the boreholes at construction and mining sites. Large-scale canal and railroad construction projects in the nineteenth century provided a huge demand for commercial explosives.

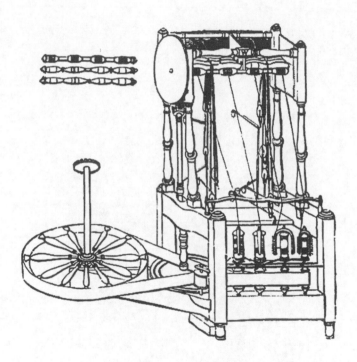

In the late 1700s, Richard Arkwright recognized the merits of a new spinning technology that had been invented by others who were not able to commercialize it. Such innovation was risky, because self-employed weavers vandalized automated operations that threatened their livelihoods. Arkwright was accused of stealing cotton-spinning innovations from James Hargreaves, who had borrowed freely from Thomas Highs. In turn, competitors freely copied his ideas after he became successful.

The incumbent technology for forming cotton thread was very labor intensive, with one person spinning a single thread. The product was expensive and had many quality issues. But, it provided an opportunity for workers to earn a living by working part time in their own homes. The Arkwright process oriented the cotton fibers, twisted them, and stretched the resulting threads into a strong and consistent product. This automated the crucial requirement of fiber spinning, which was to process the elastic thread under a constant tension. The manual technique failed to consistently fill this need.

Arkwright's new equipment was able to spin multiple strands at the same time, greatly increasing the output of the high-quality product. In addition, the process accommodated pure cotton fiber. Blending in flax was previously required for processability on the loom. Pure cotton fabric had only been available by importation from India. England became the world leader in textile production, based on the industrial-scale processes that Arkwright and others pioneered.

Red flags for the early stages of development:

- Failing to explore enough alternatives.
  - At the start of a project, every possible idea should be screened and evaluated. Sometimes, time pressure can prevent the team from giving the appropriate attention to every alternative. This is a natural result of operating in the real world.
  - The existence of an incumbent technology, solution, or process tends to discourage developers from investigating alternatives. It is quick and easy to jump to an existing piece of production equipment or utilize an existing solution. When this occurs frequently, then growth and innovation are retarded.
- Inaccurate, incomplete, and missing information after the experimentation is finished.
  - Results that fail to agree with expectations should be closely examined to understand why the theory does not match the reality.
  - Failure to record everything properly and poorly stored samples are major causes of repeating experimental trials unnecessarily.
  - Not understanding the process that the engineer is designing toward. It is often very easy on the bench scale to create interesting and wonderful prototypes that cannot be produced commercially. This is only useful if it generates information that drives the project forward and justifies the appropriate equipment purchases.
- Not realizing what factors are controlling the results.
  - It is very beneficial to hold most parameters constant while adjusting one or two. In a dynamic process environment, this is not easy.
  - Attempting to disprove your proposed solution is the best way to verify its validity.
  - Good record keeping will allow you to comb through your data to find the common factors between all the samples that succeeded or showed good results in one or more aspects.
  - Seek to replicate key test results to validate your model.

- You must recognize when a factor creates a trade-off between two properties, preventing either of them from simultaneously being at the desired value.
- Team members that do not believe in the approach being taken.
  - It is easy for participants to feel that they are chopping their way through the wrong jungle if they did not buy into the plan when they were assigned to the project. Listen to their concerns and consider their point of view.
  - Overcontrolling organizations can easily create this situation by putting too many constraints and assumptions on the work. Team members may assume the existence of constraints even when none were intended.
  - Strong-willed managers and team leaders may force people toward preferred solutions without allowing discussion or debate.
- Repeating work and correcting errors due to failures of planning and communication.
  - This can be caused by ambiguous experimental instructions. Plans are best reviewed and discussed by all involved in advance of starting the work.
  - Never give undocumented verbal instructions, especially when doing nonroutine tests. Later, it will not be obvious what the data mean or how they relate to other information.
  - Beware of time-saving, cut-and-paste editing techniques when drafting complex work plans. These detail-rich documents often retain elements from earlier experiments, which should have been deleted or overwritten.
  - When instructions undergo a lengthy editing and review, then multiple versions exist on paper and in computer drives. Further changes can be made during the experiment. Later, it can be difficult to determine which plan was followed.
  - Team members must learn from failure and not cover up their mistakes.
- Failure to protect intellectual property.
  - Record the details of all pertinent experiments. In the middle of an experiment, you never know what is essential and what is not. Log the data without premature conclusions.
  - Understand what elements are novel, which are trade secrets, and what is prior art.
  - Do not take undeveloped ideas to customers, vendors, or other outsiders before a patent application is filed or nondisclosure agreements are signed.

- Not having consistent and realistic guidelines for the time required and the deliverables.

  - Moving targets will cripple a project in a process-dependent environment.

  - Attempting to build too many features into a product early in a project, rather than concentrating on the key attributes that are important to customers.

  - When the landscape of materials and process paths is not fully explored, the results are narrow process windows, substandard FFU, and low value.

Managing the Development Organization

Never tell people how to do things. Tell them what to do and they will surprise you with their ingenuity.

George Patton

In the 1970s, an earnest young university student was working as an intern at a large industrial company. The organization was somewhat bureaucratic, with spending approval restrictions for each level of management. After years of inflation, these arbitrary limits were outdated, requiring a long chain of approvals for modest engineering projects that could extend to the upper levels of management. Rather than update the spending limits, the engineers simply worked around them by creating a series of mini requests that could be approved by low-level managers.

Everyone functioned under the system, with the exception of their temporary student engineer, who was morally offended by the deception. This individual fervently believed that every rule must be obeyed as written. After months of toiling under this ethical conflict, the student was preparing to return to university. A whistle-blower letter was written, calling the deception of the technical organization to the attention of the president. The following year, the company declined to invite this individual back for another assignment and business continued as usual. No one in the management really wanted to approve minor expenditures.

Every company has some procedures, regulations, and practices that could be improved. In fact, rules that cover every possible contingency are the crowning achievement of a bureaucratic mindset. At the other extreme, creative artists throw off all limits and constraints in pursuit of their art. Product development teams need to operate within the middle ground between control and chaos. Technical organizations must somehow grant the freedom to explore the limits of the imagination while maintaining the discipline to investigate and document every process path.

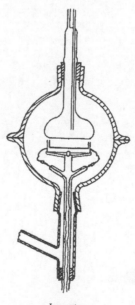

Inventor
Wallace H. Carothers

Elmer Bolton directed DuPont's dyestuffs department in the 1920s. He was aware that the company was interested in developing artificial analogs of natural latex rubber. In 1925, he attended a presentation by Julius Nieuwland, a chemistry professor and Jesuit priest at Notre Dame University. Nieuwland had extensively researched acetylene reactions, discovering a copper chloride catalyst that could produce divinylacetylene. This monomer could be polymerized into a rubbery material. While interesting, this process path was not commercially practical.

At Bolton's insistence, DuPont purchased the rights to the technology and initiated an exploratory research program to search for useful rubber properties. With Nieuwland's catalyst, they synthesized monovinylacetylene (MVA). In 1930, the chemist Arnold Collins reacted MVA with HCl, creating a liquid that polymerized into a rubbery solid called chloroprene. Additional development work led to an elastomeric material, which they named neoprene and commercialized in 1932. It was the first successful synthetic rubber. This success occurred because DuPont was willing to acquire the best tools (catalyst technology) and to dedicate a team to explore all the process paths.

Small-scale development tactics vary with the type of project you are trying to run. An exploratory project is geared to the evaluation of a new technology. Rather than trying to create a product, it should be broadly designed to exploring properties, process windows, and general capabilities versus the incumbent approaches. Well-executed exploration work will demonstrate the potential solutions and provide tools to subsequent project teams to develop commercial products.

Commercialization projects fall into three distinct categories. The first type is the most familiar one. A customer need is identified and a development team strives to fill the request. A complete process path consisting of raw materials, process conditions, operating procedures, and specifications is identified to produce the end product. This approach is common in a market-driven organization.

The second type of project works in the opposite direction. Starting with interesting process capabilities, the team seeks applications that can benefit from the new technology. This is typical of a balanced approach or a process-driven company.

The final option is a systems approach that combines different elements to provide customers with a complete solution. This is often a machine or a permanent fixture that interfaces with a consumable product. This system could be either closed (proprietary to your company) or open (multiple vendors could compete to sell products). This is most common in a market-driven approach. See the concept maps at the end of the chapter to understand the four types of small-scale development projects. "The map of small-scale development" compares the first three types.

Guidelines for managing development projects:

- In an industrial organization, it can be difficult to motivate and evaluate the people who do exploratory research. Successful outcomes may require years of collaborative effort by many participants. At other times, a single researcher stumbles onto an unexpected result such as polycarbonate or artificial sweeteners. Very few technical professionals will experience the lucky accident, but every product must be commercialized and improved with rigorous, boring detail work.

- Incent your researchers for their ability to make intuitive connections and master the details of new technology.

- Reward those who are recognized by their peers as selfless contributors. Discourage those who hoard important findings.

- The technical organization needs to continuously evaluate and weed out product development proposals that cannot be achieved with the available technology or cannot be made profitably.

  - A proposed project might deliver great things for the company. However, if the underlying science does not exist to facilitate development, then success is unlikely. Explore the fundamentals with research to establish a foundation for product development.

  - A flowchart labeled, "And then a miracle will occur" has no place in development plans.

  - Will the technology, product idea, or process path actually deliver a cash flow when completed? Is there a plan for production capacity and a sales channel?

  - Make sure that the team has the resources and expertise to accomplish the task. Ensure that low-priority projects are not inhibiting the process of the important ones.

  - Save the rejected concepts for later development, especially while doing market research, FFU development, or fundamental experimentation in the meantime. Note the limiting factors that make these concepts nonviable. Monitor new technology that might create a way around those roadblocks.

- Management should track the progress of each project against a work plan, which is subject to revision. Plans must include a list of critical questions to be answered.

  - A team can work diligently, evaluating samples and running experiments without making progress toward project goals. This happens most frequently on fundamental research.

  - Ask what key questions have been answered, rather than what work has been done. Accomplishments can include possibilities that have been ruled out. Review action plans that will achieve the next scheduled milestone. Constantly reconsider if the goals are valid.

  - Beware of planning that defers the difficult steps to the later stages. It is best to address such problems early in the program, before significant resources are spent.

- As a project nears completion, make sure that other groups within the company are prepared to implement the results. Team members are disheartened when their hard work is not moving toward a clean handoff.
  - If changing priorities halt a project, then ensure that the important findings are preserved until the project is restarted. Document the information and securely store the calibration standards, custom fixtures, and other specialized tools.
  - Evaluate the ramifications of stopping a project, such as vendors to whom promises have been made, customers who are counting on your new product, and the production facilities that have been committed to running the process.
- The development team should focus obsessively on the agreed objectives. The management should ensure that priority projects have all the necessary resources.
  - Fully resource the projects that are critical to company strategy before resourcing the less critical ones.
  - Researchers should not be punished for being tasked with dead-end assignments through no fault of their own. Neither should they receive outsized rewards if it turns out to be an idea with great merit.
  - Consider the elasticity of the project and the business model. By doubling the resources, can it be accomplished sooner? Will the additional revenue from the early product introduction pay for the extra spending?
- Projects that are leading nowhere should be killed, even if they are the brainchild of the company's founder or the head of R&D. It is very demoralizing to work on a project that seems to be doomed, yet remains active because of internal politics.
- The technical management and the product line management need to jointly administer the patent suite. Try to fully protect your future market position, not just isolated discoveries.
  - Just because you can get a patent, it does not mean that it will be useful to the company. Patents are expensive to draft, file, and maintain, especially on a global scale. Potentially patentable ideas need to be evaluated to determine if they will provide some protection against competition.
  - Occasionally, management should review which approaches need to be researched and patented in light of company strategy.
  - The easy patents to initiate and draft bear directly on the actual product under development. These require no additional

research, because most of the data and examples have already been created. However, they often neglect to cover applications that are outside the project scope and may fail to block competitive work-arounds. Full coverage may require a separate project devoted to exploring all the options, even those that are not commercially viable for you at this time.

- Keep a firm watch on the time needed to achieve project milestones.
  - Why is the team missing deadlines? Is the work more difficult than anticipated or is the team agonizing over trivia? Is there internal conflict or dissent?
  - Are the goals and constraints still reasonable in light of the team's findings?
  - Determine if resource conflicts with other projects are causing the delays.
  - Was the deadline arbitrary? Were goals changed due to shifting market situations?
  - If the group is working efficiently, but it is encountering obstacles, then the addition of extra resources or extra time might be in order if the goal is important.
  - Are there other projects that need to be adjusted based on this delay?

In the late 1930s, DuPont commercialized the cold-stretching process discovered by Wallace Carothers' team. The nylon fiber product was very successful in the market. However, Carothers had also experimented on polyesters, but this work was allowed to languish in order to focus on nylon. Several years later, DuPont and ICI cooperated, sharing fundamental research results for their mutual benefit. ICI chemists John Whinfield and James Dickson were assigned the job of researching the polyester option that Carothers had dropped. They explored the technology and found interesting properties, especially when the cold-stretching process was used to draw fiber. ICI patented polyethylene terephthalate (now known as PET or PETE) and quickly commercialized this resin for the production of polyester fibers. DuPont recognized their oversight too late and was forced to license ICI patents for their Dacron fibers and Mylar film products.

Retaining Samples and Information

Oberle's law of product development: There will always be some obscure attribute in the original sample/model/batch of a promising experimental product that will never be duplicated satisfactorily.

Corollary #1—The very existence of the lost attribute will not become apparent until after the legendary first sample has been completely consumed, discarded, mislabeled, or modified beyond recognition.

Corollary #2—The original dog-eared and coffee-stained laboratory notebook entry will not capture details of the process parameter that controls the lost attribute.

Corollary #3—The passage of time will magnify the lost attribute's importance until its reproduction becomes the key to the product's commercial launch.

Mueller's corollary—The most sorely missed sample will be the one that was discarded yesterday.

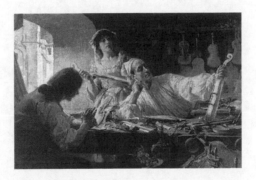

There is no more famous example of lost production techniques than the stringed instruments made by Antonio Stradivari in Cremona, Italy, during the early 1700s. Fewer than 700 instruments from his workshop survive to this day. Each piece is unique and sells for millions of dollars when it comes on the market. For centuries, the world's top violinists have preferred a Stradivarius over newer violins.

Scientists have theories to explain how Stradivari and his competitors made the best musical instruments of all time. They range from compact tree-ring growth during a solar minimum to the glue and varnish that Stradivari used.

Much controversy surrounds the question of whether even the most skilled musicians can distinguish the "voices" of individual Stradivarius violins in blind tests. However, this author and 99.9% of humanity are not sufficiently skilled to do so. That may constitute the world's toughest FFU test.

Stradivarius lessons:

- A false/deceptive memory of a fabulous performance is aggravated by a lack of objective FFU tests.
- The illusion of a flawless performance is confounded by perception, celebrity, and rarity.
- Good marketing, elegant design, and trendy fashion can influence perception more than the actual properties of a lost sample.

RULE #9 KEEP A WIDE RANGE OF SAMPLES TO DOCUMENT DEVELOPMENT WORK

No one ever knows which experimental sample or data point will represent a "Eureka!" moment in subsequent testing. Retain samples from each series of trials in the event that they are needed for comparison. Label them well. Assume that every piece of data could be the key to making the critical connection.

Eventually, the storeroom will be ransacked in an effort to find evidence from one trial, which performed better than the others in subsequent testing. Perhaps it had unique surface characteristics, trace contaminants, or a slightly different process history. It may be the same as the rest, indicating that the FFU tests were flawed, biased, or misremembered. No one will ever know, if the samples, process data, and observations are not properly retained.

Periodically, the storeroom must be selectively purged to make room for other projects. Housekeeping should be done with an understanding of how the samples fit into the project.

Tips for sample retention:

- Small samples can be saved in plastic zipper bags, labeled with a unique code number, and stapled to copies of the original notebook entries.

- Use detailed labels that will be understandable. Avoid terms such as "Last month's run."
- Larger samples should be tagged directly or wrapped in a protective covering.
- Use sealed foil bags if the material will change with exposure to oxygen, moisture, or light.
- The more nondescript the sample is, the more detail there should be on the label.
- When in doubt, collect and label a huge variety of time-stamped samples during a run and later sort out the ones that represent key transition points.
- Avoid loaning out important samples for "show-and-tell" presentations.
- Keep both the good and the bad performers for comparison. Researchers have a natural penchant for disposing of embarrassments, without understanding or learning from them.
- When reusing containers, ensure that they are clean and free of old labels.

Research notes always seem to be complete, accurate, and self-explanatory as they are recorded. Later, however, it becomes apparent that things are missing. Information about some aspect of the equipment, procedures, personnel, conditions, or materials is rarely 100% complete. These items often go unrecorded because they seemed too obvious at the time.

In 1933, Dow researchers were working on synthesis routes to make chlorinated solvents such as perchloroethylene. Someone noticed that a strange deposit had adhered to the walls of a flask while it was being cleaned. The material was unexpectedly strong, resisting attempts to clean the glassware. Ralph Wiley ultimately got credit for the accidental discovery of vinylidene chloride/vinyl chloride copolymer (PVDC).

Dow later commercialized the material under the Saran brand name and Wiley made a career of tailoring it to a range of applications. In the case of such accidental discoveries, it is critical to understand how the new material was formed. One soiled flask in a laboratory sink looks like another. Only meticulous records allow you to understand how it came to be.

Systematic procedures for recording a complete package of experimental results:

- Decide in advance exactly what information needs to be recorded. Include anything that may be relevant—equipment serial numbers, weather conditions, cycle counts, and previous materials running on the equipment prior to the trial.
  - Create prelabeled data sheets for quick reading of the entries. Group the entries logically, such as the sequence in which the instruments will be read. Decide on the frequency for collecting each dataset. Include a key with abbreviated terms.
  - Use preprinted adhesive labels for samples that match each line of the sheet.
  - Leave room on the sheets for additional information and observations.
  - Never use inappropriate sheets, designed for a different piece of equipment.
- Review the data immediately after the work is finished. Evaluate the data for completeness and add additional observations while they are still fresh. Photocopy and save important pages, if there is a risk that the book might be lost or damaged before it is full. Retire the notebook early if it contains a particularly important patentable discovery.
- When overwhelmed by the volume of information, use a data recorder to store the readings directly from the instruments. Note the time sequence of the events to correlate the data with the written or recorded commentary, if the sequence of events is highly dynamic or transient.
- Use a voice or video recorder to dictate notes for later transcription when sequences unfold quickly. You can keep your eyes on the equipment, rather than on the notes. Start the recording with a time stamp and run it continuously, even if this leaves some dead air. Later, reconstruct a time sequence. Video recorders that combine the action with an audio commentary can be useful in correlating a digital data stream with real-time events.
- Have a colleague review and sign the notes as soon as possible after the experiment. He or she should ensure that the entries are readable and complete. Correct erroneous or ambiguous data. Initial and date your corrections. Transfer the data from read/write media to a permanent read-only storage to keep a legal record.
- Take photographs or video of the equipment and the dynamic processes, especially when the configuration is frequently changed. Set the camera to record the date and time to capture a sequence. Place

known objects in the field of view for size reference in situations where dimensions may be important. Make notes to indicate what each picture means and how it fits in the experiment. Save the definitive pictures and rename them with explanatory file names or attachments. Print, sign, and date the hardcopy that documents key findings or dates.

- Devise a consistent way to cross-reference and archive related media that would otherwise be stored in different places—notebooks versus computer files versus magnetic tapes.

- Beware of data that are spread between many notebooks, which have been entered by different participants. The content and format can vary widely, unless they are discussed and agreed upon beforehand. Copy or scan all notebook entries that pertain to these trials. Add notes or links to relate them to subsequent test results. Never rely on memos with truncated data.

Alfred Wilm was a metallurgist in Berlin in the early 1900s. His goal was to harden aluminum alloys by quenching, as was done with steel. While experimenting with Al–Cu–Mn alloys in 1906, Wilm achieved excellent strength, but he could not obtain the required hardness. Just before leaving for the weekend, he and his laboratory technician quenched the week's final sample and measured another disappointing result.

The following week, the technician took the time to complete a full round of testing. An unusual discrepancy in the data was noticed. The final sample from the week before had hardened over the weekend, dramatically improving its properties. Repeated trials confirmed that this alloy underwent a transformation during 1–4 days after processing. Good experimental records and sample retention enabled Wilm and his technician to spot the unexpected change in hardness compared to the original values. This product was commercialized under the Duralumin trade name and became a key material for the construction of airplanes and zeppelins.

Later work revealed that the alloy underwent a slow room-temperature precipitation of a nanoscale crystal structure. This structure blocked slippage in the crystal structure, increasing the deformation resistance. Later still, scholars discovered that the Wright brothers benefited from the same phenomenon, without realizing it. The engine block in their historic 1903 aircraft was made with a similar aluminum alloy and displayed evidence of a similar morphology.

Scalability

> We are continually faced by great opportunities brilliantly disguised as insoluble problems.
>
> **Lee Iacocca**

Thousands of years ago, mankind learned to engineer fused clay pottery. Pieces of moist clay were molded into shape, dried, and then stacked in a kiln where they were fired at high temperatures. The inventions of the kiln and the potter's wheel improved our ability to store and transport liquids such as oil and wine. This technology facilitated the rise of trade across the Mediterranean and elsewhere. Pieces were formed from moist clay into useful shapes.

Running this process is very tricky without accurate temperature controls. Pottery is sensitive to both the peak temperature during the burn and the rate of the temperature change during heating and cooling. Trial-and-error techniques were used to develop operating conditions for each particular kiln, pot shape, and clay type. Ancient pottery production sites can be identified by the heaps of shards. Broken parts were created in abundance during the commissioning of each new production unit and when process excursions occurred.

New clay sources, pot sizes, and wall thicknesses could each dictate changes to the procedures and settings that controlled the temperature profile inside the kiln. The sizes, shapes, and finish of the product were not limited by the human imagination, but by the limitations of the process control technology. Even today, potters use caution when transitioning from one kiln size and type to another. Increasing the output capacity may require trial and error to identify the optimal conditions.

RULE #35 SCALE-UP ISSUES WILL ABSORB MORE TIME AND EFFORT THAN EXPECTED

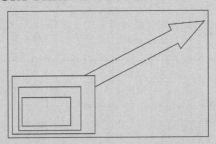

In a process-dependent environment, small-scale experimentation never simulates the commercial production operation as well as is desired. This situation must be anticipated and planned for early in the project. Much of the project work involves gathering a wider range of information to adjust the production process to duplicate promising samples made in the laboratory.

Like fired pottery, all engineered materials have process factors that are essential for consistent product properties. Every process has its own unique pressure points that can be manipulated. Sometimes, it is a matter of uniformly transferring the heat energy into and out of the material at a certain rate. Other products require controlled dispersion of additives. Achieving uniform patterns of shear stress in molten polymers flowing through distribution channels is important to plastic processing. Manipulating the size and uniformity of the crystal structures in a material is a requirement for many different industries. The critical process factors must be reproduced when transitioning from small-scale experiments to large production lines. Sometimes, they must also be fine-tuned when going from one production train to another.

Once these essential process knobs are identified for an existing application, it may be relatively straightforward to optimize the process to lock-in the solution. The problem arises when multiple products need to be run on the same equipment, each with their own customized quirks. Worse yet, subtle variations in raw materials, weather conditions, or component wear may force the production operators to hunt for the new sweet spot on a daily basis. Customer satisfaction will suffer when operators lack the means to sense whether the process is achieving the desired properties.

Product development generally begins by creating samples across a wide range of process factors. You evaluate where the process runs well and where it satisfies end-user needs. Ultimately, you need to find the intersection of those two states. The problem comes when promising bench-scale results cannot be replicated on production equipment.

Suppose that product properties are controlled by removing heat from a material at just the right point in the process. For instance, traditional black-smiths might plunge a piece of hot iron into a quench tank when they judge that the time is right. That judgment may change with the size of the part, the grade of iron, and the planned end use. The smiths have the flexibility to quench the work when they choose.

In an automated industrial process, this may not be the case. As a continuous flow of product moves through the system, the quenching step might be dictated by the physical placement, angle, and size of the cooling nozzles. If they are designed to move and adjust over a wide span, then the equipment will be flexible for processing a range of products. When the nozzles are rigidly fixed in place, then running a series of experimental variables will be difficult and expensive. When the developers can make a wide range of variables, FFU testing can be used to interpolate the best conditions for running commercial production.

Typically, development trials are intermittent as the process is repeatedly started up and shut down to make the necessary adjustments. Each sample becomes unique and distinct as a result of this process variability. Each one may be a poor representation of the products that will ultimately be generated by a steady-state operation. Such variation is one source of scale-up problems.

Dealing with scale-up issues:

- Maintain a list of factors that could create problems on each piece of production equipment. There will always be restrictions dictated by temperatures, pressures, shear rates, viscosity, corrosion, and the like. Develop laboratory tests to evaluate the raw materials or process intermediates against these known restraint limits. For instance, if a material is not capable of being molded at the maximum pressure that the press can exert, then eliminate it from consideration. Alternately, look for a way to alloy, soften, or lubricate it to achieve the processability that is needed before going to the plant with it.

- Derive correlations or dimensionless relationships that allow predictions of stable production conditions based on laboratory-scale results. Use the small-scale equipment to predict the basic process dynamics. These might include stress–strain curves, reaction kinetics, and rate-limiting steps. Granted, the heat transfer rates, residence time, and stability relationships will be different on commercial equipment. But, the small scale can give a fundamental understanding. If those relationships are established ahead of time on a technology-enabling project, then the development can proceed quickly.

- Companies with many lines of a given type should equip one line with extra instrumentation or adjustability for start-up trials of new products. This line can be used to diagnose scale-up issues and to understand how other lines need to be retrofitted.

- Use mathematical modeling to predict the properties of the materials made on the plant equipment based on the small-scale test results. Always seek to validate and confirm such predictions before making crucial decisions.

- Use the small-scale equipment and then the pilot plant equipment to narrow the range of choices down to a manageable number for subsequent production trials. Select options that have an acceptable FFU performance over a wide process range. Narrow processing windows invite scale-up problems.

- Develop equipment to predict hang-ups in the production process. Such equipment might never make prototype products—only information about their processability. A bench blender might simulate the shear rate in your production equipment, allowing an evaluation of the dispersion and particle size. An autoclave may duplicate the temperature and pressure conditions to evaluate the reaction kinetics. A small batch reactor may run a longer cycle to duplicate the residence time in a long tubular reactor.

- What is the optimal size for each production unit? Is it better to have one high-rate line or a number of smaller systems that are close in size to the research unit? A large line might have the best production rates and efficiency. However, it is the most difficult to simulate in a laboratory.

- Avoid the temptation to show off the great performance of early-stage samples to impress management and customers. The entire organization will look foolish when you fail to produce the same results commercially.
  - Beware of the researcher who builds a reputation for small-scale success that can never be scaled up to the production process.
  - A red flag warning should go up if pilot samples are hard to make. If only 1 in 10 tries yields a good result, then do not expect better results in the plant.

- Sometimes, you must reveal small-scale samples and results to customers or partners. This should only be done in an effort to explain your technology and judge its compatibility with their needs. Never suggest that the final product will be exactly the same.

Oct. 10, 1933.   E. THOMSON   1,930,327
COMPOSITE SILICA ARTICLE AND METHOD OF FABRICATING SAME
Filed May 10, 1932

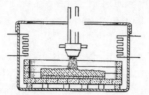

Inventor:
Elihu Thomson,

In the 1920s, work started on an ambitious telescope project for the Mt. Palomar Observatory in California. The 200 in. mirror was the largest ever constructed at the time. As mirrors increase in size, they are more susceptible to distortions caused by the thermal expansion of the material. Fused quartz was selected for this project because it had the lowest coefficient of thermal expansion of any available material.

General Electric cofounder Elihu Thompson agreed to take the job at cost and to personally lead the development team. Using an electric furnace of Thompson's design, they gradually scaled up the mirror sizes from 22 to 60 in. and finally to 200 in. The 60 in. size worked well, but nothing larger was feasible with the technology available. Tiny bubbles in the molten quartz distorted the surfaces of all their large mirrors. Thompson spent years on this effort, but ultimately failed. The telescope was completed years behind schedule with a Pyrex mirror.

Summary

- Truly novel products often spring from small-scale experimentation, which identifies critical attributes and optimal process paths for scaling up to commercial production.

- Look for ways to simulate discrete aspects of a process to evaluate key attributes.

- Cast a wide net for diverse ideas in the early stages of development, but constantly seek to identify conditions that optimize FFU and relentlessly narrow the operating window as the project progresses.

- Devise tests to overcome the known process limitations on production equipment.

- When promising results are found at one condition, explore the processing window around that point. A narrow process range is difficult to scale up.

- Make sure that positive results can be replicated. Search for those factors that inhibit good results as well as those that create them.

- Use control samples to chart the performance from one trial to the next to understand process reproducibility.

- Retain samples—both the failures and the successes. Keep copious records that document all the process aspects that created them.

- Do not assume that success on the small-scale equipment will scale up perfectly to the production line. Addressing issues of scale transition will take time, effort, and expertise. Avoid making promises to customers before scale-up issues are addressed.

- Manage exploratory development to lead toward:

  - Fit-for-use products that can be made profitably on a commercial scale.

  - Cataloged knowledge that enables further development or optimization.

  - Well-documented intellectual property that leads to useful patent coverage.

  - Motivated employees who work as a team.

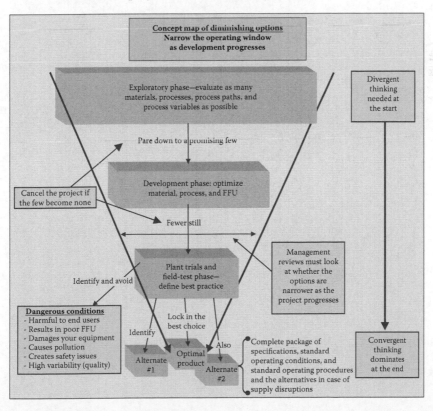

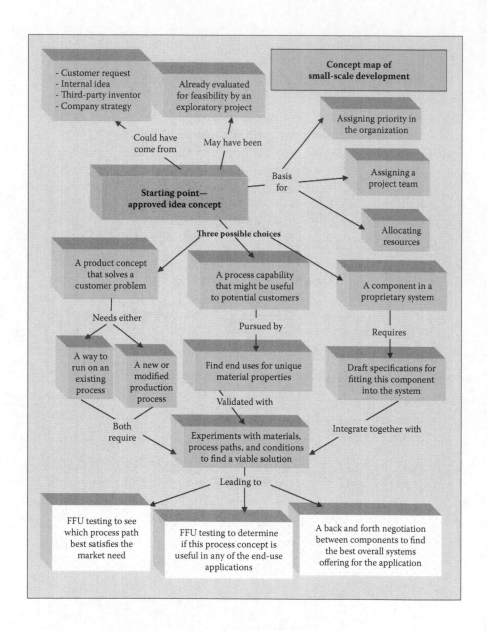

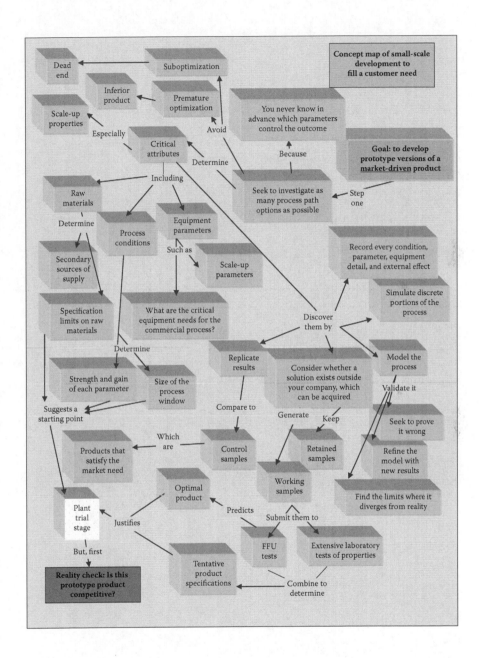

Concept map of small-scale development to fill a customer need

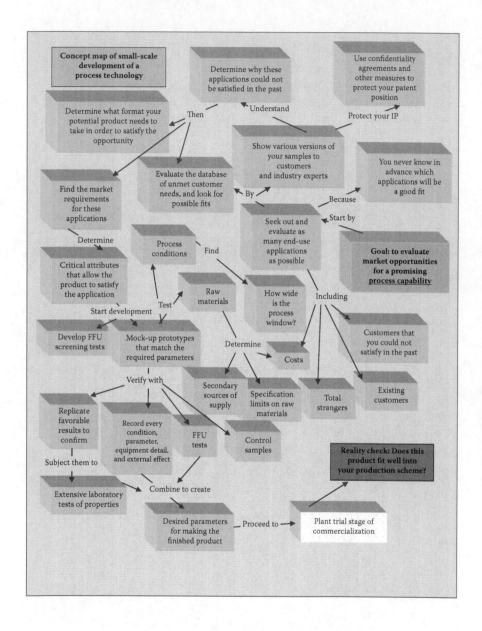

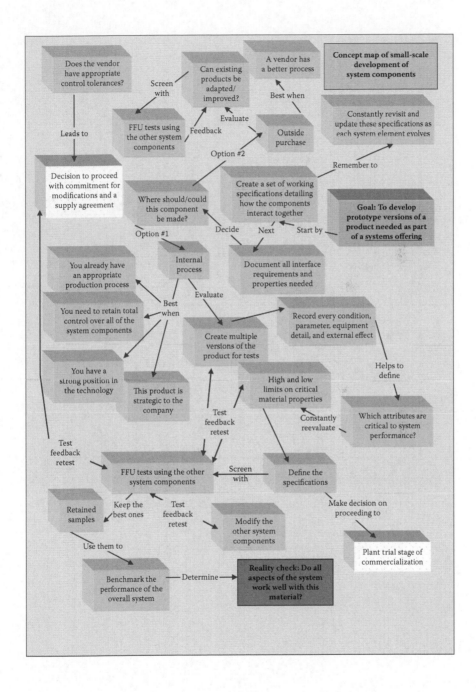

Concept map of small-scale development of system components

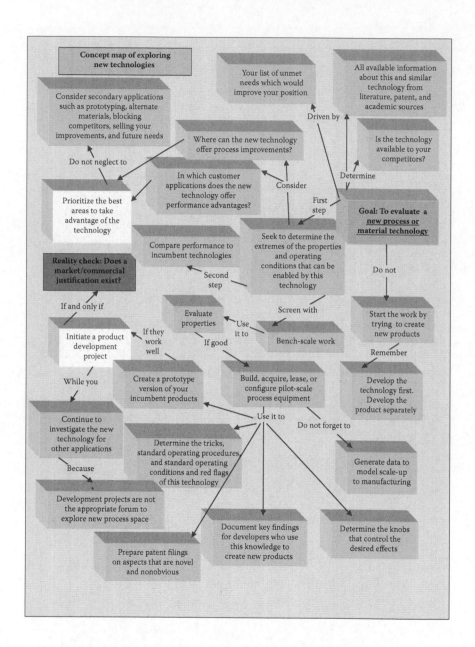

10

Development Trials on Large-Scale Equipment

How do we turn promising prototypes into commercial products? What strategy should we use for each different project type? How do trials change as development progresses toward commercialization? Why do development and production organizations view this work differently?

> If you can't describe what you are doing as a process, you don't know what you're doing.
>
> **W. Edwards Deming**

History's worst plant trial had been underway for 35 seconds during the early hours of April 26, 1986. Shift leader Alexander Akimov sat in the process control room with his hand poised over the emergency stop switch. His crew was trying desperately to maintain control over an inherently unstable situation. Akimov was only 80 minutes into his shift on the worst day of his career and had only a few hellish days left to live.

The atmosphere in the control room was tense. The engineer-in-charge, Anatoly Dyatlov, was determined to complete his experiment, regardless of the warning signs. Safety interlocks had been bypassed. The plant was running in uncharted "white spot" territory and was responding abnormally. The run plan was covered with cross-outs and handwritten additions. Because of unexpected delays, Akimov had not been briefed on the procedures. Under the original schedule, his team should have been overseeing the plant shutdown, rather than running the trial.

A historic drama was unfolding at Chernobyl Unit 4, a Soviet RBMK-1000 power plant. This reactor design had a number of flaws, which made Akimov's job both difficult and dangerous. The system was unstable when operating at low power levels and would run away in a fatal positive feedback loop if the cooling water flow was disrupted. This state secret was concealed from the plant's operators. The unit was filled to its limit with radioactive waste, which made the control response sluggish and erratic.

Ironically, Dyatlov's risky experiment was to verify a vital safety system for the plant. He was trying to demonstrate that the steam-turbine generators could provide temporary electrical power as they spun down during an emergency shutdown. This transient energy supply was needed for a critical

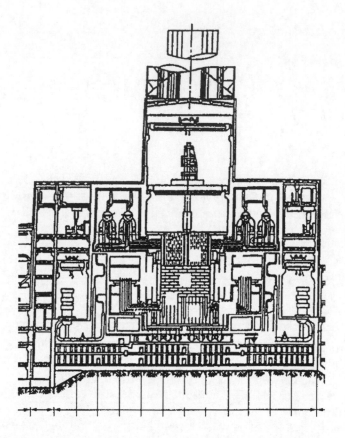

30–40 seconds until the backup generators kicked in. The central power authority required that this operation be verified with a full-scale plant test. Management bonuses were at stake.

In the run-up to the brief test, the operators needed to stabilize the reactor at half the normal power level. Control rods were lowered into the core to absorb neutrons, slowing the chain reaction. However, a strange situation developed, which was beyond Akimov's experience. The output power plunged to 5% of nominal levels. Inside the core, iodine and xenon 135 gas were collecting around the fuel rods. These reaction by-products absorb the neutrons, which sustain the chain reaction.

In this situation, the reactor should have been shut down for 24 hours to allow the iodine and xenon to disperse. However, Dyatlov was under pressure to complete the experiment and was determined to move forward. On his authority, more and more control rods were extracted to coax the power output up to the required levels.

Without moderators, the nuclear chain reaction gradually climbed out of the "iodine well" that was choking it. The reactor's thermal output increased slowly, but it was still below the planned levels. Impatiently, Dyatlov ordered the steam flow to the generators to be diverted so that the turbine test could finally begin. The big blades began to spin down and the backup generators were brought to life.

The exact details of what followed are lost to history because all the knowledgeable people either died or shaped the story to cover themselves. It is believed that the reduction in the turbine speed constricted the flow of the coolant, causing the formation of a large vapor bubble inside the pressurized chamber.

Under the original plan, this would not have been disastrous. But, lacking both water and control rods, the chain reaction suddenly accelerated out of control. Someone noticed the surge in energy output and shouted an alarm. Akimov triggered the emergency stop control, which plunged 205 control rods into the core.

Everyone expected this action to bring the chain reaction to a crashing stop. However, the process was not operating normally and the control rods had another design flaw. As the rods plunged down, a dead spot at their leading edges displaced the last margin of safety in the form of precious cooling water. Before the rods could travel deep enough to throttle the reaction, the uninhibited neutron flux briefly released hundreds of times more heat than usual. A chart recorder monitoring the power and the pressure in the system spiked ominously to the top of the paper. The remaining water in the system vaporized into a massive steam explosion.

The 2000 ton reactor lid blew off and ejected one-third of the radioactive contents into the air. In quick succession, a hydrogen explosion lit up the plant and ignited a horrendous fire. The graphite moderator burned violently for days afterwards, sending radioactive particles high into the atmosphere. Only a heroically suicidal effort by the firefighters and plant personnel brought the fire under control.

Few engineers have the opportunity to make the lasting impression that the late Anatoly Dyatlov did. He served 5 years in prison, which was longer than the remaining lives of the control room crew. His case serves as an extreme example of the consequences of a bad production test. Well-run plant trials are a key step in the development cycle. Without them, commercial production would not be feasible. Safety must be the primary concern when conducting plant trials.

Reasons for production trials:

- Plant trials are generally conducted to produce commercial quantities of prototype products for evaluation by customers. In doing so, the performance viability can be determined.
- Production tests demonstrate process scalability. Developers seek to narrow the product performance differences between the best laboratory-scale process and high-volume production equipment. Scale-up models can be validated and refined.

RULE #20 PLANT TRIALS ARE RUN DIFFERENTLY THAN LABORATORY EXPERIMENTS

The risks and requirements for production tests are unlike laboratory experimentation. Process experiments are unlike routine production operations. Production tests employ a unique set of rules. The Chernobyl experiment became an international disaster because it violated the basic precepts for plant trials outlined in this chapter.

- Plant runs ascertain the process conditions and procedures needed to produce the new product on a routine basis. No production department should accept responsibility for a new product without receiving these details.
- By demonstrating a stable and sustainable production rate and yield loss factor, the development team can calculate the cost of making a product commercially.

Rarely does a single plant trial fill all the goals listed above. You should never risk running huge volumes of product until it is clear that it can be sold to someone. The experimentation required to find the optimal production conditions will preclude the stable run needed for accurate cost calculations. Normally, a development project consists of a sequence of plant trials to develop and qualify a product.

Plant trials should be approached as a very serious business. This is not a task to be delegated to an inexperienced employee. Someone who is not familiar with the plant's processes and procedures must be oriented to them. The "concept map of plant trials" at the end of the chapter shows the progressive steps of the production experiments and how they fit into the overall development program.

Trials from the Manufacturing Point of View

Hofstadter's law: It always takes longer than expected, even when you take Hofstadter's law into account.

During World War II, the Allies initiated a crash program to build more cargo ship tonnage than existed in the entire world. Shipyards were quickly constructed and they began to turn out standardized merchant ships on an assembly-line basis. The welding process replaced traditional riveting to speed up assembly and reduce the need for skilled labor. However, the integrity of the welds was dependent on the grade of steel used, the bead uniformity, aging, the ambient temperature, and their proximity to structural features that caused stress concentration. The urgent production time line did not allow these factors to be fully evaluated in the end-use environment.

Under taxing wartime conditions, the welded seams began to fail in the frigid North Atlantic and the resulting cracks often propagated for long distances. Twelve ships broke in half without warning, often while sailing fully loaded under the strain of heavy seas. This painful experience indicated that welds at stress concentration points such as the corners of square hatchways needed to be reinforced. Mastering production operations that employ new technology always takes time. Unexpected problems need to be discovered during end-use performance and the process must be reengineered to address them.

RULE #57 HASTY IMPLEMENTATION OF NEW MANUFACTURING TECHNIQUES ALWAYS LEADS TO PROBLEMS

When new technology must be implemented rapidly in a manufacturing environment, problems are inevitable. The plant management needs to direct and understand development trials to quickly obtain the information needed to run the new process.

The plant management also needs to ensure that the trials do not damage equipment, risk their personnel, or cause disruption beyond the time devoted to the experiments. They should ensure that the end result is a good fit for their production operations. The plant managers represent a vital reality check. The optimistic economic predictions of the product manager need to be tempered with the experience of running the operation.

Uncooperative production managers can dramatically increase the time to market for a new product. This will be the case when their compensation incentives focus solely on efficiency and cost reduction. Upper management needs to ensure that all relevant functional areas are working together toward the goal of launching new products.

Plant trial considerations for production managers:

- Are any trial materials dangerous in ways that are different from the incumbents? Will they interact negatively with the existing materials? Review all material safety data sheets (MSDSs) and other documentation regarding these matters. Do not consent to the trial until the concerns are addressed. This step is important and should not be left until the last minute.

- Is the test's scope within the equipment and operator capabilities? Review the proposed conditions against safety concerns, operator ability, and equipment specifications.
    - Consider putting limits on any dangerous parameters to ensure that an ample safety margin is maintained at all times.
    - These restrictions may be enforced by the plant personnel, by limits on control systems, by rupture disk selection, by reduced feed rates, or by any means available.
    - Communicate the process boundaries to both the R&D personnel and your own people.
- Is the purpose for the test fully justified? Is this the right equipment to commercialize the new product on? Just because R&D has asked for a trial, that does not mean that their judgment is the best. Make sure that this is the correct trial to meet the stated goals.
- Are there clear ground rules on what can and cannot be done? Is everyone properly versed in plant safety procedures? Are the operators clear on who has responsibility to take action?
- Is there a system to track the development costs—line time, materials used, and extra labor?
- Does the team have a procedure to clean the equipment afterwards to get it back into production? Is there time in the schedule to perform this cleanup after the trial's completion?
- Can waste be disposed of in the normal ways or must special measures be implemented? Can recycle/reflux streams go back into the process or must they be segregated?
- How will the trial material be separated from the standard product to avoid contamination and confusion? Is there any chance that the accidental opening of a valve or chute will create a crossover? Must anything be blocked or locked out to prevent such mix-ups? Example—operators may be used to sending the product into one particular hopper, instead of the temporary bins used in the trial. Toward the end of a long shift, the tired operator or helper reflexively opens the wrong valve and sends a batch of development material into a silo of standard product.
- Can any ill effects spill over from this equipment into the broader plant? Could dust in the air, liquid spray or contamination or backflushing affect other operations? Example—a central chilled water loop that becomes contaminated with a water-soluble residue from the test product. Other lines start to suffer inadequate cooling when

foam builds up in the system. The whole plant must be shut down to flush out the cooling water lines.

- Will the trial put more demand on plant utilities? Could there be excess usage of compressed air, chiller capacity, forklifts, portable storage bins, special tools, or maintenance workers? Example— excessive use of compressed air to blow out tanks causes valves elsewhere to open when the air pressure drops too low.

- How will we know that the equipment has been sufficiently flushed out or cleaned following the development run? Measurable parameters need to be established to determine when to start collecting acceptable product again. Normal guidelines might not detect the presence of contamination or unusual properties that make a product unfit for use.

- Will the run present any extra maintenance requirements beyond the normal situation?

  - Bearings that must be repacked due to contamination from abrasive powder.

  - Electric motor brushes that are damaged by excess silicon oil from the product.

  - Dust that abrades off the product, clogging the air filters.

  - Nozzles, injectors, or flow ports that are fouled with by-product residue.

  - Control systems that need to be reconfigured before and after the run.

  - Are the standard settings on tunable control systems recorded so that the equipment can be quickly returned to normal conditions?

  - Will the appropriate people be available at the run's end to inspect for these things and address them before they disrupt the routine production?

- Is there sufficient staffing to watch the trial while supporting normal operations, especially if something breaks down or needs attention? Have priorities been established in the event of a conflict? Example—the one process engineer or lead operator on a shift has been committed full time to the development. What if significant troubleshooting time is required elsewhere in the plant?

In the early hours of February 5, 1958, a U.S. Air Force B47 strategic bomber was flying a training mission over the Eastern United States. The crew was practicing for a nuclear attack on enemy territory, with the city of Reston, Virginia, as its designated target. Following the mock bombing run, the plane turned to flee back through a simulated gauntlet of defenders. In the

darkness, it collided with an F-86 fighter, damaging both planes. The fighter pilot parachuted to safety and the bomber limped to an emergency landing in Georgia.

Throughout history, realistic training for battle has involved serious risk for everyone concerned. Members of the armed forces die every year in the course of training exercises and the testing of new weapons systems. Efforts are made to keep accidents to a minimum, but they always occur.

In this case, the B47 was carrying a real nuclear warhead. It was considered important that the crew practice with the actual device on board, because the extra weight affected the performance characteristics of the aircraft. However, the bomber pilot became concerned that the 7600 lb. weapon could break loose during a crash landing and kill the crew. So, the Mk15 was jettisoned into Wassaw Sound off Tybee Island, Georgia, as a safety precaution, prior to the landing attempt.

The air force insisted that the device was in a safe "transport configuration" and could not have undergone a nuclear chain reaction. The plutonium trigger had reportedly been replaced with an inert lead component. However, the report conceded that the Mk15 (serial no. 47782) did contain 400 lb. of conventional high explosive and a classified amount of enriched uranium. This nuclear bomb remains missing to this day, probably deteriorating under 5–15 ft. of silt and 8–40 ft. of water. When armed, the thermonuclear device had a yield of 3.8 million tons of TNT.

Four months after the accident, the Atomic Energy Commission banned the use of nuclear weapons during training missions. The B47 should have been loaded with a completely inert replica of the Mk15. However, such a dummy payload was not available at the time.

This incident represents the opposite situation from the torpedo example in Chapter 2. In that incident, money and time were saved by testing with dummy warheads that significantly changed the test results. In the B47 crash, actual warheads were employed in an effort to make the trials conform closely to reality. However, this practice created a huge risk that an accident would scatter radioactive debris across a civilian landscape. In retrospect, the cost of developing and building realistic mock warheads was insignificant compared to the jeopardy involved. *Assumptions regarding the risks associated with plant trials need to be constantly questioned.*

Preparing for a Plant Trial

Before everything else, getting ready is the secret of success.

Henry Ford

Everything that can go wrong will.

Murphy's Law

Murphy was an optimist.

Unknown

In 2006, NASA's Phoenix Mars Lander arrived on the surface of the red planet. This robotic spacecraft was looking for signs of water and alien life near the ice caps. The equipment was designed to acquire martian soil with a robotic arm and drop it into small hoppers for analysis. Unexpectedly, the soil did not slide off the collection shovel and into the selected compartments. The alien dirt tended to clump and stick to surfaces, often requiring multiple attempts over a period of days to conduct each experiment. It is not known how much of the sample may have evaporated or blown away while ground controllers ran through variations in their limited bag of tricks.

NASA engineers failed to make adequate accommodations for the adhesive and cohesive properties of the soil. Murphy's law suggests that this will always be the case in situations where you have no history to base your assumptions on. Development engineers always need to account for Murphy's interference to extract the desired samples and data from plant trials.

RULE #56 GARBAGE IN = GARBAGE OUT

The development engineer is expected to invest more preparation time before a plant trial than the time spent doing the actual work. Good results only come from excellent plans and contingencies. After the trial, even more time should be devoted to analyzing the data, communicating and documenting the findings, and testing and evaluating the products that were produced.

March 15, 1938.   T. SENDZIMIR   2,110,893

PROCESS FOR COATING METALLIC OBJECTS WITH LAYERS OF OTHER METALS

Filed July 16, 1936   5 Sheets—Sheet 2

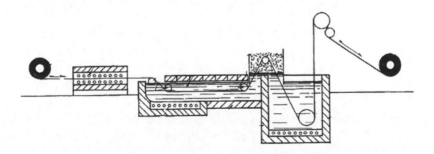

In the 1920s, a molten zinc coating process was employed to protect steel surfaces from corrosion. However, this technology was not perfect. Patches of coating often flaked off, causing rust spots. Tadeusz Sendzimir studied this problem and concluded that zinc adhesion was inhibited by an invisible layer of iron hydroxide on the surface. Sendzimir's innovation was to destroy the hydroxide contaminant with a reducing atmosphere containing hydrogen gas. This process step left a bare surface to which the zinc coating could bond effectively. With a good understanding of the mechanism that inhibited the incumbent process, Sendzimir was well equipped to engineer a better galvanizing process.

The protective atmosphere was not suitable for the traditional hand dipping of individual pieces of steel. It was perfect for an in-line process that was more efficient and consistent. This enabled the production of long coils of galvanized sheet for use in construction. Additional improvements followed quickly. The minor components of the zinc blend were adjusted to improve its processability and the finished-part properties.

Checklist for trial groundwork:

- Define and list the trial goals in order of priority. Everyone involved should be aware of the most important tasks, especially if problems scramble the work plan.

- Assign the appropriate people. Avoid pairing an inexperienced engineer with technicians/operators who are also new to the equipment.

- Plan, plan, plan. Make sure that extra materials, parts, and tools are available as contingencies.

- Put everything in writing, so that there are no misunderstandings. Have someone review the written plan for ambiguity.

- Double check. Come in early and ensure that all materials are staged properly for the run. Verify that the raw material containers really are correctly marked and staged.

  - Brief the participants beforehand and confirm their understanding.

  - Ask people to repeat significant details back to you in their own words. The terminology of the shop floor differs from the laboratory vocabulary.

- Talk to the workers ahead of time and make sure that the plan is clear to them and that it makes sense. No one should discover in the middle of a run that the equipment does not have the required capabilities. This is important for hidden components buried inside a system.

- Create a plan to deal with the health and safety issues. The Chernobyl experiment went wrong when the lead engineer opted not to follow the approved plan or safety regulations.

- Beware of recycle streams. They can contaminate development material with production scrap. The same can be said for all other hang-up spots in a process.

- Research the piping/ducts/material flow paths in complex systems. Diagram the ones that you plan to use, so that they can be visualized, especially if you need to deviate from the original plan. Know which zones are controlled by each control node. Apply temporary

adhesive labels on control points when operating in a different configuration than the staff is accustomed to.

- Allow enough time for the process to settle down after introducing
new materials or changing process conditions. The operators may
have a feel for this, based on production changeovers. However,
their experience may not be valid because the trial is different from
normal production.

- Clarify who is authorized to perform each task. Is the engineer
allowed to operate the controls, or is this solely the responsibility of
a production operator?

- Understand what risks the manufacturing organization is willing to
allow on its equipment. Debate the matter ahead of time and respect
the agreement.

- Make sure that someone is always watching the important things.
People tend to congregate around problem points during stressful
situations and take their eyes off the controls.

- Have a plan for what to do in the event of a process upset. The
plant personnel should be informed if this situation is different
from what they normally deal with. Decide in advance whether the
team will recover the process or shut it down. Agree on roles for
that situation—is the lead operator or the engineer responsible for
making the snap decisions?

- Keep the needs of the workforce in mind. They must still take
breaks and meals. They appreciate catered trials. The schedule
may require them to work harder than normal. Extra effort can
sometimes be purchased with pizza or doughnuts. Never skimp
on paying the plant to assign more operators to accommodate
the extra workload. Avoid situations where a trial is staffed
with unqualified operators or those who are tired from working
overtime.

- Know the materials that are running and the conditions that they
can withstand. Take special care during transitions.

- Be aware of the pressure, rate, and temperature limits of every component in the line. Make sure of the capacity of all subsystems, such
as the power supply, the cooling capacity, and the output capacity of
material feed systems and product haul-off equipment.

- Consider doing a pretrial if there are any major questions about the
line's capacity to run what is needed. Have all the required information to run the trial to the limits, rather than taking half measures
out of caution.

- Use a strategy for finding optimal conditions. Consider the high
and the low values of each parameter that can be safely run. Plot a

matrix of conditions around the aim point to search for an optimum. Research the approaches that have worked in the past.

- Can the control system handle the planned conditions?
  - Will alarms be going off when the process strays outside the normal production limits?
  - Will the control system trigger an unneeded shutdown if it senses unusual conditions?
  - Make sure that all the information needed to program the control system is available. This may include manuals, keys, and codes.

Critical Questions: What are the most important pieces of information and samples that we need to extract from this plant trial? What are the lesser priorities? What are the unintended consequences that we want to avoid most?

Running a Plant Trial

Without the element of uncertainty, the bringing off of even, the greatest business triumph would be dull, routine, and eminently unsatisfying.

J. Paul Getty

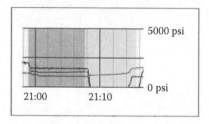

On the morning of April 20, 2010, the crew of the Deepwater Horizons drilling rig was shutting down operations at the Macondo site in the Gulf of Mexico. Engineers had just certified that the well was safely sealed. Contractors started to withdraw drilling mud from the pipe string connecting the platform to the wellhead. While doing so, it was critical for a four-member team to monitor conditions to guard against violent "kicks" of flammable gas coming up the pipe.

In the course of about 100 minutes, the drilling mud was purged from the well by forcing seawater in its place. During that procedure, several process variables—mud flow and pressure—could indicate if a dangerous bubble of gas was rising toward the surface. In that interval, a number of suspicious instrument readings were ignored, while the team dealt with a high volume of returning mud and an unrelated equipment breakdown. No one suspected that a dangerous situation was developing until they observed "a jet engine's worth of gas coming out of the rotary," according to survivor testimony. A massive explosion and fire soon followed.

Had they reacted promptly to the warning signs, the drill team could have averted the disaster. Because of operational distractions and complacency, they were not watching closely and analyzing what was going on down in the hole. In a hectic production environment, such inattention to detail is common.

When running a trial, it is not a question of whether problems, complications, or deviations will occur. An engineer must go into the trial with the mindset that things never work according to theory or that the exact results that were anticipated are never achieved. Components will break, clog, and jam. Instrument readings will be flawed. Errors will be made. The desired process settings may not result in stable running conditions. Be ready to make decisions and compromises to maximize the return on the investment in this experiment.

It is rarely possible to put the process on "pause" when problems occur on large-scale equipment. You will not have time to seek advice, sleep on it, or try again tomorrow. Be prepared to change the run plan in real time and implement alternative ideas on the fly.

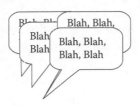

There is much concern with information overload in the form of text messages and e-mails sent to our mobile devices. The problem is severe in occupations such as the military, stock traders, news reporting, and even sports coaching. Information from a wide range of sources is constantly pouring onto the decision makers, without filtering to prioritize what is really important. The human brain becomes fatigued by this.

People who fail to make the best decisions under a deluge of information generally look like fools, because the critical clues were available in the midst of the trivial details. The same situation can occur during a plant trial. Choices must be made while looking at control screens full of constantly fluctuating process conditions. Laboratory test results arrive with a snapshot of properties made minutes or hours before. Operators request instructions and report problems, often yelling over the process noise. Information, questions, and advice arrive electronically from a management group that is anxious to optimize its investment in your trial. Development engineers must constantly filter information as they drive the process toward the desired goal. While doing so, they must be ready to adjust their plan.

Dealing with the unexpected during plant trials:

- Be flexible and opportunistic. Problems and transients represent opportunities to capture samples of "off-spec" materials for later testing. Establish which situations can be sold as "wide spec" and which cannot. Defective samples may make excellent visual examples of what not to do. They might later comprise a worst-to-best photomontage for grading different levels of quality.

- Do not panic and make a snap decision when something goes wrong. Carefully evaluate the situation and the best way to get back on plan. Evaluate the data and consider whether any of the instruments are supplying false or contradictory readings. Make sure that you understand the ramifications of a course of action before implementing it. This may go against the plant culture of jumping into action to fix problems.

- Some trial plans call for the production of different products or one product at a variety of different conditions. Do not waste much time trying to get any one of these individual objectives completed, if it will jeopardize the overall trial. Consider coming back to the problem situation(s) after finishing the remainder of the run plan. Use your priority list to determine which samples or data are most important to your needs.

- When troubleshooting, determine whether the root cause was the experimental materials and conditions or some problem with the equipment that may be completely unrelated to the trial. Consider calling in maintenance craftsmen to ensure that the equipment is working properly. Be prepared to run a known material as a control to determine if your experiment or broken equipment is at fault.

- When experimenting to find the optimal running conditions, follow a plan. Let the operators adjust the equipment in ways that they normally use to stabilize the process or fine-tune it. These adjustments are a routine practice of running the process. Only experience will tell you when to trust the operators' instincts and when to overrule them.

- If the operator fails to find the "sweet spot," systematically adjust the process conditions based on past experience. Change only one parameter at a time. Make a sufficiently bold change to ensure a significant response, but without going so far as to crash the process. Evaluate whether the change was for the better or for the worse. Let the system run long enough to reach a new equilibrium, rather than jerking it back in the opposite direction at the first indication. Learn to recognize random noise and a true process change on this process.

- When searching for conditions, be patient and do not make quick judgments. Watch long enough to get a statistically significant feel for the process response to the changes.

- Some control systems will provide real-time tracking of specific parameters. These can be charted on a screen, documenting the process shift from one condition to another. Such features can be valuable in a trial situation, since they plot the changes in transient conditions. Record the exact times when certain actions were taken so as to understand the results of process changes. This tool may need to be set up in advance by selecting the parameters to be monitored and the expected data range.

- If the line cannot plot out process parameters, consider monitoring them with a portable data-logging device. It is vital to note the times when changes are executed. Such displays can show when the process has settled into a new equilibrium after a change.

- Label each sample carefully and record the conditions that created it. Every sample must be correlated back to the running conditions and the raw materials. Never assume that anyone can remember all the information and write it down later.

- In the absence of automated features, consider having a human data recorder do all of this work. This could be an intern with a clipboard and a stopwatch. This individual would periodically record the dataset that has been designated in advance. Avoid asking for too much information. Have a backup person trained to step in when needed.

- Keep the area clean, even if extra resources are required or it takes more time than normal. Scrap may accumulate much faster during trials than in normal production conditions. Stage an appropriate storage capacity of empty bins, tanks, or totes to handle the waste.

- Maintain a separate area or containers to store any materials that are awaiting a decision on their status. Retain the process information and any test results on each container. Development trials always have different storage needs from production.

The natural response to a stressful plant trial is to unwind and move on to other pressing responsibilities. Before doing so, optimize your investment of time and money by completing the job.

Follow-Through

> The most exciting phrase to hear in science, the one that heralds new discoveries, is not 'Eureka!' (I found it!) but 'That's funny'...
>
> **Isaac Asimov**

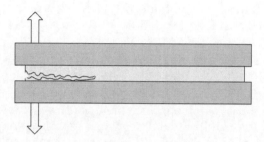

In the 1980s, a market-focused company was designing a new product to address a customer need. The design required the company to adhere two materials together with a water-based adhesive on an automated production line. All the engineering resources were focused on optimizing the properties of the two base materials, the ultimate bond strength, and the fitness-for-use (FFU) aspects of the finished product.

In the course of development, no attention was paid to the cleanup procedures on the pilot plant equipment, because it was not a squeaking wheel. The operators found that the glue application system could be cleaned adequately with soap and water. Although the glue was very fluid as it swirled down the drain, it tended to coagulate quickly. After a few months, the drain lines began to clog and the operators had to clean the components at a different sink. By the end of the project, the concrete floors around the pilot plant equipment had to be excavated to replace the nonfunctioning drain lines.

At the same time, the commercial process was started up and debugged in a distant production facility. Once again, a new set of operators created their own cleanup procedures. The production plant's wastewater volume was much greater than the pilot plant's wastewater volume. Eventually, the company had to pay to have the surrounding streets excavated to unclog the city's sewer lines. In retrospect, no detail is too trivial to investigate and document.

Actions to take following a plant trial:

- Make sure that the finished products and samples are labeled, packaged, and stored appropriately. Get everything out of the production area and secure it. The exceptions will be products that need to be stored under special containment conditions or in an inventory warehouse.

- Document and verify all data as soon as possible. Transcribe handwritten notes and annotate them appropriately, so that information will be in a context that can be understood and interpreted by others. This author finds nothing more difficult than going through old notebooks and trying to read the handwritten scrawl of someone who is working under pressure in a plant test. This includes entries made by the author himself.
- Submit selected samples to the testing laboratories. Keep retained samples, in case the laboratory needs more samples or new questions arise.
- Report the results back to everyone involved, especially the manufacturing people.
- Check with the plant for feedback on any effects that the trial had on its process. Give the plant any comments or suggestions for improvements.
- Be sure to thank everyone who participated. Make a note if there are individuals who should or should not be included on future trials.
- Review the costs that have been charged to your budget versus the plan or assumptions.
- Reflect on what happened in the trial and what should be done differently in the next effort. Were there unexpected results or discrepancies? Reevaluate your model of the system.
- Issue a comprehensive report, complete with conclusions and recommendations. Plan for internal tests and customer trials.

Types of Plant Trials (From the Plant Perspective)

The status quo is the only solution that cannot be vetoed.

Clark Kerr

A man rarely sees more than he is looking for.

Old saying

Good judgment comes from experience. Experience comes from bad judgment. Make your mistakes in test tubes and make your profit in vats.

A. C. Gilbert

The Airbus A320 airliner was introduced to commercial service in the mid-1980s. The automated fly-by-wire control systems promised to be much easier to fly than the earlier generation of large jets. This was a key selling point

when marketing the planes globally. Many of its potential customers lacked the pool of skilled ex-military pilots that were available in the United States and Europe. There was keen interest in automating pilot decisions.

Aircraft enthusiasts crowded the Mulhouse-Habsheim airport in Eastern France on a pleasant summer day in 1986. The highlight of the air show was an A320 overflight. This particular aircraft had been delivered to Air France only days earlier. The plane was to make two low passes over the field to permit the public a good view of the latest technology.

The demonstration started well, until the flight crew tried to climb back up from its first low pass over the field. The plane hit a grove of trees beyond the field and crashed, killing three people. The control system was later blamed for being too slow in responding to the pilot's frantic attempts to gain altitude.

In this unusual maneuver, the control system may have considered itself to be in a normal flight configuration rather than in "takeoff" mode. It did not engage full engine power because this seemed to conflict with its directive to conserve fuel. The situation may have also been influenced by a faulty altitude setting that had been entered into the computer. Unprepared engineers can also find themselves in conflict with control systems when conducting plant trials.

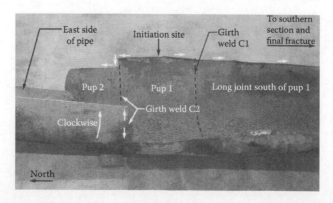

On September 9, 2010, the Pacific Gas & Electric Company (PG&E) was performing maintenance on an uninterruptable power supply (UPS) at one of its natural gas terminals in California. The UPS failed and power was lost to critical control systems that governed gas distribution within the network. This caused a regulator valve to open, sending a pressure spike down a 30 in. diameter pipeline. Immediately afterwards, the line burst in a residential section of San Bruno, killing 8 people and destroying 40 homes.

A subsequent National Transportation Safety Board (NTSB) investigation indicated that the 1956 vintage pipe failed at a junction point where numerous short sections were welded together on-site to dip under a roadway. PG&E records indicated that this leg of its distribution system was composed of seamless tubing. In fact, the failed section had been formed from flat stock, which was bent into a round cross section and welded together. The dynamic pressure spike caused by the control system failure would put additional stress on the line compared to a static loading situation.

Similar conditions can occur when development trials are conducted on production equipment. Control systems are designed with a margin of safety appropriate for the state of the equipment at the time of installation and the process conditions that were originally expected. When significantly

RULE #12 CONTROL SYSTEMS HIDE BOOBY TRAPS THAT ARE REVEALED DURING PLANT TRIALS

Production equipment is often fitted with alarms, controller limits, and physical restrictions to prevent them from being driven into potentially dangerous parts of the process window. Operators may not even be acquainted with these limitations, if they never encounter them during normal operation. When running a plant trial, everyone must appreciate these restrictions and understand which ones can be safely overridden and which ones must stay in place.

different materials are run at new operating conditions, the equipment may fail in ways that the designers never expected.

These decisions will differ from one situation to another, depending on the type of trial being conducted and the objectives. From the production department's point of view, there are different types of plant trials. Each type should be approached differently.

The Break-in: This type of trial is done in the middle of a long production campaign of a standard product. It employs the same equipment configurations as the conventional process, but may have different raw materials, running conditions, size, color, and the like. Minimal changes are needed to transition into and out of these trials.

Select a time when a similar product is being produced. When the commercial run is complete, then the process is transitioned into the experimental grade. By doing so, the new item is run the way that it will be produced as a standard product.

Special considerations for break-in trials:

- Make a decision on who should staff the run. Is it more appropriate to have the most experienced personnel or the least? The most skilled operators will provide the best results. However, their ability to overcome problems may conceal the difficulty that their less experienced coworkers will have.

- Be wary of accepting a process that is running poorly before the start of the trial. If there are problems during the test, then manufacturing will be less than anxious to take on the new product. You may spend a lot of money on a trial that does not give an acceptable product.

- Plan the transition and the return to standard conditions. Time is critical in break-ins.

- Record the process conditions on the production run just before the changeover. Collect samples to act as a control for the trial.

- Transition from the standard conditions and materials to the test product. Record times and the sequence of the steps that are performed. This is helpful in troubleshooting, in writing procedures, and in estimating costs.

- Segregate the first batch or the first few "good" parts after the transition. Compare its properties and running conditions with the following batches to identify any contamination or carryover effects from the previous products.

- At the trial's end, repeat the process of recording times and sequences of actions. Again, take transition samples and verify the time needed to change over to the next item on the schedule. Save samples of that next product to determine any effects of contamination from the experiment. Consider quarantining the first batch to ensure that it works properly.

Line Start-up Run: This is always the most challenging type of trial. It will be remembered with either fond nostalgia or great regret. It requires the team to start up a new piece of equipment and run a new product. The task calls for a methodical approach. Many questions need to be answered and numerous running conditions need to be evaluated. Hard work and long hours are required.

Before the process start-up begins, the line must be fully checked out by the installation crew. These are capital engineers that work for the equipment vendor, a consulting firm, or the company's own engineering staff. They certify that the equipment has been built to the prints and that the controls function appropriately; the motors have to turn in the correct direction, the gearboxes must hold their lubricant, the pressure vessels are certified, and the like.

The first priority is to get the line running with a standard product or a dummy material. The kinks need to be worked out of new equipment before anything challenging is attempted. Experimentation can only begin once the system has met predetermined performance goals. Days may be spent flushing the iron filings, lubricants, and cigarette butts out of the equipment's nooks and crannies. Installation errors must be identified and corrected before you can hope to run developmental trials.

Leaks will develop, filters will clog, breakers will trip, joints will vibrate loose, and a hundred other problems will be discovered. The idea is to shake out all the problems and document them on a punch list for corrective action. A lot of scrap will be generated in finding the problems. It is much cheaper to identify weaknesses at this stage, rather than later when customers start to complain.

The next step is to make one of the easier standard products or sizes. Experiment with the key variables to seek stable running conditions that meet quality targets. Samples from these conditions should be quickly evaluated in FFU tests. It is important to know if the process output is close to the desired target or not.

Gradually expand operations to additional products in some logical order. Test the size of the process envelope to make sure that stable conditions exist for production. Make sure that transitions from one product to another are problem free.

Clean Start Run: This is a plant trial beginning with a cold start on a process train, which has been out of service for overhaul or lack of orders. This situation is favorable when the new materials are not compatible with the standard product or because a significant equipment reconfiguration is required. This could be the checkout of a line for sale or rent.

Special considerations for clean start runs:

- A clean start situation usually requires staffing with experienced technicians and senior operators. The skills to properly bring up a line from a cold start may not be routinely used by development personnel. But, the normal production procedures need to be adjusted for your trial. Question the procedures ahead of time to make sure that they are appropriate.

- Ensure that clean really means clean. A production operator may have one standard for periodically cleaning out the production equipment. However, the new raw materials or end product may be incompatible with the residue from standard production. Flush out any raw material inlet streams, drain lines, air intakes, vents, sight glasses, transducer taps, and any other point where traces can be retained. Completely change out any components that are notorious as hangout points, which cannot be manually purged.

- Consider replacing any components that are questionable. Look for corrosion or pitting on thermocouples and transducers.

- Conduct a complete safety review. Procedures that were appropriate for the standard raw materials or process configuration might not be safe for your trial situation.

- Survey the equipment to make sure that everything is configured in accordance with the plan. This can include inspecting the important dimensions on components that influence the process. For instance, the plant may have replaced the standard one-half inch valve with a three-quarter inch valve to increase the production rate. If the team is depending on the published pressure drop across the original equipment, then it will be disappointed with the lower pressure drop from the larger valve.

- As you discuss the start-up steps, stress the differences from standard procedures.

- Leave specific instructions for any special warm-up, gas flushing, sterilization, or conditioning that needs to be done in the hours prior to the start-up.

- Following the run, discuss the cleaning and maintenance needs to return the equipment to production use. Highlight all of your configuration changes that need to be undone.

- A team member should inspect the removed equipment afterwards to ensure that special parts and leftover materials have been correctly removed, cleaned, labeled, packaged, and stored.

Absent Trial: This is a straightforward test that does not require the physical presence of the project team. This might involve the production of a new product, which has been run multiple times in the past, but which has not yet been accepted as a standard product. It could also entail the production of a sample with a slight variation over a standard product. It is more common in plants that the development team cannot easily travel to.

Special requirements for absent trials:

- Every step must be documented in writing, highlighting any deviations from normal procedures.
- Complete standard operating procedures (SOPs) and standard operating conditions (SOCs) must be provided.
- Product labeling, packaging, and disposition should be fully specified in advance.
- Special forms may be provided to record the conditions, quality control (QC) tests, and observations, if standard production reports are not adequate.
- Detailed instructions should be written for dealing with problems.
- Team members should be on call in case of questions—provide contact information.

Tail-End Test: This is a case where manufacturing allows experimentation on a continuously running process at the end of a production cycle or wheel. Following this trial, the equipment will be shut down for routine cleaning and maintenance. Scheduling a trial last assures the plant that a test will not jeopardize production commitments. The plant assumes that any harmful residue from the trial can be scrubbed clean during the scheduled overhaul. This is often an acid test. If a product can run at the tail end of the production wheel, then it can probably run at any time.

Special considerations for tail-end trials:

- Retain the option of canceling the run if the equipment is not up to the task.
  - Review the production records to ascertain what has been running on the line.
  - Look out for any upsets, unscheduled maintenance, or mystery events.
  - Quiz the operators to fully understand what the equipment has been through.
- Check the personnel schedule to see who will be running the equipment. This is a must if the run is to be made just before a holiday shutdown. Vacation commitments may divert the appropriate talent.
- Study the process data and the current instrument settings and readings. Make assumptions about which readouts may be out of calibration or problematic. Have them replaced if possible. Consider bringing in portable or temporary instrumentation if some vital piece of information will not be measured. Perhaps an armored data

sensor can be run through the process to capture information that cannot be read from the outside.

- Run an extra-long purging/transition cycle to flush out as much contamination as possible. Antioxidants, lubricants, surfactants, or scavengers should be on hand in case they are needed.

- Run a control especially if the instrumentation is questionable.

- Make sure that the samples and products are labeled, packaged, and safely stored away. Ensure that all data are recorded and archived safely. Do not assume that anything will survive the cleaning and overhaul processes.

- A team representative should be present to inspect the equipment during the overhaul. Look for anything that might have compromised the run and any damage that might have resulted from it. Take the opportunity to measure and record any dimensions that will affect the process performance.

Idle Equipment Trial: This is a trial done on unused process equipment that will be idle before and after the trial. On the upside, this provides the flexibility to schedule and run the trial when needed. The team can modify the process to suit its needs and presumably run as long as needed.

Special considerations for idle equipment trials:

- Idle equipment is frequently robbed of parts and is generally dirty, unlubricated, and unloved. The experimenters are responsible for getting it running correctly.

- The operators may not be up to date on all the procedures for running this line. It may not have the latest control system that they are accustomed to. They may take more time to do normal tasks than is the case on production equipment that runs every day.

- Ask for operators and technicians with the longest seniority. They may have a better feel for running the less automated equipment than their younger coworkers who were only trained on the latest control systems. Do not confuse the computer phobia of the older workers with lack of process knowledge.

- Ensure that all the necessary supplies and tools are available. If the equipment is out of date compared to the standard production units, then it may need different consumable parts, which could be out of stock in maintenance stores.

- Perform tests of all components and utilities prior to starting any trial work. Just because there are cooling water pipes running to the equipment, that does not mean that these lines are not clogged with rust. Flush everything and evaluate to ensure that utilities deliver the proper pressure and flow rate. Do not rely on the assurances of people who have not been using the equipment.

- Ensure that ducts, pipes, tanks, bins, chutes, and conveyor belts that will be in contact with your product are clean and serviceable. If not, your samples will be contaminated with residue or scale.

- Running a control is an absolute requirement, rather than a luxury. Without the verification that the known material can be run successfully, there is no basis to make judgments on the results obtained in running a new material.

- If you are conducting a series of discrete trials over some period of weeks or months, then there will be stretches of idleness when the equipment is not running. Take measures to ensure that the equipment does not deteriorate in the meantime. These might include filling the interior with inert gas, blocking off pipes, or wrapping sensitive components with a protective covering. Put locks or seals on control panels or other access points to ensure that important parts are not scavenged for adjacent lines. Some plants erect chain-link fences around idle equipment.

Types of Plant Trials (From the R&D Perspective)

What the people believe is true.

Native American Proverb

There are not enough Indians in the world to defeat the Seventh Cavalry.

General George Custer

Like an experimental product being run for the first time in a plant trial, every new aircraft design must undergo a first test flight. In the early days of aviation, these experimental missions were often fatal. Test pilots were not ideal customers for life insurance companies. It is traditional for the designers to place their car keys and wallets onboard the test models. This is a tangible demonstration of their confidence, as well as a personal investment in the success of the first flight.

Development engineers must have the same degree of assurance and preparation before their experimental products go to the plant for their test flights. Understanding and preparing is an important requirement. There are many types of production trials, depending on what stage the project is at:

The Exploratory Trials: This is the first attempt to run a new product on large-scale equipment. The team will be feeling its way to establish stable conditions by extrapolating from small-scale trials. It will be experimenting to find the impact of process conditions on the product's end-use attributes. Problems and opportunities come from process paths that have not been tried previously.

Do not expect to sell the material made in an exploratory run. It will be evaluated for FFU. Usually, the need for experimentation means that significant quantities of any one variation will not be produced.

The cost of this trial will largely be written off as an R&D expense, rather than being funded by the sale of the product. This trial might be the biggest single unrecoverable expense in the entire development project. It should be meticulously planned and executed to get the most data.

Maximizing the results of exploratory trials:

- Avoid focusing too narrowly on the targets that were identified in the small-scale studies. Seek to find the limits of processability. Evaluate a range of process conditions to bracket the optimal end product. It is always better to interpolate, rather than to extrapolate.

- Given the first trial's importance, do not skimp on the manning of the equipment. Allocate sufficient manpower to ensure success or to learn from failure.

- This trial is ideally run on idle equipment or in a break-in situation.

- An exploratory trial puts a premium on quick decision making to exploit opportunities. Material property data may be needed to quickly adjust the process conditions when searching for the optimal conditions. Dedicate test equipment to this trial for quick feedback.

- At this stage, only rough estimates of the production costs are feasible. Look for factors that could push costs higher than what your model predicts—low rates, high scrap levels, or instability.

- Do not expect this first run to produce the final product. Large numbers of samples will be collected. Testing will identify which ones to focus on in the next run. The exploratory trial usually generates a punch list of issues to work through before good samples can be made.

The Field-Test Trials: This step follows the experimental run(s) and assumes that some ground rules have been established for satisfying end-user needs. This run will yield product to support significant tests at customers' facilities. The output need not be refined in every way. There will be some rough edges and unanswered questions about the final design, packaging, look, and feel. Ensure that the customers and the salespeople understand that this is a test. The product evolves based on plant trials and field-testing.

This trial may require making the many different permutations that individual customers require. This is a good opportunity to demonstrate that the operators can control the process to make all the necessary variations. It allows a determination of the time and the yield loss involved in product changes.

Guidelines for plant trials to make field-test product:

- Choose process conditions and procedures based on the best samples from the exploratory trials. Use the less successful samples as a guide for where not to run.
- Freeze the process conditions for the duration of each batch. Isolate samples made during process excursions. If you must experiment with conditions, make sure that those samples are uniquely labeled and tested.
- Calculate and record the production rates and yield losses for the periods of stable running of each product. This will refine the preliminary cost calculations. At this stage, you must pay close attention to the production costs in order to predict profitability during mature production.
- Make process capability runs during these trials. Hold all conditions constant and measure the variation in the important attributes. Calculate the manufacturing specification ranges for these parameters. Capture the high and low samples and subject them to controlled FFU tests. It is vital to ensure that the process's day-to-day variation is narrower than the end user's FFU requirements.
- Make small experimental batches during the trial to evaluate the effect of minor parameters on the important product properties. If the results are good, then the samples can be shipped along with the field-trial materials. If the results are bad, then an important piece of information can be added to the punch list.

The Troubleshooting Trial: This type of trial resolves serious problems uncovered during other types of plant trials. Perhaps too much product was downgraded from the field-test trial. Perhaps the product is functional, but it has defects or excessive variability. Maybe the rates and yields must be increased to hit the cost targets. Sometimes, you just need to generate data to support the process equipment upgrades and debottlenecking or the design of a new line. This run will primarily generate data and samples. Some salable product may be produced, but troubleshooting is the primary goal.

Troubleshooting guidelines:

- Idle equipment is the ideal situation when many different equipment configurations need to be evaluated. This allows time to repetitively set up the equipment, run the process, shut down, and install the next variation.
- Break-in runs are better for modifying blends, formulas, or alloys on existing products. You transition the process from some stable production situation to the new set of raw materials. You run your product(s) and then transition back to the next product in the production wheel. The next week or month, you come back for another try, based on your findings from the previous trials.
- Arrange to have immediate test feedback when evaluating numerous options in a limited time.
- Prior to the trial, create a list of potential solutions and evaluate the best options first.
- This type of plant trial is a pure experiment. Treat it as one. Run a control. Hold nonexperimental conditions as constant as possible.

The Transition Trial: This trial demonstrates the commercial rates and the yields that can be expected for normal operations. The product must be finalized and have the look and feel of a commercial offering. All of this material will be sold to customers without shortcomings or apologies.

Prior to this trial, a complete package of SOPs, SOCs, formulas, specifications, and quality tests should be submitted to the manufacturing group. The performance of field-test products at customers' facilities must have been satisfactory. There must be no mystery factors that cause product performance to be suboptimal. In short, all aspects of production must be understood and documented, at least for the products that you will release to manufacturing. Issues with other grades or applications should not cloud this situation.

Transition trial considerations:

- The equipment should be configured in accordance with the written recommendations in your documentation package. Hesitate to accept compromises.
- Run with proven SOPs and SOCs determined from your previous work. Do not try to push the envelope to increase the production rate

or improve the properties. These actions put the demonstration at risk of failure. If the manufacturing manager insists that you test the higher rates, do so at the end. Separate that portion of the run from the rate and yield data collection.

- This run should be self-financing with the sale of product to customers or commercial inventory.

- Transition runs will highlight any organizational preparedness issues for the product launch. The management should monitor these trials as closely as the process team. They ought to be looking for breakdowns in the handoffs from one functional area to another. This is the proof run for financial, logistical, information, and management systems.

- The manufacturing manager might prefer a clean start or a tail-end run. This demonstrates that the product runs well in the worst conditions.

- The development team would like this trial to be a break in, following a smooth-running compatible commercial product. This will generally give a smooth transition with fewer start-up headaches.

- Involve the production management as much as possible. Communicate your wants and needs, but then let them do their jobs based on your technical guidelines.

- Do not even consider doing anything experimental during this trial. A stable production run with no problems is the only objective. Do not even talk about any experimental things. It will only confuse the issue with the plant management and the operators.

See the flowchart at the end of the chapter that describes the relationship between these trial types.

Equipment Considerations for Plant Trials

Well, that's what it was bloody well designed to do, wasn't it?

Frank Whittle

Well, it was kind of an accident, because plastic is not what I meant to invent.

Leo Baekeland

Chance favors the prepared mind.

Louis Pasteur

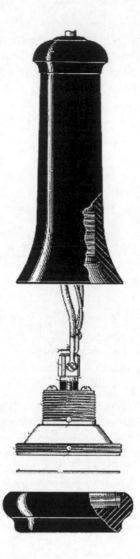

Early in the twentieth century, the chemist Leo Hendrik Baekeland was researching synthetic alternatives to shellac, a natural resin derived from Asian beetles. This material was widely used in electrical insulation applications. However, some 15,000 beetles gave their lives for one pound of resin, so the supply was limited and the cost was high. This was a big problem for the electrical industry.

During the course of his experiments into synthetic shellac, Baekeland discovered phenolic plastics by reacting phenol and formaldehyde together. When the reaction was carried out at the proper temperatures and pressures, the mixture hardened into a rigid thermoset polymer. The process could mold identical, mass-produced product shapes, which was important for commercial applications. When combined with the appropriate fillers, the resulting products were ideal insulators with cost and performance advantages in numerous applications.

Baekeland patented his Bakelite technology and hoped to collect a healthy revenue stream by licensing third parties to employ his invention. However, most companies lacked the process knowledge needed to control the FFU attributes with this new technology. So, in 1910, he formed the General Bakelite Company to commercialize the required processing techniques. The company created specialized plastic molding equipment and fine-tuned each application. Under these conditions, Bakelite was able to achieve widespread success in a range of products.

The successful commercial production of engineered materials requires more than an understanding of the basic chemistry. The combinations of raw materials, equipment, and process conditions must be evaluated for their suitability in each end-user application. Flexibility is the key to solving these diverse problems.

RULE #40 HIGHLY EFFICIENT PROCESS EQUIPMENT IS OFTEN NOT FLEXIBLE FOR RUNNING NEW PRODUCTS

The flexibility of your process equipment and the breadth of your organization's process knowledge in running or modifying that equipment will dictate how plant trials are run. The more a production line has been specialized and optimized to produce a particular product, the harder it will be to satisfy new applications.

Plant trial strategies for various equipment types:

- Large output equipment is slow to equilibrate to new conditions.
  - Obviously, you need an outlet for the changeover scrap. Reconsider the project viability if the transition material cannot be sold, recycled, or otherwise disposed of at a viable cost.

- Carefully consider how to staff the trial. Does someone need to watch the process during the changeover? Is the time required for equilibration predictable? This may be a good opportunity to spend time observing the process and learning how it works.
- It may be necessary to collect many discrete samples as the conditions slowly approach your desired target. Set standards for when the product is "close enough" to the target to create salable "in-spec" product.
- Determine how to "tune" the process to reach the target quickly without overshooting.
- Management must understand that trials on this equipment will be costly.
- New "one-of-a-kind" line that has been custom built.
  - First, determine if this is the right line for the product or if design changes are needed.
  - You need to write the book on how to run the equipment. There will not be suitable SOCs and SOPs at the start of the project.
- Process with high variability from one run to the next.
  - Extensive data from long runs are needed to establish process capability and FFU.
  - Test customers must see and evaluate the whole range of product variability. If not, they could experience mysterious quality problems at the extreme ends of the process window.
  - Intensive data collection may be required.
  - Determine if products that are run earlier in the schedule have an impact on subsequent runs.
  - Identify if other products that are run after this one will be affected by your test material.
- Lines that feed semifinished product to a second step that is located in a different plant.
  - You will need to retain data on each discrete unit of production (roll, box, or drum of product) to correlate with the process results on the second step and the FFU at the customer's facility.
  - Monitor all steps as intently as the first operation in the sequence, even if those later stages are considered to be routine. Watch the performance and quality to better understand the effect of the process knobs in the original step.
- Equipment that is run by "feel," skill, or intuition.
  - Try to staff the trials with multiple operators or technicians who have experience with the operational techniques. Cross-train them and try to document the procedures, judgments, and rational for their control abilities.

- Record all available data on the operation to correlate them with the operator's actions.
- Develop and record a vocabulary of perceptions that the operators employ to monitor the process and what actions they take in each situation.
- Try to understand if some of these control knobs are better than others and rate them in terms of how they manipulate the process.
- Never commercialize a process that can only be run by one or two skilled operators.
- Many different lines, all a little different in design.
  - Try to make the first few trials on only one line—establish viability and capability.
  - Alternately, try different lines in succession and identify which one best meets the product needs.
  - Identify the significant equipment features and attributes that affect the new product for better or for worse. Rank the lines for suitability to run the new product.
  - Understand if it is okay to only run this product on the best line or if it must run on every line. Identify what must be standardized in order to make it run everywhere.
- Different geographic regions, each with its own market needs and equipment configurations.
  - Start the development with the one region that has the greatest urgency for the new product. Spend more time in this plant trying to understand how to control the process and meet the FFU requirements for this material.
  - Understand which requirements are process needs and which are local regulatory or legal requirements. Do not blindly assume that one region's "must have" attribute is common to all.
  - Treat each successive region as a new project that builds on what was learned before. Retain experienced members from the original team for subsequent trials. Do not assume that the needs are common or that the process works exactly the same way everywhere.
  - Carefully consider whether each region is allowed to do development independently.
  - Start work in each new region by exporting product from a plant location that you already understand well—even if this is not commercially viable. Use the well-known process to test the FFU requirements for the new regions, before you undertake to experiment with the local process.

Summary

- The manufacturing group must ensure that development trials will not damage process equipment, injure employees, or disrupt the production that follows the trial.

- Running development trials on production equipment requires far different procedures compared to running laboratory-scale or pilot-scale experiments. The risks, rewards, and costs are all greater for large-scale development.

- Product development trials should be run in close consultation with the production management to ensure that the process will be commercially viable. They must ensure that their organization is prepared for it (QC, operator training, production control).

- Plant trials require detailed preparation by the development engineers to get the most out of the expensive line time. When running the trials, be opportunistic and calm in the face of the inevitable surprises.

- Much work is required after the trial to ensure data retention and the proper labeling and retention of experimental materials.

- Plant trials are conducted in many different production situations. Each requires the development team to employ a different strategy.

- Each stage of a development project will require different types of production trials. The first production attempt to make a promising prototype will be very different from the final run prior to turning responsibility over to the manufacturing group.

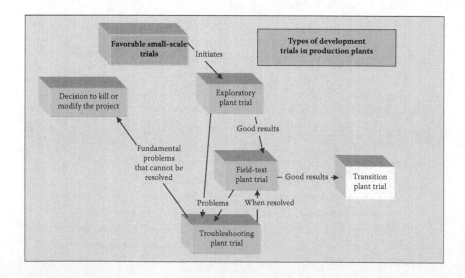

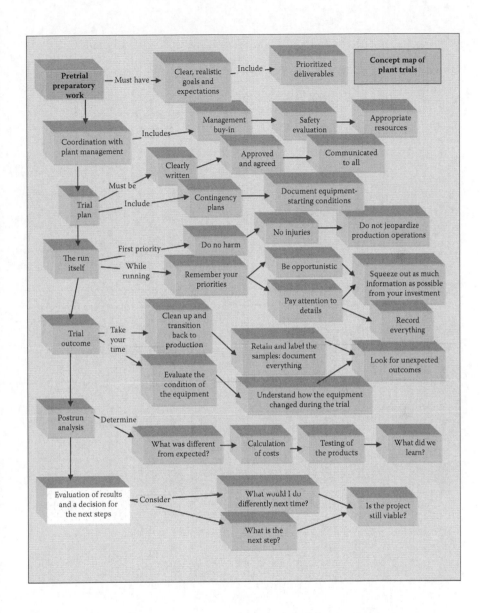

11

Managing and Controlling the Process

How do we manage manufacturing operations for engineered materials? Why do products process differently from one line to the next? Why are job responsibilities different in a dynamic process situation? What questions should we be asking?

> Technology is dominated by two types of people: those who understand what they do not manage, and those who manage what they do not understand.
>
> **Putt's Law**

In 2010, the Campbell Soup Company recalled 15 million pounds of SpaghettiOs with Meatballs, a canned pasta product. A routine warehouse inspection had turned up cans that were bulging, due to a bacterial growth inside them. An investigation blamed a faulty retort step in the packaging process. This heat treatment should have killed all bacteria in the cans. However, the malfunctioning equipment was running at a lower temperature than mandated. Lacking periodic validation of the unit's performance, the company had no way to determine when the system started to deviate from its target temperature. As a consequence, the company was compelled to recall every can of product in its supply chain.

The venerable brand had been on the market for 45 years. Food safety issues were extremely distressing in this case because the product is often purchased for consumption by children. The technology for the sterilization of canned products had been well understood for a century when the problem occurred. In this industry, drastic measures are often taken to prevent the product from crossing over to the sterile side of the facility without passing through an autoclave.

It was difficult to spot the problem because the sterilization step was running just outside the range of acceptability. The system appeared to be operating normally and the cans were hot when coming down the line. The vast majority of the containers were safely sterilized, but a small percentage of them were not. The problems did not become evident until after the product was packed in shipping cartons. These issues went undiscovered because no one asked, "How can we be absolutely sure that every can is being adequately sterilized?" Instead, plant personnel trusted that the control panel was displaying accurate temperature conditions.

RULE #32 EVERY PART FAILS EVENTUALLY—YOU MUST BE PREPARED

Every process system will eventually deviate from its optimal state. The effective management of a dynamic process is all about building circuit breakers to ensure that any divergence is not disastrous. Variability could originate with suppliers, individual workers, systems, or machine components. You cannot anticipate and prevent every possible failure. But you must be able to detect problems before they impact your products.

Detecting Process Variability

- Experienced operators may have the best feel or touch for running the process. Everything tangible that they know should be documented in writing. Document procedures for each task in a format that can be used for training new personnel and can be audited by the appropriate experts.

- Plant management should constantly evaluate the effectiveness of each operator and each piece of equipment to ensure that they are working correctly.

- When the product is process sensitive, management should be especially diligent that the specifications are followed as written. Changes cannot be made without communicating a revised specification. Verbally authorized deviations are often at the root cause of process disasters.

- Do not allow critical procedures and tests to fall into disuse simply because they have not detected problems in the recent past. Operators on different shifts must not be allowed to gradually evolve different habits and technologies to suit their own wants and needs.

- Encourage employees to harmonize their procedures for consistency in both the process and the resulting product. New best practices should be uniformly applied across the workforce after they have been proven. If not, the variation in the product will gradually increase with time. When the variability in the properties is not apparent to the workforce, then they will have no feedback on their actions.

State of the Process

When the well's dry, we know the worth of water.

Ben Franklin

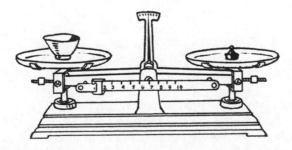

In mid-2006, Nova Chemicals experienced an unscheduled shutdown of a polymerization unit in their manufacturing center in Ontario, Canada. Upgrade work was being done in parallel with routine production operations. In theory, both could be performed safely. Unfortunately, a contractor activated the wrong circuit, initiating a crash shutdown of the entire plant. Several weeks were needed to repair the resulting damage before the unit could be restarted. An 11 million dollar charge was taken against corporate earnings to cover the resulting losses.

This is a situation where a dynamic process is balanced on a knife-edge. Unexpected inputs could send it spinning out of control. In such cases, the sensitive points must be isolated from dangerous external factors. These include utility upsets, airborne contaminants, temperature changes, and vibrations. Not every dynamic process will suffer weeks of downtime if the wrong switch is thrown. It could just as easily encounter weeks of substandard performance or process fluctuation before the root cause is identified.

It is not uncommon to encounter problems when running new products on existing equipment. Occasionally, it is difficult to run standard products that have not been scheduled for a long time. These problems could manifest themselves in any number of ways. Imagine running your massive paperboard machine to produce a slightly different product grade—a heavier thickness and a wider width than usual.

The equipment was designed to handle the product in these dimensions, so this should be a routine order to run. However, numerous problems crop up during the changeover. The vacuum boxes have trouble feeding wet pulp uniformly onto the mesh belt. The pumps are overheating and the winder cannot maintain good edge alignment. The thickness control at the edges is out of specification, because the oven is not drying the board uniformly.

In short, a modest deviation turns into a fiasco. Things go so badly that this nonstandard order becomes a money loser. It seems as though the equipment has a "groove" in which it wants to run and any variation from the routine is not tolerated.

Invariably, the line is modified over time to optimize the production of routine products. In this case, the baffles in a drying tunnel were redirected to maximize the airflow onto the normal web width to boost the production rate. When a wider sheet was run, the heat and the airflow were not uniform across the machine's width. Motors that normally run in their most efficient torque range respond poorly at a different speed.

It is vital to recognize every trade-off decision as you maintain and control the production of the grade slate. Efficient plant operations and extreme flexibility are often incompatible in an industrial environment.

A dynamic process can be in one of three states of readiness:

Intelligently controlled: The process can run all product types and configurations that have been developed and proven on it. The operators can readily transition from one product to another. The engineers know how to add new products to the process by changing components and conditions.

Evolved control: Routine products can be produced efficiently. Infrequently run items take longer to set up and may run with high yield loss. Considerable time and scrap will be required to get an unfamiliar item to run within specifications. A learning curve must be overcome to make even modest transitions.

Uncontrolled: The process gradually loses its ability to run even the routine products over time due to wear, corrosion, or lack of maintenance. The ability to meet specifications is gradually lost, resulting in a diminished level of quality. The staff is reluctant to make changes out of concern that the process cannot be returned to its previous conditions. The equipment requires constant maintenance attention to run at all.

Critical Questions: What state of readiness is our process operating in? Could we improve operational efficiency or product performance by spending the time and money to upgrade? Does it make good business sense to do that?

Upgrading a process to a higher state of readiness is not just a matter of getting smarter about how you run the equipment. Some processes are just inherently hard to control and transition. It may be appropriate to have lines that can run only one product and nothing else, in order to be efficient. If so, it is necessary to find just enough business to keep this line running on that product.

On some equipment, it will be necessary to changeout key components when transitioning from one product to another. If this can be done quickly and easily, then the line may be very flexible. Some equipment will have diverter valves to allow the flow stream to be rerouted along a different path to switch on the fly. Other lines must be shut down, cleaned, and rebuilt to make such a change.

Market dynamics dictate which approach makes sense for your business. It can be very gratifying to have sophisticated process equipment that can rapidly switch from one product to another without scrap and quality problems. However, if your low-tech competitor is able to make an equivalent product significantly cheaper on an inflexible line, then you will be uncompetitive.

The scheduling and operation of a production plant must be based on the control state of your equipment. A financial analysis may suggest reductions in inventory levels. However, if your process is not flexible, then this strategy could increase changeover scrap as you make short runs with frequent process changes.

Even the decision to offer a new product will be driven by the production equipment's ability to transition to the necessary process conditions and configuration. If this product is expected to grow into a big sales volume, then any short-term inefficiency will be offset by the long-term potential to add dedicated equipment. If it remains a niche product, then the lack of efficiency will be a great burden.

Asking the Right Questions

> A process cannot be understood by stopping it. Understanding must
> move with the flow of the process, must join it and flow with it.
>
> **Frank Herbert, Dune**

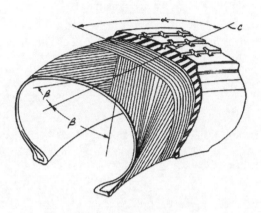

In the early 1970s, market forces were pressuring the Firestone Tire and Rubber Company to sell steel-belted radial tires. For several years, the company had resisted this move, relying on a bias-ply design as a transitional strategy. Customers were cool to this compromise and they were increasingly switching brands. Once Firestone's board had finally voted to offer radials, it was necessary to reconfigure the entire process very quickly. To do so, the existing production sites were converted to produce the new product, rather than building new facilities.

The company's factories were retooled and the workers were retrained. The new brand was launched as the 500 Series. In their haste, Firestone was not able to fully validate the new production process with life testing across the spectrum on environmental conditions. Initially, its bold strategy seemed to work. But, reports of premature tire failures began to trickle in. Soon, the company was deluged with negative publicity over the fatal accidents linked to its products. By 1978, the company was subjected to the largest tire recall in history.

The 500 Series radial tires were prone to have small points of delamination, where the tread has poor adhesion to the carcass of the tire. The damage slowly builds up over years of use, often after prolonged exposure to sunlight, hot weather, and high stress. The eventual failure is often sudden and catastrophic while the vehicle is running at highway speeds. Tires produced at the Decatur, Illinois, plant were especially problematic. Product defects were blamed for 34 deaths. The company paid a record fine and incurred huge losses both financially and in consumer trust.

Tire experts suggest that a number of production factors must be optimized to avoid delamination problems. Batches of raw rubber must be homogenized to a high degree to prevent the ingredients from separating out into discrete domains. These batches need to be vulcanized within strict time limits or oxidation will cause problems. While the different materials are being assembled into a tire, great care must be taken to avoid cross-contamination between the components of the sidewall and the tread. During assembly, it is crucial not to contaminate or touch the exposed surfaces that are about to be laminated together. Even a fingerprint could later become a failure point.

RULE #24 INSIGHTFUL QUESTIONS ARE KEY TO PROCESS OPERATION

If managers ask the right questions and require proof to document decisions, then they will be well equipped to oversee the care and maintenance of a dynamic process. Accurate feedback from end users is essential to manage the operation. Invariably, product performance is the final arbiter, rather than opinions, theory, and operational convenience.

Questions for engineers and product specialists:

- What factor might affect the quality of raw materials?
    - Time in storage—do materials need to be labeled with an expiration date?
    - Type of packaging—are the materials sensitive to oxygen, dirt, or moisture? What about biological factors such as bacteria, insects, and rodents?
    - Shipping method—does vibration in freight transport cause particles to segregate?
- Things that a product might be exposed to over its life.
    - Heat—should we stabilize the product? Label it for maximum storage temperature? Attach temperature indicators to the product?

- Moisture—protective packaging, encapsulation, desiccants?
- Light—what frequency of light? Light-blocking wrappers? Sunblock additives/coatings?
- Vibration or G-forces—cushioning material? Harden the product to resist the effects?
- Insects, mold, bacteria, bird nesting—packaging? Protective additives? Date coding?
- Freezing—ship the product in a temperature-controlled truck? Pack in an insulated container? Special labeling?
- Environmental factors that influence the process:
  - How do the neighboring processes interact?—utilities, contaminants, vibration?
  - Night/day and hot/cold cycles—do we need to isolate the process from these effects?
  - Component wear—how do we monitor and detect when components should be replaced or refurbished?
- Variation between operators:
  - Training—what operational factors are different? How do we train to eliminate the effect? What is the best practice that every operator should follow?
  - Complacency—how do we keep the operator engaged? How do we prevent the defects that inattention causes?
  - Physical strength or dexterity—how do we change the job to eliminate these requirements? Do we create special requirements and personnel performance testing for this job?
- How can we ensure that the operators will perform the essential tasks the same way every time?
  - Create smart checklists for starting up and at the beginning of each shift.
  - Provide a place where the operators can find the most current specifications, standard operating conditions (SOCs), standard operating procedures (SOPs), and any special alerts/quality concerns.
  - Create menus in the control system which configures the line to the specific conditions needed for each product.
  - Provide fixtures and guides to ensure consistency and check periodically to ensure that they remain in tolerance.
- How can mixtures differ?
  - Settling or segregation of the components during storage or handling?
  - Particle size variability from one batch to another or within a batch?

- Continuous or discontinuous phases?
- Phase changes due to temperature, pressure, or humidity?
- Volume shift due to thermodynamic effects?
- Need for surfactants to compatibilize incompatible materials?
- Preferential adhesion of one component to the walls of a container?
- Static electric attraction of fine particles?
- Reaction of one component with the environment?
- Does the product depend on something that vaporizes away during the process, during storage, or when used by the customer?
  - What factors or conditions can block or encourage the escape of that volatile from the product?
  - How can the volatile be measured or controlled?
  - Where does it finally recondense and can any of those places suffer harm?
  - Is there an odor associated with the volatile?
- Is there some surface property that dictates the product's utility? (Optical properties, cleanliness, adhesion to something else, frictional properties, marring of something it contacts, visual appearance.) How do you ensure that the surface will retain the properties that you want? How can you check to see if the properties are still there?
- Are there morphology issues that influence performance? (Changing from one crystalline structure to another, phase changes.) How can these be measured or detected?
- What can be done to make the process scalable—how far can you increase capacity in increments as opposed to building a whole new unit?

Production management questions:

- What process components could break down at inopportune times? Remember Oberle's third law—every component fails eventually.
  - Gradual failures of subsystems are not immediately obvious. The line does not crash to a stop. Are the operators trained to recognize failure and act on it?
  - Do we have trained people and spare parts available to make repairs on a timely basis?
  - Can we isolate and test the products that were affected by the breakdown?

- How can we prevent unpredictable failures from causing harm to people, the environment, and the equipment?
- Can we implement preventative maintenance procedures to replace components before they start to affect the process?
- How do we encourage the workers to act in the company's best interest?
  - What are the most critical things that we want them to focus on?
  - Does management behavior send the wrong message?
- What is Plan B? The real world is never static. Regulations, competitors, price fluctuations, and raw material shortages periodically take away our options. Proactive managers are always looking for the most probable sources of trouble and planning contingencies.
- How do we run the process to optimize profit? What are the most important factors that will advance the company's competitive position?
  - Will running the process faster reduce the unit production costs or will it just harm the quality, equipment reliability, and fitness-for-use (FFU)?
- What are the bottlenecks and trade-offs? How do we remove these restrictions?
- Which individual products are most sensitive to these limits?
  - Will more economical raw materials save money or cause problems elsewhere?
- Are longer runs of each product going to affect or help costs?
  - Longer runs mean more inventory-carrying costs.
  - Short runs mean that set-up costs are spread over less production.
- Will a new product be incompatible with the existing product mix?
  - If the manufacturing plant is too accommodating of inefficient, customized, or short-run items, then production efficiency will suffer.
  - If the production manager is too pessimistic, then new products will never be introduced.
- How long can a continuous process run between preventative maintenance shutdowns?
  - The dirtier the line becomes, the more time that is lost cleaning between batches.
  - The longer the line runs, the greater the product variability and FFU problems are.
  - The shorter the campaign is, the less production that is made between downtimes.

- How much housekeeping should be done in the production area?
  - Clean and orderly work areas are safer and encourage people to take pride in their work.
  - Cleaning is a boring activity that takes time away from important responsibilities.
  - Cleaning should be clearly focused on things that make a difference to the process and to safety. It should not be limited just to areas visible to passing management personnel.
- How much raw material should be kept on hand?
  - A lower inventory conserves working capital.
- What happens to the total operation if you have a supply disruption?
  - Keep just-in-case stocks of materials that have long lead times.

Functional Responsibilities

Criticism from hindsight should not be confused with wisdom, but a skilled sycophant can fake it.
Reacting to random events does not constitute strategy, although it sure reads that way in the annual report.
Any operational fiasco can be addressed by firing the appropriate scapegoat and issuing stock options.
Staging meetings and issuing edicts is not leadership, but an expensive suit helps you to look the part.

Law of Organizational Indifference

Early in the twentieth century, British shipbuilders were making the transition from iron to steel rivets. Twenty years earlier, steel had replaced iron in structural components, such as hull plates. But, it took longer to make the same change in rivets.

While the Titanic was under construction in 1911, both the new and the old technologies were employed. Steel rivets were inserted into the central hull section by large riveting machines suspended from overhead gantries. However, the sloped bow and stern sections were hard to reach, so this work was done by hand, using iron rivets. The rivet quality has since been cited as the weakness that caused the massive vessel to sink quickly the following year. An exploration of the wreck site indicates that the hull was not ripped open by the iceberg. Rather, some rivets popped under the force of the impact, allowing seawater to enter through the open seams around the plates.

The wrought iron rivet production process was performed by blacksmiths in the area surrounding the Belfast shipyards. Their capacity was taxed by the huge volume of rivets needed to finish the Titanic and her sister ships on schedule. The FFU requirement for this application dictated that the slag content should have ranged between 2% and 3%. Slag is a nonmetallic waste product of smelting. Slag-free iron is too soft, allowing deformation over time, eventually resulting in leakage. A higher slag content results in a brittle rivet, which fails easily during an impact, especially at the cold temperatures of the North Atlantic.

Rivets recovered from the wreck site were found to have an unacceptably high slag content. This impurity was driven out of wrought iron by laborious heating and hammering. As the blacksmiths were pressed to supply more rivets, it stands to reason that the number of hammer strokes was reduced.

The ironworkers used their experience to judge the slag levels, but they lacked qualitative test methods. Apparently, no one at the Harland and Wolff shipyard was looking after the end-use performance of the purchased rivets. Ten years earlier, the British Board of Trade stopped testing the iron used in ships because it considered the technology to be mature. During the intervening time, ship designs were steadily growing in size, putting additional stress on the fasteners.

In a process-dependent environment, the participants must have clear responsibilities to ensure that every aspect of production is under control. The obvious factors are usually discovered quickly. If the process runs poorly or the output fails to pass basic quality control (QC) tests, then action must be taken. The key is to ensure that these QC tests capture every aspect of the end-use performance.

In May 2010, Johnson & Johnson was forced to recall 43 different Tylenol products. These popular pediatric medicines lacked proper quality safeguards and the company failed to follow up on customer complaints. But the most troubling finding in a Food and Drug Administration (FDA) inspection report concerned uncontrolled process factors. Excerpts from the report are shown below. The government redacted confidential information from the original document prior to its public release.

OBSERVATION 2

There are no written procedures for production and process controls designed to assure that the drug products have the identity, strength, quality, and purity they purport or are represented to possess.

Specifically,

Lack of process validation for the manufacture of Infant's Dye-Free Tylenol Suspension Drops, Cherry, Formula (b) (4) 80 mg/0.8 mL. The compounding and transfer of the (b) (4) batch size suspension to the (b) (4) hold tank is not in a "state of control". The firm did not effectively evaluate the change in the manufacturing process (agitation and tank level time to shut off of agitator)

OBSERVATION 4

Control procedures are not established which monitor the output and validate the performance of those manufacturing processes that may be responsible for causing variability in the characteristics of in-process material and the drug product.

Specifically,

Control procedures used did not validate the manufacturing processes that caused variability in the characteristics of the drug product. For examples, the agitation speeds and time to reach (b) (4) in the hold tank during processing of the super potent batches that failed APAP (end of run) assays, released batches, and the demonstration batch. The firm did not demonstrate the adequacy of mixing to assure uniformity and homogeneity for Infant's Dye-Free Tylenol Suspension Drops, Formula (b) (4) using a (b) (4) batch in a (b) (4) hold tank. Agitation and tank levels with (b) (4) the amount of liquid) in a (b) (4) hold tank were evaluated with one demonstration bulk batch, lot

| SEE REVERSE OF THIS PAGE | Anita R. Michael, Investigator
Matthew R. Noonan, Investigator
Sharon K. Thoma, Investigator
Nala L. Whitesell, Investigator | 04/30/2010 |

The report indicates that the company neglected to properly monitor and control key steps in the manufacturing process. Specifically, uncontrolled factors resulted in unacceptably high and low levels of individual ingredients in some batches. The net result was a blow to consumer confidence, especially among parents concerned about how to treat their sick infants. In March 2011, the situation became more serious when the federal government implemented a consent decree, which set a strict timetable for the company to bring its plants into conformance with drug regulations.

Both the Titanic and the Tylenol examples suggest a failure of production management to assign the appropriate people to monitor critical aspects of their operations. In modern facilities, no one person can be responsible for all the details that control variability and FFU. Problems will eventually occur when crucial process factors are left to chance.

Special considerations for each job function:

Plant Process Engineers: These individuals are responsible for finding ways to run the process more efficiently and/or with less variability while not impacting FFU attributes. They should question undocumented notions and cautiously experiment with parameters that are normally kept constant, to evaluate their effect. When comparing two sets of conditions, every possible process impact should be recorded. Make very sure that the effect that you are seeing is really the result of your adjustments. Repeat all trials to prove the effect's strength. Beware of pushing conditions beyond current experience without validating the end-use performance.

Development Engineers: Obviously, development engineers must be knowledgeable about the process. They understand the effects of each knob on the process and end-use performance. The trade-offs presented by the various raw material options should be documented.

It may be less obvious that these individuals must be willing to embrace uncertainty. The complete path from start to finish is never clear at the start of projects involving dynamic processes. Engineers must be flexible because the process does not always act as predicted. They need to learn from the exceptions and document why critical process factors have been selected.

People who write and maintain the specifications, SOCs, and SOPs: These individuals must understand and communicate what is really important for each application. Be aware that standard document formats generally fail to capture the complexity needed to understand the interactions in a dynamic process.

Purchasing Agents: Purchasing agents do more than negotiate for the lowest price. Look for the best value for the process, even if it has a higher unit price. Find vendors whose products are consistent and uniform in the properties that are important for the end use. When trying to make consistent products at widespread locations, look for suppliers with a global reach. Getting the same feedstock for each location ensures a uniform and interchangeable product.

What do your vendors think of you as a customer? Are you a tier-one customer, who is worthy of support and is viewed as a long-term partner? Are suppliers willing to tailor products to your needs? Or do they view you as a price whore, who is always trying to wring out concessions and never pays on time? You need not be best friends with every supplier, but you do need to understand how raw materials affect process consistency. Build relationships that will provide access to essential information and insights about your important feedstocks.

Production Managers: Production managers should not have compensation incentives that are at odds with market-driven product goals. If a minor process deviation could cause major problems for the customer, then the manager should not compromise quality by speeding up the process. Manufacturing should be punished, rather than rewarded when it makes ill-advised "process improvements." The situation will be reversed if the product is a commodity where profitability is driven solely by manufacturing costs. Operations that include both types of products can be very confusing.

Product Line Managers: These individuals should strike a balance between the sales and manufacturing organizations, which naturally tend to have opposing views of how the business should be run.

Salespeople: Salespeople should communicate problems that the customer is having, particularly if a shift in performance is encountered.

These first responders need to obtain samples of problems along with a complete explanation of what happened. They need to be open to experimentation by the development teams and know when changes and tests are not desirable for a customer.

The people in direct contact with the end users should constantly be looking for new places to employ proven solutions. These are applications where the process has delivered value to other customers. Finding similar end uses allows you to grow sales without a long project to develop new solutions. The sales staff needs case studies and examples to identify similar applications.

The Old Timers: The most important people in the company are those who know why certain control variables should never be touched. Their knowledge needs to be passed on before they transfer, retire, or die. This process knowledge was expensive to acquire and is invaluable to running the process properly.

Top Management: These individuals need to periodically examine the validity of all their paradigms about the process and the products. They need to ask hard questions and not take the slick answers at face value.

The authority to make decisions is the ultimate source of power in an organization. The enthusiasm with which these decisions are carried out by engineers is directly related to how well grounded they are in the process reality. The workforce can make any decision maker look foolish by executing their strategy in a haphazard manner.

The rank and file become disillusioned when their ideas, opinions, and needs are ignored. This can easily happen, even in a responsive organization. It happens when communication across functional lines is poor and people are spread over great distances. The reasons for decisions and new policies need to be communicated, especially when there is a large gap between the process and the end use. The direction from the top must be clearly focused and unwavering.

See the "concept map of manufacturing goals" at the end of the chapter. The character of the production organization must match with the character of the larger company that they serve. New initiatives cannot constantly change the focus of the group. The company must select one overriding mission goal for the production group above all other options.

Critical Questions: Is there a fundamental disconnect between the operating philosophy of our production group and the rest of the organization? What kind of manufacturing organization is right for meeting the goals of our company/division/product line?

Dealing with Trade-Offs

All movements go too far.

Bertrand Russell

In the race to provide more compact and powerful electronic gadgets, batteries are a limiting factor. Consumers want lightweight devices to run for the longest possible time and recharge quickly. These conflicting desires create stress when pushing capabilities to the limit. Process variability in this situation results in spectacular failures. In 2006, fires in laptop computers sparked massive recall operations due to a few Sony lithium-ion batteries at the extremes of the bell curve.

In this technology, a thin composite sheet contains all the essential elements of the battery. A 25 µm porous insulator film is sandwiched between a graphite anode sheet and a layer of lithium cobalt oxide, which acts as the cathode. This composite is rolled up into a tight cylinder, creating a battery cell immersed in a flammable lithium salt electrolyte solution.

Problems occurred when tiny shards of metal were created in the cutting/ crimping operation that completes the process. These particles can embed themselves in the porous film, resulting in a short circuit. Large specks of metal can be rejected by QC inspections in the factory. The smallest shards go undetected for years. They might cause the individual cell to fail late in its life, degrading the performance of older devices. Medium-sized particles can trigger a violent reaction when vibration and constant charge/discharge cycles breach the insulator. These cases result in energetic chain-reaction fires, which spread from one cell to the next.

This defect class was difficult to detect in the production plant. During development, it is hard to evaluate metal contamination, because problems are infrequent and take years to occur. A successful experience with one application may not be applicable to the next, due to differences in the end-use environments. The commercial pressure to supply the lightest product at the lowest possible cost will tip the balance away from the safest options. Cost concerns force companies to "leverage the technology" and upgrade the original production line to provide solutions to many customer applications.

Everyone has their own perspective on how a process should perform, how flexible it should be, and how much redundancy is needed to ensure product consistency. The product manager wants production lines with infinite flexibility, to make every possible version that the customers might request. Of course, this should done at the lowest possible cost and the product should never fail. The manufacturing manager would like to restrict the line to making the bare minimum number of different variants. This will mean fewer changeovers, lower training burden, and less opportunity for operator error.

RULE #61 PROCESS FLEXIBILITY IS EXPENSIVE

Very flexible processes allow for multiple products to be run on the same equipment, but they will cost more than lines dedicated to a single product. Versatile plants are more expensive in terms of the capital, the operational complexity, and the technical investment to control them. Adjustable processes are more maintenance intensive and require more personnel training. The more flexible a production system is, the greater the chance of errors.

Building flexibility into a process:

- Small lines can be optimized and run in a particular configuration for long periods, rather than being reconfigured constantly. Each line has very little flexibility in itself, but the total plant can handle many different combinations, depending on which lines it chooses to run.
    - Small lines can be added in smaller increments of capital when capacity is needed.
    - These lines are convenient for development, as there are fewer scale-up problems.
    - Labor costs may be higher to run many small lines versus a single large one.
- Consider how often the equipment needs to be changed from one orientation to another. If the changes are infrequent, then you can afford to shut it down and rebuild it into a new configuration. At any given time, the process equipment is inflexible, but has the potential to be changed.

- Considerable maintenance costs may be required to perform changeovers.
- Unanticipated changes in customer demand will upset this approach.
- Some operations need many similar pieces of process equipment to serve their markets. Such companies should build at least one very flexible line with many potential process paths that can run many different product configurations.
  - Use this line to create products for new applications. Once the process needs for those applications are understood, then efficient, dedicated equipment can be built for them.
  - This line allows for technology options to be compared in a production environment.
- Flexibility might represent the ability to use many different feedstock operations when making the same end products. This is "back-end" flexibility. It allows a process-driven company to choose the best-value raw materials and to switch from one to another as prices fluctuate.
- "Front-end" flexibility might represent the ability to blend, convert, and package the finished product in many different configurations, ranging from tiny, sterile vials to bulk tanker ships. The dynamic process step might run continuously, but its output is sent to where it is needed.
- A company needs to have a consistent approach when selecting these options. Understand that company strategy must guide decision making on matters of production flexibility versus efficiency. The manufacturing operation must be structured to service customers.

Technical Decision Making

No plan of battle ever survives contact with the enemy.

Heinz Guderian

Ötzi the iceman was entombed in the Tyrolean Alps some 5300 years ago. He carried a copper axe head, which must have been a rare and valuable tool among his people. For centuries, copper was mined, smelted, and cast in many different countries. However, it is not a very good material for this application. Some early process genius recognized that adding tin-stone to the copper ore could produce a bronze alloy that was significantly stronger than pure copper. This discovery must have resulted from intentional process experimentation, because the two metals are not found together in the same ore deposits.

This ore-blending process was not optimal for achieving a consistent performance, because bronze properties are dependent on the ratio of the two components. A later process innovator thought to smelt the two ores separately and then melt them together into a finished product. This control over the alloy composition allowed the tailoring of product properties. High tin levels created shiny jewelry and mirrors, which were brittle. Alloys with less tin produced stronger weapons and tools with a dull finish.

Still later, lead began to appear in bronze alloys. The lead content gave the molten metal a lower viscosity, which results in better definition in the casting of intricate parts. The invention of the bellows facilitated larger batches of metal, which could be cast into big parts in a single pour. The advances in processing technologies enabled the creation of materials that were optimized for specific applications. Early civilizations guarded these discoveries to keep advanced weapons out of the hands of their enemies and to improve their balance of trade. Tin mines and the trade routes that delivered the ore became strategically important.

Archeologists can date bronze artifacts by analyzing the maturity of their metallurgy. Their places of origin can be traced to the ore deposits that provided the raw materials. Trade routes can be uncovered by looking at the dispersion of the product from one particular production center. The proprietary processes of each culture are still frozen in the metal, thousands of years later.

In the late nineteenth and early twentieth century, foundries carefully guarded their bronze formulations while competing for prestigious contracts to cast sculptures for famous artists. A century later, the history of some artwork is missing or incomplete. Works are occasionally uncovered, which bear a style reminiscent of one of those masters. Using a mass spectrometer, it is often possible to evaluate the composition of the metal to determine where and when the statue was cast. The foundries' trade secrets are fingerprints that reveal if the artwork matches known works of an artist or if it is a copy.

Today, companies must make decisions on complex matters that are strongly influenced by factors that they cannot fully understand. They must reproduce and improve upon products made by competitors without seeing their production techniques. New scientific knowledge must be turned into a well-controlled process that can be manipulated to solve customer problems. It is crucial to make good decisions when attempting to pursue these commercial goals.

Decision-making procedures in materials technology are often confused by poor communications, false claims of expertise, overconfidence, and interpersonal conflicts. Outside experts in the field and internal specialists all compete with one another for the attention of the decision maker. In the worst case, executives are reluctant to admit that they need help to choose the best path. In other situations, the technical experts try to protect their privileged positions by not freely sharing knowledge. It is important to remember that no one geek has the answer to every specialized question. The biggest challenge is to determine the validity of the conflicting opinions and missing information.

RULE #17 FIND OUT WHAT INFORMATION MAY BE MISSING

KNOW/DON'T KNOW

The first obstacle to making decisions on complex issues is the natural tendency for people to make their judgments based only on what they know. In every situation, there will be things that are known and things that are unknown. The first group is the one that gets all the attention and consideration. The second group is much harder to comprehend. Most decision makers may not even understand how large the unknown group is.

Late in the evening of December 29, 1972, Eastern Airlines Flight 401 was on its final approach into Miami International Airport. The pilot aborted the landing because the instruments suggested that the front landing gear was not properly locked in position. The crew had to determine if the mechanism was actually broken or if the indicator was faulty. They began to troubleshoot the problem while orbiting on autopilot. At one point, the flight engineer crawled into a narrow equipment space below the cockpit to visually inspect

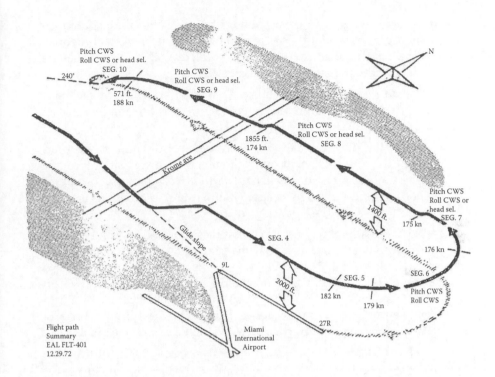

Pitch CWS
Roll CWS or head sel.
SEG. 10

Pitch CWS
Roll CWS or head sel.
SEG. 9

240°

571 ft.
188 kn

Pitch CWS
Roll CWS or head sel.
SEG. 8

1855 ft.
174 kn

Krome ave

N

Pitch CWS
Roll CWS or head sel.
SEG. 7

1400 ft.
175 kn

176 kn

Glide slope

SEG. 4

9L

SEG. 5

2000 ft.
182 kn

SEG. 6

Pitch CWS
Roll CWS

179 kn

27R

Flight path
Summary
EAL FLT-401
12.29.72

Miami
International
Airport

the problem. The pilot and the copilot were occupied with reinserting the indicator light lens back into the instrument panel after having taken it apart.

The flight crew was tired and impatient with the unexpected problem. Because of the crew's distracted frame of mind, they did not notice that the autopilot had put the plane into a gentle decent. Within minutes, the Lockheed L1011 plowed into a swamp at 227 mph, killing over 100 people. The accident investigation blamed the flight crew because they had failed to monitor the instruments. This is a case of an experienced team becoming narrowly focused on one task at the expense of what was most important.

Dealing with the known factors—first, evaluate the data quality:

- Review the evidence. Experts should explain their proposals or theories in a concise summary. Seek the group's opinion to determine if the expert conclusion is compelling. Are there alternate explanations for the facts? Is the logic flawed? Does it pass the "sniff test"?
  - Some managers feel that they look weak and indecisive when they give others input into their decisions. In fact, the staff is often delighted when the management asks its opinion. Seek the widest possible decision-making base and let them review the evidence and evaluate the conflicting models. Guard against the group-think mentality that comes from a show-of-hands vote.

- Avoid relying on one trusted advisor to provide all the guidance in a particular field. This is a comfortable trap for both people. The expert will always have an answer, even if he or she lacks the data or the perspective needed to be sure.

- Beware of the people who only confirm your existing beliefs. Such positions are more persuasive than those that contradict your paradigms.

- After long experience with a process, people may develop an intuitive feel for what will happen in a given situation, but they will not be able to articulate their reasoning. These opinions should be weighed against other data, but neither method should ever be the sole basis for a decision.

- Avoid "over-weighting" newly discovered information. In the airliner crash, an autopsy revealed that the pilot had a brain tumor. Much speculation went into whether this one factor explained the whole accident. For instance, a gradual deterioration of his peripheral vision might have prevented him from noticing the movement of the altimeter. Eventually, it was determined that the medical condition could not explain the poor performance of the flight crew as a team.

- Watch out for individuals or groups who hoard their information and only present a recommendation. Use a peer review technique to verify the validity of their conclusion.

- Test proposals by trying to falsify the claims. Look for exceptions, contradictions, alternative explanations, and discounting evidence.

- Put more weight on proposals that have been proven by experiments on the particular process or materials that are pertinent. Always remember that interpolation is superior to extrapolation.

- Beware of explanations that depend on the reasoning that the competitor is doing the same thing. No one can know everything that goes on at the competition. Examples given in their patent may not represent their actual production process.

- Distrust reverse explanations. It is quite easy to turn flawless diamonds into worthless carbon. It is the reverse process that is difficult.

- Be wary of simplistic reasoning based on stories and analogies. Entertaining presentation techniques are much more interesting, but no more valid than boring ones. Look deeper into the reasoning before investing money.

- Additional information often confuses the decision-making process. Complex issues can be boiled down to a handful of pertinent factors that control the outcome. It is always desirable to identify these parameters. Seek to build a decision tree and test the validity of each node.

Dealing with the unknowns—first, evaluate the importance of missing information:

- When faced with a significant decision, lay out a series of questions that need to be answered. Do not take action until these questions are answered or a good reason is advanced for why they do not matter.

  - This list should not be limited to technical issues. Trends in the marketplace, industry, government regulations, and the patent landscape are also important. Killer variables are often recognized well in advance even if you cannot predict when they will strike.

  - Forecasting the market situation 2 years down the road will give a picture of what the customers will need when the work is finished. This approach works well on fundamental megatrends such as energy costs, environmental restrictions, and a shortage of skilled labor. It is not good at predicting changes in consumer tastes, public opinion, or social hot buttons.

- In the annual report of an American corporation, a summary is required detailing the adverse situations that could impact the business in the future. Investors should be informed of potential problems for their investment. Apply the same reasoning to market changes or competitive moves that could undermine a new product line. Assign realistic probabilities to the various scenarios.

  - Never apply this thinking in isolation. A risk-adverse analysis can easily kill any plan by overestimating the potential problems. Use this approach to compare different options and choose the one that has the highest risk-adjusted return on investment.

  - Adjust the strategy to minimize the risks. It is easier to address concerns at the beginning of a project, rather than reacting to them later.

- Always remember that there are many possible solutions to any question requiring a decision. The job is to find the one that is best for today's situation.

  - What is optimal for the competitor is not automatically right for you.

  - The customer's preferred solution may not be best for your profit and loss. Do not surrender decision making to your customers.

  - New solutions will become available in the future. Few decisions will stand the test of time as more options become available. Constantly reevaluate your position.

- A key decision concerns the trade-off of spending resources to find and verify superior solutions versus continuing to milk the best technology available today.

  - Today's technology is obviously the known information. This is the knowledge base that nourishes the business. There are often more project requests than there are resources to pursue them. Diverting technical assets to the pursuit of the unknown will delay the development of urgently needed short-term development efforts.

  - The superior solutions are the unknown factors. These options are not unknown to science. Often, the theory was advanced years ago. However, their economic exploitation for the application is the unknown. Millions of dollars could be spent on the technology, only to discover that it is less profitable than the existing process.

Anticipating Risk

> It is a capital mistake to theorize before one has data. Insensibly one begins to twist facts to suit theories instead of theories to suit facts.
>
> **Sherlock Holmes**

In May 1845, the British Royal Navy dispatched Captain Sir John Franklin in one of that century's many efforts to discover a Northwest Passage from the Atlantic Ocean to the Pacific Ocean. Up to that time, his two sailing vessels were the best equipped to be sent into the Canadian Arctic. Unlike the previous expeditions, this effort ended in disaster, with the loss of both ships and their entire crews.

Subsequent investigations raised serious questions about the men's mental state and the poor decisions they made in trying to survive. With food stocks running low, they tried to hike south to settlements on Hudson Bay. The survivors overloaded a small boat with unnecessary items and tried to drag it across the frozen landscape. This effort soon exhausted and killed the remaining crew members. The search parties were shocked by Inuit reports that these men resorted to cannibalism at the end.

Modern researchers have investigated the causes of this behavior. They discovered excessive levels of lead in the frozen remains of the expedition. A subsequent analysis suggests two sources for heavy metal in their diet. Over 16 ton of canned meat was supplied to Franklin's ships by the Goldner firm. This was one of the company's first contracts to supply tinned meat to naval ships. Empty cans found in the ice had sloppy lead seam welds. Goldner's production process apparently had serious issues in its weld quality at this point.

Other investigations suggest that the water purification equipment on both ships was a major source of heavy metals. Before sailing, the vessels were fitted with distillation systems to desalinate seawater. Lead was the most convenient material to fabricate both the pipes and the fittings for the system. Due to the high operating temperatures of the distillation process, significant quantities of lead leached into the water that was used for drinking and baking bread. This is a case where a serious problem had multiple factors that combined to turn a bad situation into a disaster.

RULE #52 THERE ARE ALWAYS HIDDEN PROCESS PROBLEMS WAITING TO SURFACE

The problems that result from a lack of process knowledge are not always visible at the time of production. However, uncontrolled attributes eventually find a way to make themselves evident.

Ideas for controlling minor components:

- Suppose that a product could contain one or more minor components that fluctuate from one batch to the next. However, it is difficult to test each batch to measure the levels of trace contamination. Spike some samples with higher levels of these materials than any one production batch is ever likely to contain. Evaluate the product with FFU tools at various concentrations to determine where the product performance starts to deteriorate. Identify the process condition envelope that is likely to generate undesirable levels of by-products or defects. Compare this to the envelope of conditions that will produce the product at an acceptable profit. If there is too much overlap, consider redesigning the process or dropping the business.

- Install safeguards to ensure that production never ships a batch that was made during an excursion into a dangerous part of the process window.

- Automatically record and chart the process conditions on a continuous basis. Require a review to generate a certificate of compliance to ship each lot.

- Periodically screen for these by-products to ensure that they are always below the danger levels that have been identified, especially after the process is altered.

- Run the process at proven conditions, process configurations, and raw materials. Do not allow the operational staff to get creative with new rates and conditions. Take the line down for routine overhauls well before routine wear and tear pushes the process out of its comfort zone. Qualify any modifications to the process and the materials for dangerous interactions or problems.

- If the contaminant or the defective formations cannot be reliably predicted by the process conditions, then measure them in real time with in-line sensors that constantly sample the process. Divert the product stream into a holding tank or bin when dangerous levels of minor components are detected.

  - This is not limited to chemical systems. The same technique is performed with a computerized vision system on molded parts. Any deviation in the appearance or dimensions of even a single piece should alert the production staff.

  - Packaged food products routinely pass through metal detectors that sense the presence of even small amounts of metallic iron. No one wants to bite down onto a machine screw that worked loose from some component in the process.

- Look for ways to scrub, neutralize, or buffer the dangerous factors out of your product.

  - For many years, the Perrier mineral water production process had included a charcoal filter bed to remove naturally occurring aromatic contaminants. This routine worked well until the plant neglected routine maintenance and failed to replace the overloaded filter pack. A massive product recall was needed after traces of benzene were discovered in the bottled product.

  - This approach is best accompanied by a redundant monitoring system and an aggressive preventative maintenance program.

- A product might develop dangerous characteristics as it ages or interacts with the environment sometime during its life span.
  - Use forced-aging to evaluate possible degradation products. For instance, store the product under elevated temperature conditions in the presence of a range of different materials that may cause issues. Correlate the forced-aging test results with real-world conditions.
  - Perform flame tests to determine the combustion products produced in oxygen-rich and oxygen-poor atmospheres.
  - Sample the air around the customer process that uses the material to detect and quantify fumes that may be given off. Compare them to the same operation when idle and when using alternative products. Specify the required ventilation standards or other measures.
- Assess applications for risk, evaluating how property fluctuations influence the problem.
  - A critical performance part in an automotive brake system should get more attention than a part that does not have dire consequences upon failure.
  - Do not sell into applications that cannot accept failures from your worst-case levels.
  - Pay attention to items that may be very difficult and expensive to inspect and replace once they are in the customer's product.
- Establish the point where product failure will cause real danger. Measure the long-term standard deviation of these attributes for the product and the process. Make sure that there are a lot more than three standard deviations between the mean and the danger point.
  - Continue to improve the process to reduce the variation over time.
  - Research the assumptions about the danger point and the safe operating range. Look for changes in product use that will invalidate the earlier assumptions.
  - If the above is not cost effective, then exit the business or get a new process.
  - Remember that standard deviation only measures "natural variation." Process instability that is not statistically "normal" (broken components or process excursions) will make the standard deviation predictions invalid.
- Buy premium equipment for key process functions such as metering in additives and controlling temperature.
  - Use redundant instruments to ensure that the process is running where it should.

- Frequently calibrate redundant monitoring systems to ensure consistency.
- Beware of situations that can fool both monitoring systems, such as software errors.
- Conduct rigorous preventative maintenance to ensure that the components are within tolerance.
- Chart the control parameters to ensure that the process is not drifting over time.

- Beware of new applications where products like yours have never been used before. There will be no basis on which to make judgments about the relative importance of one attribute versus another.
  - Be very afraid if the product is expected to continue to function for years into the future. It is problematic if there is no way to monitor the product in the end use or to model changes.
  - Ensure that the customer is aware of your uncertainty on these matters.
  - Beware of applications where the customer is not clear about the risk of failure or how the product will be used.
- Remember that the biological response to a product is rarely linear or predictable.
  - Individuals can have lethal allergic responses to a material that is safe for the majority.
  - Some people can work happily with a material for years and then become sensitized to it. Both urethane and epoxy products can trigger negative reactions in sensitized individuals.
  - Bacteria that are normally killed by a chemical, a process condition, or a cleaning procedure may develop an immunity or resistance. This could create a contamination problem, where none existed before.
  - A chemical with low toxicity in laboratory rats might be dangerous for humans or other species.
  - A material is odorless to most people, but can be offensive to some people or animals.
- Suppose that your company produces a wide variety of wax formulations. A customer proposes to attach feathers to his or her arms using one of your adhesive products. These human powered flight systems will allow a person to fly like a bird. The customer is Greek, hoping to escape from an island prison. In knowing the product's softening point, limits on the maximum safe altitude should be proposed. Advise the consumer to avoid flight problems due to close proximity to the sun.

Failure Analysis and Corrective Action

From disaster good fortune comes, and in good fortune lurks disaster.

Chinese Proverb

An ounce of prevention is worth a pound of cure.

English Proverb

It is better to be a mouse in a cat's mouth than a man in a lawyer's hands.

Spanish Proverb

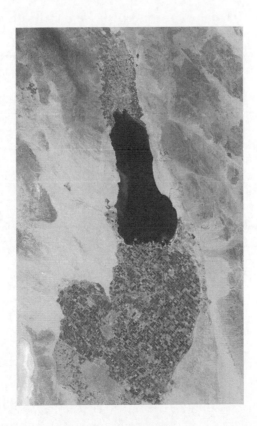

Early in the twentieth century, Robert R. Rockwood's California Development Company (CDC) was in the business of diverting irrigation water from the Colorado River to farms in Southern California. There were numerous problems with the company's process for delivering water over long distances. The river carries a great deal of sediment as it cascades down from the mountains. This mud tended to settle out in slow-moving water and clog the CDC's canals. The company's dredges were operating full time, but they could not remove the silt as quickly as it built up. Combined with the low water levels on the river, Rockwood was not able to deliver enough water to satisfy his customers.

In 1905, the company cut a temporary canal just south of the Mexican border. In the interest of expediency, they did not install any type of flow control system on the new works. Had the weather conditions been normal, the canal would have silted over before the end of that year's dry season. Instead, there were heavy rains that year and a vast torrent of water poured into the canal. The surging water flow enlarged the channel width from 60 to 160 ft. Several earthen dams were thrown up to control the flow, but these were washed away before they could be completed.

The 900,000 ft.^{3}/s deluge soon surpassed California's distribution capacity. The overflow gushed into the lowest point in the area, which was a dry rift valley over the San Andreas Fault. Several desert settlements and a salt-mining operation were quickly inundated by the rising water level.

The following year, the CDC went bankrupt as the channel grew to 600 ft. in width, carrying 80% of the Colorado River's flow. It took years of work for Rockwood's successors to reroute the river back into its original channel. To this day, the Salton Sea (pictured in the satellite photo) remains the largest accidentally created body of water on the planet.

RULE #16 THEORY ≠ PRACTICE IN A PROCESS-DEPENDENT ENVIRONMENT

Dynamic processes rarely conform to the theoretical predictions on the first trial. Models must undergo constant refinement based on actual performance. Things can and will go wrong. You need to anticipate and prepare for those eventualities. Process problems can manifest themselves in a number of different ways.

Types of process disasters:

- An industrial accident that causes facility damage, loss of production, and/or risk to personnel.
  - The lucky fact that no one was standing adjacent to the autoclave when it exploded is no reason to skimp on corrective action. Guilt prompts us to go overboard in reacting to human tragedy. Relief at our good fortune should be an equally powerful signal to take corrective measures.
  - The fact that the fire department was not called and that the media did not report the incident is immaterial to your corrective action. If the process puts people, the environment, or equipment at risk, then it must be addressed. Serious accidents are often preceded by predictive events that were ignored.
- Quality problems resulting in the wide-scale recall and the extensive reinspection of product.
  - Such incidents will prompt even the most loyal customers to spend time and money to seek alternatives to your product.
  - Death or injury to consumers will put the incident in a whole other class.
  - Again, do not defer the investigation because no one got hurt. Such inaction will count against you the next time.
- An uncontrolled release of regulated materials into the environment.
- A mistake or process excursion that causes a large batch or lot of product to be rejected.

Launch a thorough investigation to determine the problem's root cause and the contributing factors. Carefully considered corrective actions must be implemented to prevent a reoccurrence. Note the plural "causes" and "actions." The discovery of only a single root cause can indicate a shallow investigation. It is rare that only a single corrective action is appropriate.

The reasons that employees resist investigations:

- People want to put mistakes behind them. "The past is past. We need to move on. That was a fluke event. We know better now. That will not happen again."
- The search for the root cause can require a long and meticulous investigation. Few companies have the appropriate internal resources to devote to such an endeavor. This is especially true when everyone is dealing with the disaster—inspecting products, supporting the families of those injured, finding replacement production capacity,

repairing the damage, and dealing with the government and the media.

- The investigation must be balanced. This is difficult when everyone who understands the technology is either responsible for the problem or stands to benefit from some outcomes.

- The investigative process requires contemplative, analytical thinking. Most managers are action-oriented people, especially when faced with a crisis.

- Major incidents are rare; therefore, few employees have the experience and expertise to conduct a methodical investigation.

- People assume that the effort to find the cause will be used against the company in court later. In fact, victims are more likely to seek legal action if they perceive an inadequate reaction by the organization responsible.

Options for implementing a proper investigation:

- Employ an engineering firm with experience in root cause analysis. Detail a group of knowledgeable employees to participate in learning the technique and to provide support.

- Bring in process specialists from another facility or division who are not vested in the problem.

- Train a plant safety team to conduct a root cause analysis of routine accidents.

- The management must communicate that correcting the root causes is their top priority.
  - Employees should understand that it is in their vital interest to prevent reoccurrences.
  - People should not be withholding information for fear of the consequences.

- Involve upper management or directors in the final review and in the implementation of the corrective action. Verify that the measures will eliminate all the root causes of the incident and any other serious safety concerns that are discovered. Audit the unit to ensure ongoing compliance. There should be zero tolerance for backsliding at any level.

- Senior management must get its own house in order before dealing with actions of the workforce.
  - If pressure to meet production targets or reduce costs was the root cause, then this should be admitted and addressed.
  - The incentive and compensation of managers should not encourage them to condone lax practices and shortcuts.

- The management should ensure that any truly negligent or dangerous behavior is addressed, whether it results in an accident or not.
  When bad practices are routine and widespread across the organization, it makes no sense to punish the minority who happened to be
  performing them when the incident occurred.

Major events rarely have a single root cause. Situations triggered by a single factor are generally known as accidents. These may be serious problems,
even causing loss of life. But, they are not disasters. To generate a true disaster,
there are often multiple failures in the safeguards, which could have contained
mistakes.

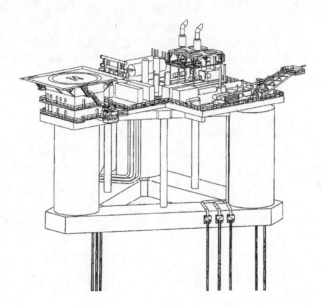

In the late 1980s, a large fraction of North Sea gas and oil production off
the coast of Scotland was processed through the Piper Alpha platform. This
massive facility was anchored in the middle of deep-water oil fields, where
it consolidated the output of nearby wells and pumped the product ashore.
The equipment was designed for crude oil, but was later upgraded to also
accommodate natural gas.

On July 6, 1988, one of platform's two large gas compressors was down for
a scheduled overhaul. Because of inadequate procedures, the equipment was
not locked out in such a way that would prevent it from being remotely activated from the control room. During the night shift, the primary compressor
failed and could not be restarted. The operations staff rushed to reactivate
the off-line unit, on the mistaken assumption that the maintenance work had
been completed. In fact, the housing had not been resealed. The resulting gas
leak caused an explosion, which immediately killed two workers and started
a localized fire.

At that point, a serious accident should have been contained by the automated fire suppression equipment. A massive volume of seawater would have extinguished the fire before it could spread. However, the pump system was in manual mode because divers had earlier been working beneath the platform. In that configuration, the pumps required activation from a control room. Due to a design fault, that control room was not adequately isolated from the effects of the initial explosion in the adjacent compressor room. The firewall was designed to deal with the hazards of crude oil fires. It was not adequate to protect the occupants and equipment from natural gas explosions, which are much more destructive.

In the resulting cascade of events, the fire spread throughout the platform. The flames were fed by additional gas and oil that were forced to the platform from huge pressurized pipes. The intense blaze destroyed the facility, including the crew quarters above the production modules. The dense smoke hampered a helicopter rescue and the fire prevented survivors from descending through the production area to the lifeboats. Only 62 of the 229 people on the Piper Alpha survived the night. The United Kingdom's national oil production was significantly reduced for over a year.

Critical Questions: What could go wrong with our process? How could these potential events affect operations, people, and the end users? Would this disaster be immediately obvious or would it only become apparent afterwards? Are we well equipped to deal with the results?

Summary

- Process equipment will eventually suffer component failure, which will affect output and quality. The worst case problems are subtle and will not be immediately obvious to those running the process.
- No process decisions can be made in isolation from the downstream functions that depend on product performance.
- Give the operators the training and the incentive to watch carefully for changes in the process, even if the product continues to meet specifications. Delegate the authority to take action.
- Products that are not run frequently are often difficult to process due to operator unfamiliarity and unfavorable equipment conditions.
- Seek the right balance of flexibility versus efficiency in the process equipment to suit the company's business model.

- Be aware of what critical process information is known and what is unknown. Beware of assumptions. Ask a lot of questions. Confirm and document the answers.
- Decide what the operation's top priority should be. Do not try to implement initiatives that are at cross-purposes to the overriding goal.

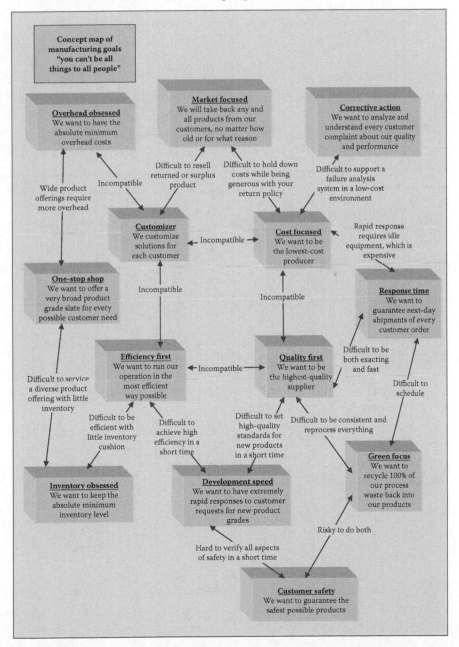

12

Controlling Raw Materials

Which raw materials offer the optimal balance between cost and performance? How do you control and monitor feedstocks to ensure consistent performance? How do we specify raw material attributes and qualify second sources?

> Quality in a product or service is not what the supplier puts in. It is what the customer gets out and is willing to pay for.
>
> **Peter Drucker**

In the 1970s, a packaging company had been buying a particular grade of polyethylene from a large petrochemical vendor for several years. The polymer acted as a hot-melt glue to hold two substrates (paper, foil, or plastic film) permanently together. In this process, a thin layer of polyethylene is extruded into a nip point where all the materials are squeezed together by a large set of rollers. This operation worked well until one day when the finished product failed to meet a particular quality test. The bond strength between the layers was not up to specification.

Production was halted and engineers investigated the situation. The surface properties were scrutinized for contaminants that could impact the adhesion strength. The laminator was inspected for damage and wear. New and old polyethylene batches were compared to one another in a series of laboratory tests and production trials. The vendor was contacted for answers because the old batches worked better than the newest ones.

Eventually, the embarrassed resin supplier admitted to causing the problem. It seems that the supplier had "improved" its resin in the course of a major reactor upgrade. Previously, this grade of polymer was considered to be very sloppy. It was composed of a wide variety of different molecular weight strands of polyethylene molecules. Every pellet of resin was a mix of polymer chains ranging from low molecular weight waxes up to strands that were millions of links long.

By improving its polymerization process, the supplier was able to greatly reduce the variation around the target, producing a more consistent product. In doing so, it kept the same grade designation and data sheet, despite the drastic changes in the molecular architecture.

It turns out that the original version, with its broad molecular weight distribution, was ideal for an extrusion lamination process. The supplier's customer had selected it after evaluating a large number of alternatives.

It ran well in the process and had the most consistent adhesion strength across a range of applications. However, no one fully understood which material properties were important for the application. And no one anticipated that the supplier would undertake such a radical "process improvement."

The resin vendor was tired of defending its "sloppy" product against complaints from customers, who were using the material in completely different applications. The vendor upgraded the catalyst system, making the false assumption that the satisfied customers would remain satisfied.

The story highlights the nightmare scenario for any engineered material operation. If some aspect of a unique raw material changes without warning, then your product may fail in unexpected ways. In this example, the weak bond strength during lamination was immediately detected by quality control (QC) test results. A worse situation would have occurred if the deficiency had not been found until the products had been delivered to the customers. *It is critical to know which feedstock attributes control process functionality and product performance for your customers. There is nothing more dangerous than an unexpected new twist on a process feedstock.*

Ideally, the victims of an invalid assumption crisis will discover something from the experience. The packaging company learned which properties of polyethylene were important for the extrusion lamination process. It developed reliable tests to evaluate incoming batches and to screen new candidate materials. The resin supplier found an end-user application that could appreciate the wide molecular weight distribution that was considered deficient by most customers.

See the "concept map of raw materials" at the end of the chapter for a visual understanding of how suppliers, feedstocks, and raw material specifications fit into a process-dependent environment.

> *Critical Questions*: What factors could vary in our raw materials resulting in serious fitness-for-use (FFU) problems for end users? How can we detect those changes? Is our span of knowledge wide enough to understand these issues?

Material Guidelines in the Development Phase

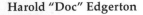

That's the nature of research—you don't know what in hell you're doing.

Harold "Doc" Edgerton

Thousands of years ago, our ancestors first learned to ferment agricultural products into alcoholic beverages. Ingredients such as ripe fruit, grain, and honey could be transformed into beverages, which had a mysterious and addictive effect on the human brain. Perishable calories could be stored for months, providing valuable energy during the lean times. With centuries of experimentation, a range of different recipes and process techniques were developed to produce ethanol. Some scientists argue that this quest may have prompted mankind's shift from a hunter/gatherer lifestyle to a more settled agricultural existence. Most agree that the development of clay pots and amphora was closely tied to the production, storage, and transportation of alcoholic products.

While this was underway, early humans employed animal parts such as bladders and stomachs as a means to carry water and other liquids on long journeys. Following the domestication of cattle, milk became available as a nutritious beverage. When milk was carried in a container fashioned from a calf's stomach, a magical transformation occurred. A natural complex of stomach enzymes called rennet transforms the milk into cheese curds and whey. This discovery allowed people to produce and store an energy-rich food that could last much longer than milk. Over the years, this simple product has proliferated into thousands of different types of cheese. A large portion of these cheeses are still made with rennet extracted from calves' stomachs. Most of the differences between the varieties are linked to process factors.

RULE #42 KEEP YOUR EYES OPEN. MATERIAL INTERACTIONS ARE OFTEN DISCOVERED BY ACCIDENT

Curious people wonder how to take advantage of the unusual material properties and process interactions that they observe. Over time, experiments and continuous improvements transform curiosities into materials that suit a range of needs.

	SPODE & COPELAND SPODE, SON &
SPODE	COPELAND
1780-1790	1797-1820

1800-1820	1805-1830

| 1805-1833 | 1815-1830 |

In the eighteenth century, English potters were struggling to duplicate the fine porcelain ware made in China and Germany. Their raw materials, refining methods, and processing techniques were seriously deficient compared to the Asian technology. In the early 1790s, Josiah Spode built on the earlier work of Thomas Frye and perfected a technique for making fine porcelain products.

In reading the accounts of a Jesuit missionary, Frye was puzzled and inspired by a quote attributed to a Chinese merchant. This man symbolically referred to kaolin clay as "the bones inside the porcelain body." This was just a metaphor. Traditionally, the Chinese never incorporated bone in any form. Mistaking the reference for a literal formulary guideline, Frye incorporated the ash of cattle bones into his ingredients and found promising results. He patented the concept, but failed to achieve commercial success.

Spode experimented with Frye's ideas for several years, adjusting both the formulation and the process. The result caused a sensation in the marketplace and it was soon copied by other producers. Bone china is whiter, more translucent, and stronger than porcelain that incorporates only the kaolin clay and powdered rock of the conventional Chinese approach. Spode's discovery increased the utility of the porcelain technology by creating more process options to solve customer problems.

General ideas for raw material qualification:

- Try to determine why the promising samples bring the benefits that you prefer. Understand the failure mechanism of unsuccessful candidates. Confirm and verify your results with similar materials and different lots. Minor components, additives, and contaminants invariably confuse the data, especially if the vendor does not certify their presence.

- Record everything. Keep samples, especially the defects. Major production problems are often encountered and ignored as minor issues during the development phase, especially if the team is under pressure to qualify alternate feedstocks quickly.

- Information about every factor that affects the process is crucial for the makers of engineered materials. Some of this knowledge is held by your suppliers. Cultivate relationships to mine this experience.

- Never close off communication channels by demanding that vendors should never change a thing about their product or process. This will discourage them from telling you about things that are going on in their operation.

- Resist the urge to narrow the choices down to one material too early in the project. Try to keep several options alive, in case the preferred solution develops problems. The second-choice material may eventually become the backup in the event of supply problems with the best material.

- Choose to work with suppliers that are appropriate for your needs. Avoid commodity suppliers when the precision and repeatability of a specialty product are required. Never commit to a material based solely on a favorable price quote, if its long-term cost structure is highly dependent on spot market conditions.

- Show the suppliers your failure samples and the drawbacks of their offerings for your process. Graphically demonstrate what you love and hate in the alternates. The vocabulary used across different companies and industries may not accurately describe your requirements as well as tangible examples do.

- Make sure that your application is one that vendors want to support and service. Are you too small, too large, too risky, or outside their comfort zone? Excessive secrecy at the outset may mask these concerns.

- Do you feel comfortable and have experience with these vendors? Do they understand applications such as this one? Are they aware of all the requirements and are they willing to accept the responsibility of meeting them?

- Are the candidate's benefits really worthwhile? A researcher can easily become enthralled with the interesting properties that a new material brings to an application. However, if customers find no value in those attributes, then the effort will be wasted.

Finding alternate materials for existing applications:

- List the required attributes, the aspects that need to change, and the factors you want to avoid.

- Why are different materials really needed? Research the incumbent materials and their benefits and limitations for your process and end uses. Study the records of what has worked on this equipment in the past and what has failed. Understand the reasons behind its success or failure.

- Collect as much information about alternatives as possible. Probe the suppliers for their knowledge. Be open about your needs and the perceived shortcomings of the alternatives.

- Is a new material needed to accommodate new process conditions? Or is it required to give a different end-use performance? The end result may be the same, but the two situations are very different from a development and qualification standpoint.

- In what ways can your process be reconfigured to take advantage of alternative materials? Decide if the current equipment constraints should limit your scope or if they could be changed.

- Keep the total process economics picture in mind. If a dramatically different feedstock cannot pay for itself in better value for the end product, then avoid it.

Finding materials for new applications:

- Cast a wide net in the search. Never let the incumbent materials limit your thinking. All types of technologies should be investigated, rather than just versions of the existing raw materials.
- Be honest and upfront with the vendors regarding volume and price requirements. Do not exaggerate your potential in order to induce them to customize a product to your needs. It is not a good basis for a long-term supplier relationship.
- Look for ways compare options fairly and consistently. It is difficult to evaluate different technologies directly. Some are easy to experiment with; others are less convenient for researchers. This can lead to a limited scope of innovation unless you procure better tools.
- First, compare different families of raw materials. Evaluate the generic properties of the competing groups to the potential process economics. Choose the most appropriate class and then evaluate the assorted choices from all the materials in the selected category.

Early civilizations used copper to make their first metallic tools. While copper was easy to process, it had its drawbacks, such as lack of hardness. We now know that this problem can be overcome by alloying copper with secondary metals. One of the first choices was arsenic, which tended to be found in conjunction with copper ores. However, this poisonous ingredient causes serious occupational health issues.

Early metalworking innovators overcame the arsenic side effects by switching raw material suppliers and using tin as the alloy component in their bronze. However, tin rarely occurs in close proximity to copper ore deposits.

So, long-distance trading was often required to bring the two components together in one production site.

A material may be perfect for the application, but if it brings stringent regulatory baggage, then its commercial use may be problematic. Before going too far, ensure that the selected material is really suitable for the application, in all respects. Researchers may be comfortable in working with hazardous materials in a laboratory setting, but the company may be less willing to commercialize them.

Ensuring material suitability:

- Does it meet the regulatory requirements of the customers, governments, and industry standards?
- Does the use of a raw material dictate special transportation requirements for your finished product? Must it be specially labeled as a hazardous material or is it restricted from air shipment? Is special packaging required to ship the material?
- Are the potential raw materials interesting to drug dealers or terrorists?
- What about foreign regulatory requirements? Even though a product may be deemed safe in your primary market, it may not be acceptable in all other countries. Are there patent restrictions that prevent its sale or use in those other jurisdictions?
- Even if the material complies with today's laws and regulations, what about the future? Are environmental or consumer safety groups lobbying to have it banned?
- What about disposal limitations? Will the use of one material make the end product hard to dispose of? Will it prevent recycling of the product?
- Is there emotional baggage with the material that is incompatible with your company's values? Is the supplier using any morally questionable practices that will reflect badly on your company's image?— hurting some endangered species; sweatshop conditions, convict labor, or child labor?
- Will the material cause your production site to have unacceptable air/water/waste emissions?
- Study the material safety data sheet (MSDS) to ensure that it does not dictate special handling procedures or precautions that the production plant or the warehouse is not prepared to handle.
- Will the product require special labeling for sale in locations such as California that have greater restrictive regulations to protect consumers? Will those regional requirements extend to all of your production because you need to minimize the number of discrete inventory items? What new requirements are on the horizon?

- Is the product available in a form compatible with your material-handling systems?

The many roadblocks just mentioned can be overcome, if the need for the product is sufficiently compelling. Plants can add engineering controls and emission scrubbers. Personnel can be trained to use new protective equipment and procedures. Special labeling and shipping containers are routinely employed to handle regulated materials. The key question is whether these measures are confined to a small subset of your operation or if your customers must accommodate them as well.

There can be tremendous resistance from the organization in implementing new materials that require stringent safety measures. It is far easier to evaluate the impact of a questionable material at the outset, rather than reformulate the product after the various interest groups in the company have rejected it. In large organizations, these interest groups may not proactively puzzle over every potential component in the development pipeline. Ask questions and discuss alternatives if a hazardous material has positive results early in a development project.

Material Rules for Industrial Scale-up

A rule to live by: I won't use anything I can't explain in five minutes.

Philip Crosby

June 24, 1947.   G. M. EISENBERG ET AL   **2,422,777**

METHOD FOR PRODUCING PENICILLIN BY MOLD CULTURE

Filed Dec. 6, 1944

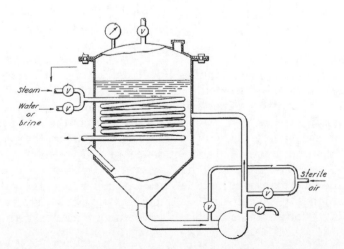

When chronicling the history of penicillin, it is impossible to omit the story of Alexander Fleming. In 1928, he returned from a holiday and discovered that an airborne mold had colonized a Petri dish on his workbench. Surprisingly, the *Staphylococcus* bacterium that was growing in the culture had been killed in a circle around the fungus colony. Intrigued, Fleming cultured the mold and speculated that the active agent could have therapeutic applications. Apparently, this minimal effort is all that is required to win a Nobel Prize, once someone else does the development work to commercialize the solution.

In the following decade, little progress was made toward the commercial production of a functional antibiotic. Tiny quantities of penicillin were produced on a small-scale process. The few experiments conducted on laboratory animals demonstrated that the new material was not toxic. Only by 1941 was enough antibiotic available for clinical trials on human patients.

The process limitation was that the Fleming strain of mold would only grow on the surface of culture media. This growth mechanism was not suitable for industrial-scale production. From the mold's point of view, it had other priorities in addition to producing an antibiotic. In fact, the fungus will only generate penicillin as a defensive measure when its growth cycle is inhibited by stress. Early researchers did not know that closely related *Penicillium* mold strains would produce different versions of the penicillin molecule, such as 2-pentenylpenicillin versus benzylpenicillin. These are not interchangeable. Each type is unique in treating the spectrum of the disease bacteria. This sample variability made it difficult for doctors to predict the optimal dose and the course of treatment.

In 1942, an intense effort was launched to research the penicillin process technology at various locations, including the USDA laboratory in Peoria, Illinois. Located in the leading region for corn (maize) production and processing, the laboratory chose a readily available corn by-product mixed with lactose as the growth medium. It screened thousands of mold samples to find the ones that produced antibiotic toxins and could grow while submerged in an aerated fermentation tank.

By manipulating and mutating the mold strains, optimizing the growth media, finding the most favorable conditions, and scaling up the tank sizes, it was possible to increase production exponentially in just 2 years. The yields were found to be highly dependent on the temperature, pH, lysine level, phosphate level, and oxygen availability. By the end of the war, production was taking place in large 40,000 L vessels and 80%–90% of the batches met or exceeded specifications. This performance was only possible by sourcing new mold strains and growth media.

One of the most ambitious material scale-ups in history was conducted by Howard Hughes in the construction of his "Spruce Goose" seaplane in the 1940s. This 200 ton aircraft still holds the record for the longest wingspan.

The Duramold plywood process was a clever method for making smooth, light composite structures, especially when aluminum was in short supply. Hughes had previously used it to fabricate prototype aircraft, but nothing on this scale had ever been attempted. Thin layers of wood veneer were layered into curved mold forms with an alternating grain direction. The wood was bonded together with liquid resins under heat and pressure. The resulting structural surfaces could take almost any curvature needed for airframe components.

Howard Hughes' flew his creation in 1947, primarily to answer his critics. The plywood laminate performed to design specifications and has held its shape to this day in a museum setting. The wide-scale aviation applications of composite technology eluded material suppliers for 50 years. Recently, composite airframes have started to displace aluminum in order to reduce weight and reduce fuel consumption.

Ideally, the small-scale development phase identifies a range of material options. Evaluate candidate raw materials to determine how a batch-to-batch variation affects the production rate, the scrap generation, the changeover, and the cleanout time. Are the materials sufficiently consistent that the product can always be run at a single set of "cookbook" conditions? Or does the process have to be constantly adjusted to accommodate each batch of raw material?

When changing from one raw material batch to another during a run, could anything bad happen? Will the process be upset by sudden swings in the material properties? Will there be radical process variations in the pressure, temperature, reaction rate, process stability, emissions, and process yield? These situations are most frequent when processing unrefined minerals, agricultural feedstocks, and external recycle streams.

RULE #43 THE OPTIMAL RAW MATERIAL IN DEVELOPMENT ≠ THE OPTIMAL RAW MATERIAL FOR PRODUCTION

When new products are developed or reformulated, the biggest hurdle to commercialization is in implementing a full-scale production process. Process conditions, equipment parameters, operating procedures, test methods, quality guidelines, safety issues, and environmental concerns must all be optimized for commercial production. It is often necessary to change feedstocks as part of the process optimization. Raw materials that made great prototypes are often discarded for more process-friendly alternatives.

Process engineers can intentionally introduce a significantly different raw material batch during a closely monitored production run. This allows the careful monitoring of the feedstock transitions, with the appropriate QC tests. Product samples can be taken for later evaluation of the downstream effects. Evaluate the FFU performance on samples from before and during the changeover. If the worst case batch-to-batch difference gives a stable transition, then the lesser differences should be no problem. Remember that some processes and applications may have more than one kind of worst case situation. Never assume that stable transitions for your process will always be transparent to the customers.

If batch change effects are detected, then several different strategies can be employed:

- It may be possible to purchase only special, certified batches from the vendor. This requires that the suppliers have some outlet for the materials that cannot be fed into your equipment. Some vendors are anxious to customize their offerings to suit end-user needs, especially if it locks the customer into using their product.

- Batches of like material can be staged together and run in sequence through the process. Manufacturing discipline is needed to ensure that all the containers of a batch are stored and used together. Someone needs to know which lots are similar and can be inserted into the production scheme without upsetting the process. Pallets of raw materials cannot simply be randomly shelved in rack spaces. Forklift operators find it convenient to access the nearest pallet, rather than searching for a specific lot number.

- Add equipment and procedures to deal with the problematic factors. Homogenization, drying, grinding, or buffering may be needed to minimize the impact of lot changes.

- When dealing with minerals taken directly from the ground, processors must rely on their own analysis, record keeping, and experience. Some storage mechanism must be introduced to segregate, store, measure, homogenize, or transform the different batches to accommodate natural variability.

Dealing with raw material variations within a batch:

- What happens to a container that has been sampled or partially used and then saved for the next run? Does the material oxidize, dry out, absorb water from the air, or change in any way? Is there a special procedure describing how these partial containers are resealed or repackaged for storage? Are the lot number and product identification always preserved during storage? Is there a system to use these partial containers first in the next run opportunity, or do they collect in some dark corner?

- Can storage conditions within the inventory affect the properties due to temperature or moisture variations from one location to another? If the entire lot is not shipped together, can variations from one shipping mode to another cause problems? Do you need to establish shelf-life limits and a system to use the oldest stock first?

- Is there any tendency for a material to settle or segregate in a container or a tank? An alert engineer can monitor the process effects of the first material dispensed from a given tank, silo, bulk truck, or railcar compared to the middle and end material. He or she may also do aging tests to simulate the impact of leaving the material in the tank for long periods of time. Segregation in shipment may depend on the amount of vibration that the container is subjected to on each trip. It is often possible to model these effects on smaller containers mounted on shaker tables.

- Being physically present to watch all phases of an operation is vital to a full understanding of the sources of variation.

Material Considerations During the Specification Writing Step

Colours, sounds, temperatures, pressures, spaces, times, and so forth, are connected with one another in manifold ways; and with them are associated dispositions of mind, feelings, and volitions.

Ernst Mach

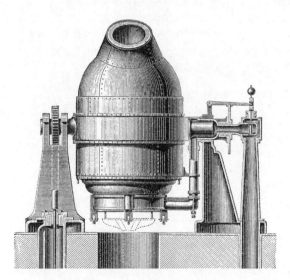

In the 1850s, Henry Bessemer demonstrated a new process for converting brittle, high-carbon cast iron into malleable steel. The Bessemer converter forced air through molten iron to oxidize the carbon content. Based on the demonstrated results with medium-sized crucibles, several British companies licensed the technology. The process was compelling because the incumbent steel-making technology required weeks of heating in a small crucible. The Bessemer technique was both more energy efficient and worked on a much larger scale.

However, once the licensees had built full-scale production equipment, they found that their end product had poor properties and lacked the expected commercial value. Bessemer was compelled to repay some license fees. It turned out that his successful experiments had been conducted with iron smelted from premium Swedish ore. Steel made from most English ore deposits failed to work as well. The local raw material's high phosphorus content had a negative impact on the steel properties. The process also oxidized most of the carbon content, while steel needs a specific level of carbon (0.2%–1%, depending on the grade) to obtain optimal properties. By the 1860s, solutions had been found. The addition of chalk stripped out the phosphorus into a slag layer, which floated to the top. Adding minerals such as Spiegeleisen improved the properties by introducing controlled amounts of manganese and carbon.

In one sense, Bessemer was fortunate when he selected a low phosphorus material for his first trials, allowing the concept to be proven. However, he did not evaluate a wide enough range of raw materials to fully understand the process requirements.

The Bessemer process had widespread success supplying steel tracks for railroads, which were expanding during this period. The open-hearth process replaced it as the dominant steel technology later in the century. Open-hearth furnaces allowed more flexibility to monitor and adjust the composition of each batch. This permitted the mills to control the product properties for each application.

When drafting a new specification, there is a tendency to look up a similar material and then cut and paste its attributes, tests, and specification limits. This is a good starting point, but it is crucial to identify the critical properties that control process stability and end-use performance for each situation.

Work with the supplier to ensure that your requirements are within their production capability. Perform "round-robin" evaluations of test methods so that both parties agree on the results and the error range. Calibration standards and techniques should also be harmonized. Do not set the specification limits unnecessarily tightly unless that is profoundly important to the process. Tight specification ranges are problematic if the vendor keeps the price high to meet unrealistic requirements. Only essential properties should be tested with each delivery. When too many attributes are evaluated, the odds increase that at least one result will be out of bounds. This can lead to a routine practice of waiving negative results, which undermines the organization's respect for the specifications.

Organizations can waste resources by conducting raw material tests that have no bearing on performance. Most vendors publish specifications for their products, which list the high and low allowable values on selected properties. This does not mean that you as a customer are compelled to perform the same tests to verify that each delivery is "in-spec."

RULE #38 IDENTIFY WHICH RAW MATERIAL PROPERTIES MUST BE ROUTINELY MEASURED AND CONTROLLED

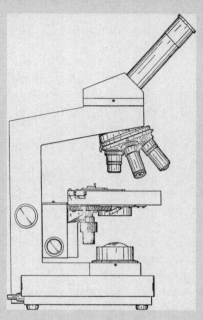

The development project must identify the beneficial and harmful feedstock properties. Create raw material specifications that totally delineate the process requirements for each end-use application, along with test procedures for incoming batches.

In fact, those specifications may have no impact on the fitness of your product.

When dealing with a trusted vendor, many companies rely on a certificate of compliance (CofC), rather than testing the product themselves. This is especially useful when you lack the test equipment necessary for all the important parameters. This approach may not be prudent if product failures pose some significant danger to end users.

During the summer of 2010, Kellogg's Cereal Company was compelled to recall 28 million boxes of breakfast cereal, due to customer complaints. According to some consumers, the product had an objectionable odor and taste. The chemical 2-methylnaphthalene was identified as the source of the problem. This contaminant was believed to originate in the liner packaging.

The liner serves to protect the product, keeping it fresh and dry during shipment and in storage. In this case, it had the opposite effect. Normally, the operators running the packaging machines would be expected to notice when one batch of material has an unusual odor or other troubling characteristics. Any deviation from normal behavior should elicit comment from the people who handle it on a daily basis.

However, this safety mechanism can be short-circuited by commercial factors. When the company finds a much less expensive raw material, there can be a tremendous incentive to make it work. Production workers may be told to use the new vendor's product, no matter what extra effort is required. Operators are often resentful when this occurs, especially when the new option runs poorly on their equipment. Management may ignore the workforce's observations, believing that complaints spring from a negative attitude. When that happens, an important QC feature is eliminated.

Do not limit the specification to standard tests:

- Keep retained samples of materials for visual, odor, and tactile comparisons with the new batches. Make sure that the material-handling people will check the product codes and the appearance of all containers that they receive into inventory. Even a certified vendor, freight company, or distributor will make mistakes and deliver the wrong material, or damage the protective packaging.

- Everyone needs a means of catching these blunders before they crash the process. Be vigilant if the supplier packs incompatible materials in similar packaging to other applications within your operation. Do not allow temporary or new employees to load the wrong material into the process.

- Keep "no-go" retained materials that should be specifically excluded from the process. These may be examples of specific raw material defects that are not acceptable for the process.

- Be careful of containers that have been reused, relabeled, or repackaged. Watch for shipments with containers from many more lots or batches than the normal orders. Test each discrete batch. Do not assume that one randomly selected sample will give a snapshot of the entire shipment.

Qualifying Second-Source Raw Materials

Every time man makes a new experiment he always learns more. He cannot learn less. He may learn that what he thought to be true is not true.

Buckminster Fuller

Like many American food producers, a major retail donut chain felt pressure to change its formulations to eliminate trans-fatty acids from its products. This food ingredient has been linked to an increased risk of heart disease. Consumer brands are eager to label their offerings as free of trans-fats.

Alternative fats and oils are widely available, often at a higher price than the incumbent shortenings. However, trans-fats were originally selected to optimize the properties in the fast-food marketplace. Loyal consumers crave the taste and texture of these unhealthy choices. Trans-fats are highly useful to food processors because they resist rancidity, extending the shelf life of baked goods and dairy products. Frequent heating and cooling does not induce the thermal degradation of products with trans-fats as quickly as that of natural products. In short, trans-fats are engineered to be more process friendly than healthy oils, so eliminating them is not a painless change.

The donut chain experimented with alternatives, trying to retain the same taste, appearance, and texture. This was difficult because alternate fats process differently, resulting in undesirable product properties. For instance, one raw material gave the donut a very smooth outer crust. The icing slid off the edges because it did not adhere properly to the slick surface. Other companies have had similar problems in reformulating their products without sacrificing consumer preferences.

Good businesspeople always look for alternate sources for their important raw materials. Suppliers' plants occasionally blow up or shut down due to earthquakes, hurricanes, or other unpredictable issues. Tariff duties rise and fall, as do currency exchange rates. Multiple sources of supply allow for more leverage in price negotiations and more flexibility in dealing with supply disruptions.

When looking for alternate materials, first consider how difficult it was to qualify the incumbent material. Was a special grade developed exclusively for your needs? Is there a key ingredient that is crucial to your process? Does it have exacting specifications to achieve FFU goals? If any of these are the case, then a well-defined project is called for.

When evaluating new sources, the developers must discard assumptions and start with the basics. Evaluate which attribute is crucial to performance. The fact that the current raw material has no problems with lot-to-lot variation is no guarantee that alternatives will have the same consistency. Exercise caution with cheap "wide spec" materials or products derived from sources that are completely different from the incumbent source.

Never trust salespeople who claim that their product is an exact copy of the incumbent. Just because other industries use them interchangeably, there is no guarantee that your process will be so forgiving. Make sure that everyone is talking the same language and that all the current supplier's positives and negatives are understood. Use this opportunity to better learn which properties influence your process and the end-user FFU. Find out what aspects of the supplier's process create adverse outcomes. Compare the properties of the prospective options with the existing material. Ask to see a history of vendor QC tests. What is their quality plan and do they really follow it? Determine what will be done with nonconforming batches, especially if you will be the only customer. Both parties need to agree on test methods and return policies for nonconforming batches.

Also, ask the incumbent vendors the same questions for comparison purposes. Is their production equipment similar or are the processes completely different? What other products or materials do they run on the same equipment? Will any prospective trace contaminants cause problems for your process?

Think about the endgame for this qualification. The ideal situation is to have two or more qualified sources available for your ongoing needs. Do you have enough order volume to support both sources? This might not be important if the materials are standard items for both vendors. You can go a long time without ordering from one of the suppliers. But, products that are customized for your needs will not be available in a crunch if they are not purchased regularly.

Joseph Henry was an electromagnet pioneer who created the most power-ful magnets of the 1830s. His technology allowed the iron works at Crown Point, New York, to significantly improve the feedstock consistency of their smelting operation. Initially, iron particles were separated from rock and sand by an inefficient water-washing process.

Henry improved the process efficiency with a rotary magnetic ore sep-arator that attracted iron particles out of the ore stream. As the wheel turned, the pure iron was brushed off and collected. The system mimicked the cotton-ginning process documented in Chapter 2. These improvements allowed the smelter to better accommodate the natural variability of the ore beds.

Implementing raw material changes:

Follow a defined plan to implement the routine production of new feed-stocks. Try to get material from several different batches, lots, production lines, or campaigns. If possible, work with the vendor to select batches that are as different from each other as possible in terms of the QC results or the production dates. The sample size should be sufficient to allow the new material to completely displace the old material in your process and allow it to settle down to new conditions. Each discreet batch should be thoroughly tested by all appropriate QC tests that would normally be employed on this type of feedstock or incoming part.

Often, a single lot will be sufficient to qualify a material if you know the process to be insensitive to lot-to-lot variation. Such confidence only comes from having evaluated materials from different vendors and understanding their impact on your product. Carefully record all process conditions while running a standard batch of the old material. Ensure that the line output is completely within the desired specification limits and is stable. Make sure that all equipment is in good operating order.

Each sample batch should be of a sufficient size to allow for an exhaustive test in simulated use and in real-world applications. It should be labeled and stored in such a way that it will not be mistaken for the standard product and sold to an end user before it is fully qualified. Look for any property differences between the samples made from different lots. Are the effects more significant than the current material's lot-to-lot variation? Is it possible to distinguish which source was used when doing blind tests? Will the customers be able to tell?

Document the specific situations where the material has and has not been qualified. The number of exclusions may decrease as the plant gains more experience in running with it. As positive results come in from the field tests, order larger batches and use them in standard production. Continue to monitor for any adverse results on the process, the QC results, and the performance at the end-use applications. Look for unusual build-up, corrosion, or wear on the process components after long runs.

Summary

- Determine which attributes of your raw materials influence the process consistency, the product properties, and the FFU performance.
- Cultivate free and open communication with your vendors. You will discourage information transfer if you are never open to any changes on the vendor's part.
- Evaluate batch-to-batch feedstock variability during development. Do not wait until the product is in production to find that a raw material is too inconsistent for the application.
- Consider all the impacts of a raw material on your product over its entire life span.
- Spend time selecting tollgate tests for incoming materials that will predict their performance in your process. Preferably, your suppliers should be using the same tests. Ideally, they will do the testing and provide you with the results.
- Keep retained samples of the good and the bad examples.
- Develop methods to detect blunders and shipping errors in the critical feedstocks.
- Qualifying the alternative sources may require as much effort as the original development.

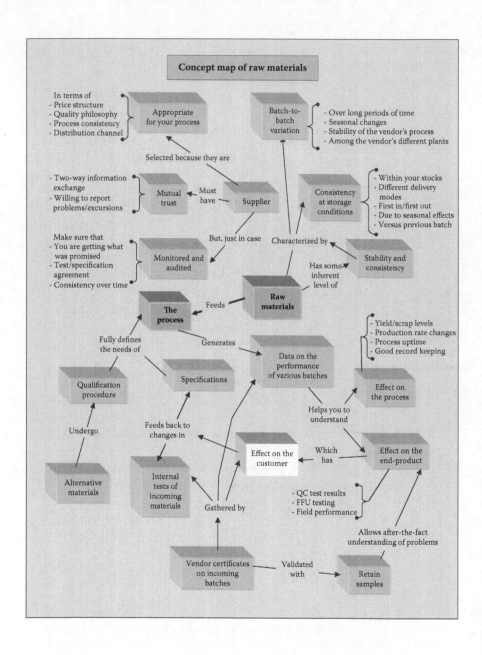

Concept map of raw materials

In terms of
- Price structure
- Quality philosophy
- Process consistency
- Distribution channel

Appropriate for your process

Batch-to-batch variation
- Over long periods of time
- Seasonal changes
- Stability of the vendor's process
- Among the vendor's different plants

Selected because they are

- Two-way information exchange
- Willing to report problems/excursions

Mutual trust ← Must have — Supplier

Consistency at storage conditions
- Within your stocks
- Different delivery modes
- First in/first out
- Due to seasonal effects
- Versus previous batch

Make sure that
- You are getting what was promised
- Test/specification agreement
- Consistency over time

But, just in case

Monitored and audited

Characterized by

Stability and consistency

Has some inherent level of

Raw materials

The process ← Feeds —

Fully defines the needs of

Generates

Data on the performance of various batches

- Yield/scrap levels
- Production rate changes
- Process uptime
- Good record keeping

Qualification procedure

Specifications

Effect on the process

Undergo

Feeds back to changes in

Helps you to understand

Alternative materials

Internal tests of incoming materials

Effect on the customer ← Which has — Effect on the end-product

Gathered by

- QC test results
- FFU testing
- Field performance

Allows after-the-fact understanding of problems

Vendor certificates on incoming batches — Validated with → Retain samples

13

Complex Manufacturing Situations

How do we optimize production schemes containing a series of dynamic process steps to ensure consistency and fitness-for-use? How do we control product performance on a process that we do not own? What should we look for in a production partner to balance quality with flexibility?

> It is not a question of how well each process works, the question is how well they all work together.
>
> **Lloyd Dobens**

In the 1970s, an American rubber company developed a weather-resistant polymer—an alloy of acrylonitrile, ethylene–propylene rubber, and styrene. This material had improved properties compared to PVC, but it also had its own unique process idiosyncrasies. Premium exterior doors were identified as a market opportunity. No single company could bring this product to the market. The rubber company produced the polymer, which was then extruded into sheet stock by a second firm. A third company thermoformed the sheets into panels, which were assembled into doors by a fourth company. Finally, a national retailer marketed the finished product.

Problems were immediately evident. Mysterious blemishes were found on the door panels, which had not been noticed during the earlier steps in the process chain. Some colors had a high frequency of defects while others were problem free. Occurrences of the problem were random and erratic. Measures were taken to protect the affected surfaces from contamination, marring, and abrasion throughout the processing sequence, yet defects continued to appear. Special inspections were implemented and passed, but still the blemishes appeared days later, as if by magic.

A careful investigation eventually revealed the source of the problem, which was unique to this polymer. Even though the finished surfaces were protected, it was not uncommon for the sheets to be bumped or dinged on the hidden face. No damage was visible, so the semifinished parts passed inspection and they were shipped to the next processer. However, the impacted area carried invisible stress lines. Only after aging did a flaw appear on the opposite side of the panel from where the compression had occurred. The visibility of the damage was enhanced by some pigment colors and was concealed by others.

Resolving this problem was difficult due to the disjointed production process and the many companies involved. Every handoff situation is a potential source of problems. The number of transition points in a process

should always be kept to a minimum, even when one company is doing the entire job. The ideal situation is to have a single production line that runs the process from start to finish. However, this is often impractical. Even if the sequence of production steps must be physically separated, the system should be organized as a seamless whole.

RULE #26 DO NOT SUBOPTIMIZE YOUR PRODUCTION SCHEME WITH AN INAPPROPRIATE THEORY

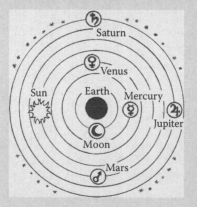

In a well-run organization, the semifinished product flows smoothly from one step to the next, even if it is stored in a warehouse or shipped across the world between steps. Optimizing the consistency, quality, and cost of the end product must be the one overriding goal.

Indications of a smooth-running process:

- In the best-managed situations, the scheduling of an order is done once. It is not repeated at each stage. Each production node is aware of when work will arrive and what production load to expect.
- Quality assurance is done continually. Each unit ensures that its work product is ready for the subsequent steps. The semifinished product should not arrive at a downstream operation with unexpected errors.
- Workers in one department are aware of how their performance will affect downstream operations. Everyone in the sequence works to ensure that a good-quality product comes out at the end.
- Management can see the costs of the production problems, administrative errors, faulty specifications, and special customer requests.

An integrated textile company was in the business of producing colorful bedsheets with printed patterns. The process started with the spinning of thread from raw fibers and ended with the packaging of the folded sheet sets. This technology was well understood by both the management and an experienced workforce. One day in the 1980s, the printing department abruptly started having problems with the dye uptake by the cotton fabric. The cloth was not holding the color as cotton normally should, resulting in a product that could not be sold as a first-quality item. An investigation was launched, gradually working its way back through the plant to the first process step.

In the spinning room, it was discovered that a significant quantity of polyester fibers was contaminating the incoming stream of raw, natural cotton. The equipment that conveyed the fibers to the spinners was designed to blend together a variety of different materials. A broken valve allowed a polyester staple stream from an adjacent process to be drawn into the cotton-spinning operation. What should have been 100% cotton thread was actually a cotton/polyester blend. The dyes that were selected for their adhesion to pure cotton were not suitable for the synthetic fibers.

The broken component was hidden deep inside the process, which was sealed to protect the worker's lungs from airborne cotton dust. The thread's polyester content was not routinely measured, because it was not supposed to be there.

Indications of a poorly organized production scheme:

- Each department considers the job to be done when its work is completed.
- In any given order, there tends to be a wide variation in either the production quantities or the amount of semifinished goods that are left over when the final step is completed. Production planning is not an exact science.
- It is difficult to forecast a ship date until the order is in the finishing department. Its prediction will depend on the pressure to ship quickly.
- Area managers, operators, and engineers are focused on optimizing the output and the yield of their individual operations. There are incentives for sending slightly out-of-spec products downstream.
- Managers find it difficult to estimate their workload in the near future.
- A high percentage of orders are processed on a priority basis due to delays in earlier steps or the need to rerun or rework a defective product.
- Extra costs result from meeting urgent ship dates that should never have been issued.
  - Pressure from sales causes the scheduler to agree to unusually short lead times.
  - Mistakes in estimating the schedule or the availability of materials necessitates a rush job late in the process.
  - Expedited orders have higher defect rates than normal.
  - Significant administrative effort is put into juggling the schedule.

- The plant may have a hard time in calculating the true cost of any given job.
  - The financial impact of errors and inefficiencies cannot be assigned to a specific order.
  - The management knows the cost of everything, but the value of nothing.
  - Small orders for nonstandard items are priced the same as routine items.

Discrete production departments take on their own personality, which springs from the individuals involved and the type of work that they do. Some say they do the best job possible with the variation that comes from elsewhere. Some workers resent mysterious specifications and cryptic requirements dictated by people who do not understand the impact of their requests.

The production schemes become flawed when the management measures each department in isolation. Suboptimization results when each production node is encouraged to maximize its own metrics, rather than the system as a whole. All the intangible quality and fitness-for-use (FFU) concerns are allowed to slip, because they cannot be measured. Under this system, marginal quality is likely to be sent on to the next operation, because that is easier than correcting the problem.

The alternative is to focus holistically on the larger process. True value can be achieved only when the workforce has an incentive to minimize the total cost of completing an order and maximize product functionality. They need to see the actual costs and appreciate how the product works for the end users.

Integrating the process into a seamless system:

- Audit the actual workflow to look at handoffs. Zero is a good target, but it is not always practical. Reorganize the system and cross-train the workforce to eliminate delays caused by "it's not my job" situations. Look for role conflicts where conflicting duties are assigned to a person or a group.
  - The maintenance craftsman who must be called to the line for a set-up or changeover task, but who is too busy dealing with other problems.
  - The forklift driver who has to transport the semifinished materials to the next department, but who is, instead, looking for misplaced inventory.
  - The supervisor who must sign off to release a job, but is in a meeting.

- The only QC inspector on the shift, who is reinspecting a rejected order, in the hope that it will pass the second time.
- Look for sources of yield loss and process inconsistency. Monitor the mass balance of the material in and out of each node. Evaluate the effect of batch-to-batch raw material variation.
- Be prepared to understand every cost associated with each order. This will allow the management to see which special products or process steps are creating inefficiencies. (Warning—will the work needed to record detailed accounting information cause more problems than it fixes?)
  - If the raw material for an order must be expedited, add the extra costs for quick delivery.
  - If extra craftsmen are needed to run the job, then allocate their hourly billing rate to the order. Include the time that they or the equipment sits idle in preparation for the task.
  - Consider the working capital for inventory sitting in the warehouse waiting for this order.
  - Add the shipping costs to get the goods to the customer, especially if production delays dictate the use of express freight to meet the promised delivery dates.
  - Consider product spoilage due to process variability that creates out-of-tolerance material.

Some extra costs will be unavoidable one-time charges. These could include inevitable equipment breakdowns that add extra machine hours and maintenance charges. There is a temptation to subtract these out, because they are unusual events. The trick is to identify the real cost to make the material, so that the opportunities for improvement are obvious. The spending might reveal that this product (or the one before it) is stressful on the equipment, requiring time to return the process to normal.

Involve all personnel in the solution:

- Rotate workers between production functions to experience the entire process.
- Train operators to perform duties that are currently done by specialists. Consider basic maintenance tasks, quality inspection, production reporting, and movement of goods.
- Use special pallets or containers to move products through the process to prevent damage and to reduce handling. Adapt a production scheme that is appropriate for your product.
- Employ special marking to flag unavoidable defects so that they can be removed downstream. For instance, a huge roll of paper may have

print defects deep inside the roll. These should be clearly flagged to indicate which parts should be sliced out later. Known flaws should never require rediscovery in subsequent steps.

- Create separate functions to deal with the recovery of damaged material or the processing of small batches of leftover materials. These special activities should not delay orders or affect the consistency of a shipment. The extra costs will be more visible.

- Identify problems that result from false assumptions. The customers might want an entire order shipped on a specific date to minimize transport costs. However, they will consume the material over the course of months. It could be cheaper to release the order in increments to have a longer run on a single piece of process equipment, rather than to have short runs on multiple machines. The reduced set-up costs and improved consistency might defray the increased freight and delayed payment.

- Feed individual parts to a finishing machine in the same order that they were made on the upstream process, rather than in a random or reverse order. This may help to identify periodic property variations that influence the product quality.

In-Line Process Trains

It is the function of creative man to perceive and to connect the seemingly unconnected.

William Plommer

The production of plastic shopping bags is performed in three discrete unit operations: film extrusion, label printing, and bag sealing/cutting. The maximum output rate of each operation is different. In isolation, the specialist in each operation might purchase the fastest, most efficient equipment for each step.

Instead, it can be beneficial to size and operate each unit at a rate that allows a continuous operation. One extruder, one press, and two bag machines might be colocated in an area. The extruder continuously runs the particular thickness, width, and color of film needed by a large retail chain. As rolls come off the extruder, they might only wait a few minutes before the press operator loads them onto the adjacent printing press. The printed film rolls run through one of the two bag machines to complete the process.

Problems with one unit operation will immediately become evident on the next operation. Scrap film from the converting steps can be recycled back into the extruder. There is no inventory of semifinished material in the warehouse. The entire process is run at the speed of the rate-limiting unit operation. If the press can run twice as fast as the extruder, that is okay. The downtime might allow the press operator to change the print plates or clean the ink pans in preparation for a new order. Or, the press operator might be asked to man the extruder or a bag machine while another operator is eating lunch.

The entire team bears joint responsibility for their work. Each member is vested in the efficient operation of the adjacent machine. No one can suboptimize his or her own function or blame problems on another department. Defective film will not be delivered to the press because problems in an earlier extrusion run went undiscovered.

The most efficient way to run a multistep process is to physically connect the various unit operations into a continuous, in-line process, when possible. Raw materials are fed into one end of the train and the finished product emerges from the other end. Each production step is synchronized to run at exactly the same rate as the others.

In the early 1920s, the U.S. automotive industry was anxious to buy wide widths of high-quality sheet steel to form body panels. However, the hot-strip sheet mills of that time employed an inadequate process. Workmen gripped hot bars of steel with tongs and fed them into massive roller assemblies multiple times, tightening the gap after each pass. The adjustments were made by sight, based on the worker's experience and skill. Each piece was handled as many as 22 times during the process. The semifinished sheets cooled between each step, introducing differences between one piece and the next. Gruesome accidents were common while performing this exhausting work in hot, noisy conditions.

> No two lots of hand-rolled sheets were exactly the same thickness and temper, and it was these variations which the automobile makers complained most loudly about. Lack of uniformity in steel sheets became a serious impediment in stamping body parts, where nothing caused more failures and rejections than variation in gauge and stiffness.*

The rough steel surfaces had a poor finish, even after painting. Due to the width limitations, the body parts often had ugly weld seams. John Butler Tytus recognized these limitations and spent the better part of 3 years and millions of dollars to improve sheet metal production. In place of manual handling, long strips of steel were pulled directly from one station to the next, running thinner and faster with each step. Tytus discovered that perfectly flat rollers were not the ideal means to squeeze hot metal into thin sheets. His train of nip rollers started out with convex rolls to do the initial compression, followed by a series of increasingly flat rolls at each subsequent station. This allowed his new sheet mill to produce a much wider product with a far better surface finish.

Within a few years, uniform sheet metal was flying through continuous rolling machines at speeds of 2000 ft./min. The automotive and appliance industries were transformed by the improvements. Metal stamping systems grew in size to form larger pieces. Car body designs changed to take advantage of the deep drawing ability of the new materials. With surface finish improvements, car bodies became an art medium.

---

* Douglas Alan Fisher, *The Epic of Steel*, Harper & Row, New York, 1963, p. 142.

Today, consumers expect glass windowpanes to be crystal clear and available in any size. However, until the middle of the twentieth century, most windowpanes were produced by hand-blowing molten glass into a cylindrical shape. These tubes were then cut in half, reheated, and pressed flat. The blown-glass process was cost effective, but it was limited in panel size. This is why windows were traditionally assembled from small panes set into wooden frames. The glass had wavy imperfections as a result of process inconsistencies. Plate glass had a much better clarity and was available in larger sizes, but was much more expensive. It was cast as a wide, thick sheet followed by a laborious finishing process of grinding and polishing.

This changed with the introduction of the float glass process in the 1950s. A continuous ribbon of glass is cast onto a bath of molten tin. The liquid surface is perfectly flat, resulting in a flawlessly smooth pane. By manipulating the temperatures of sequential bath zones, melting and then refreezing the surface, a glossy "fire finish" can be achieved.

The British glass producer, Pilkington Brothers, demonstrated the process on a pilot plant scale. This costly stage of development took over a year to produce good product. Their experiments revealed that the tin had to be very pure and it had to be isolated from oxygen to function properly. After overcoming the technical issues, large-scale production equipment was authorized. However, the initial performance was disappointing. Months were lost while process adjustments failed to reproduce the original product of the pilot process.

An investigation revealed that there had been a damaged component in the original pilot plant configuration. This one deviation from the original design made the difference between a high-quality product and unusable scrap. Once the production line was modified to reproduce this process fluke, it too began to turn out top-quality glass. Afterwards, the discontinuous methods for making windowpanes became obsolete. The time needed to identify this factor caused many sleepless nights of financial concern for the owners of this family business.

In 1800, paper was made laboriously, one sheet at a time. The Frenchman Louis-Nicolas Robert sought a better solution. Under the financial backing of his employer, Saint-Léger Didot, Robert was encouraged to develop a continuous process for making paper. After a long series of experiments, he created a successful working model, which demonstrated the validity of his ideas. In 1799, the French Ministry of the Interior rewarded Robert with a patent for his invention.

Despite this recognition, commercial success was difficult in the unsettled French political environment of that period. Didot took the technology to England and sought the backing of Sealy and Henry Fourdrinier. Under their sponsorship, full-scale equipment was developed and put into commercial production without Robert's help. This production technique spread rapidly and was named after the brothers who financed it, rather than its developer. The Fourdrinier process continues to be the primary source of paper in the world today. Modern production lines run continuously to make a very uniform product that can be smoothly converted by large presses and high-speed photocopiers.

Reasons to use in-line processes:

- An in-line process requires less labor because workers are shared between the various steps. Intermediate materials need not be collected, stored, and reoriented into each subsequent process step.
- A problem in one operation is immediately evident in downstream operations. There is less opportunity for defective intermediate materials to build up in inventory before their shortcomings are discovered in later steps. Intermediate quality inspections are reduced.
- The combined process takes up less room in the plant. Eliminating the storage of in-process intermediates frees up space and working capital.
- Product quality may be more uniform because every component is running continuously at the same speed, without the need to ramp each discrete operation up or down.

Some products and product slates are better suited for discontinuous production schemes. It is rarely possible or economical to combine every multi-step process into an in-line operation.

Reasons to avoid production on an in-line process:

- In-line operations benefit operational efficiency, at the expense of flexibility.
- The discontinuous process allows time for aging between stages. Curing, off-gassing, crystallization, and molecular relaxation can occur at their natural pace.
- In a continuous production scheme, downtime on one unit will curtail the operations of all the process steps. The failure of one component might shut down the entire train. Frequent failures will cause excessive start-up and shutdown waste.

- Short runs can generate excessive changeover scrap while each separate operation equilibrates to new process conditions. This is especially problematic if one operation has long residence times. Continuous process operations are best suited to limited numbers of high-volume products, rather than a large variety of short-run items.

- The optimal production rates of each operation may not be compatible. It may not be possible to establish stable process conditions for each unit at the common output rate that is dictated by the line's bottleneck.

Considering Outsourced Production

I have a hunch that the unknown sequences of DNA will decode into copyright notices and patent protections.

Donald E. Knuth

The Boeing Company attempted the most ambitious toll-manufacturing experiment in history with the production of their 787 Dreamliner. The initiative was unique in that they delegated significant development and engineering to component vendors, rather than providing a finished design package. The company was shifting from being an integrated manufacturer to a program manager overseeing a massive collection of collaborators, many of whom were subcontracting to second-tier and third-tier vendors.

This approach opened the insular business to new resources and ideas, but it caused problems that would have bankrupted a smaller company. In particular, the global collaborators were isolated by distance, language, and the customer/vendor relationship. Boeing's own employees were demotivated by a program that clearly aimed to reduce their contribution to the project.

The vendors were overwhelmed by modification requests that were often unclear and contradictory.

At one point, Boeing had 37 boards, which could dictate design changes to the network of vendors on the program. Such adjustments become a serious issue when process dependency is involved. Significant process work or simulations might be required to determine the impact of each change.

A dynamic process requires two-way communication to evaluate how change requests affect the performance of the total product and to identify optimal process conditions. Boeing could not possibly anticipate the impact of all these process-dependency issues across their complex web of associates. They discovered that large assemblies did not perform as originally modeled when the process dependency of many vendors was scaled up into an integrated whole.

RULE #46 YOU CANNOT MANAGE TOLL MANUFACTURING THE SAME WAY AS INTERNAL OPERATIONS

The biggest mistake in outsourcing is to assume that the process of your contract manufacturing vendor will run identically to your in-house operation. It will be different in a hundred subtle ways that cannot be anticipated. Only careful commercial and technical monitoring of the toll relationship can ensure that products will meet end-user requirements.

Reasons for using toll vendors:

- Seasonal or transient demand exceeds the internal production capacity.
- Sales exceed capacity and capital is not available to expand production. Expansion/debottlenecking are underway, but are not complete.
- The new product might not be successful in the marketplace. The management is reluctant to add internal capacity until market acceptance justifies the investment.

- An outside vendor may have more expertise in a specific production technology or a significantly lower unit cost due to high-volume equipment. This is especially attractive for market-driven companies that cannot pretend to be experts on every different process.
- The production of a niche product that is not strategic for the company, but which must be tightly integrated with a system solution to work correctly.
- An internal production line is currently out of service for an overhaul, upgrade, or relocation.
- The market manager wants to begin supplying a product in a new geographical region, but does not have the appropriate supply chain.
- The product is needed for a short-term opportunity or promotion.
- The plant is optimized for the mass production of standard products. Special requests for low-volume specialty items tend to disrupt the production scheme.

Ways a toll-manufacturing contract differs from a vendor/customer relationship:

- The supplier is making a product to a single customer's specific requirements. The customer is required to specify those details in writing and assume all responsibility for FFU.
- No salesperson, broker, distributor, or middleman should be involved. The supplier saves the cost of sales, advertising, marketing, and distributor margin.
- The supplier is entrusted with the customer's internal knowledge regarding end-user needs.
- The customers always have the option to add internal production capacity and pull the business back from the toll vendor. Conversely, the vendors may elect at any time to terminate the agreement if they find a more profitable use for their process capacity. In the worst case, they may cut out the middleman and service the end users directly.
- The production vendors guard their process secrets and the customers hide market information.

Qualifying a Toll Vendor

In the beginner's mind there are many possibilities. In the expert's mind there are few.

Shunryu Suzuki

When outsourcing production, take nothing for granted as to what the vendor might do wrong. Process conditions, equipment configurations, raw materials, and operating procedures must all be explored, as you would do with an internal operation.

When providing specifications to your own plant, the most basic things might not need to be called out. These matters are implicitly understood by a workforce, a QC team, and a management structure that fully comprehend the end-use requirements. In addition, the manufacturing process might be tuned to make this specific product and nothing else. Many potential errors are impossible, because there is no "knob" to adjust that attribute. Toll manufacturers will not have an understanding of your intangible needs and may have more flexibility to do things wrong.

Imagine a facility that grinds certified organic grains into flour and meal as the first step in the process for making specialty food items. Its products are sold to consumers with special dietary needs or an appreciation of artisanal quality. When upgrading its milling equipment, the grinding process is temporarily outsourced to a commodity flour-processing mill.

New procedures must be put in place to ensure that the toll-grinding operation adheres to all the appropriate requirements to maintain strict standards. The internal operation has fewer safeguards because it normally processes materials from controlled sources. The commodity facility does not employ the same controls and procedures because its customer base is not concerned with those issues.

The vendor's standard operating procedures (SOPs) and standard operating conditions (SOCs) must be verified for every item in the grade slate. Unusual grains do not process into flour in the same way as the wheat that the vendor is familiar with. Even if the flour processes well, it may have variations in its particle size, microtexture, or heat history, which will affect the downstream operation. These subtle differences may not be evident until the finished product is baked and tasted.

Technical oversight of outsourcing:

- At least one person dealing with the vendor must be knowledgeable about the process, the products, and the specification requirements. All the aspects that impact the product need to be understood and locked-in. It is never advisable to simply provide the current specifications and ask, "Can you make this for us?"

- Outsourced production of your particular product will often call for changes to the operational procedures at the toll processor. The operators may have completely different expectations of quality, based on their experience with other customers. Not all of the differences will be revealed in an inspection tour or a brief, monitored trial. The QC inspectors may pay too much attention to attributes that are not important to you. The process management group may try to run this product at very high output rates, to make money after it unwisely bid too low to get the business.

- The toll process will run differently from your operations, even when using the "same" equipment. Each company adjusts and modifies things to optimize for its unique needs. These subtle adjustments are difficult to detect and appreciate.

- Your "process expert(s)" must have a broader base of knowledge than is required to do the same function in a captive production environment. Experts may work for you permanently or as consultants. Never rely solely on the vendor's people in this role. These specialists must be aware of all the process paths that the vendor may employ. If the operators have more or fewer knobs than was the case on your process, then the experts must understand what the impact will be.

- In sensitive situations, process experts may need to stay on-site to provide the same level of control that you have in your own facility. Over time, this will evolve into a firm set of operating guidelines supported by periodic inspections visits. Less important applications will only need a detailed specification. In all cases, your requirements need to be boiled down to a system that allows the product fitness to be evaluated in an unambiguous way.

- In working with outside suppliers, the depth of the experts' knowledge can only grow in comparison with what they can learn inside their own operation. Insights and ideas will be filed for future use when the time comes to buy additional equipment and change your internal process paths. This learning opportunity is an unrecognized benefit to your span of process knowledge.

- The toll manufacturers must clearly detail their trade secrets under the terms of the confidentiality agreements between themselves and the customer. Both parties must be prepared to document the specific information that they want to protect.

- The vendors may not have the same test equipment in their laboratories or it may be calibrated differently. They may use unique procedures, terminology, and interpretations.

- The vendor's raw materials may come from a different source, even though they are certified by the global supplier as being interchangeable. Initial trials should be conducted with a small batch of your standard feedstock shipped in for this evaluation in comparison with the equivalent grade from a local source. Any differences in processability must be evaluated.

- In running other products, the vendors' production line may be contaminated with residues of raw materials that are detrimental to your needs. Even when using a dedicated line, the air in their plant may be laced with dust that becomes part of the product. To protect the proprietary formulations of their other customers, the vendors may not be willing to disclose all the potential contaminants. Even if they did, your product specialist may only be able to speculate on the affect that traces of these materials may have on the end users.

July 11, 1967　　　R. J. EVERITT　　　**3,331,002**

SEALED CAPACITOR HAVING HYDROGEN ABSORBING MATERIAL THEREIN

Filed Oct. 10, 1963

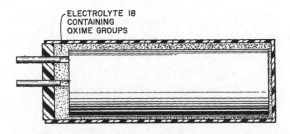

ELECTROLYTE 18
CONTAINING
OXIME GROUPS

Personal computers and mobile devices are often assembled by specialized vendors. They, in turn, purchase circuit boards and other subassemblies from second-tier suppliers. Those companies may procure individual components such as chips from still other sources. This complex chain of responsibility offers many opportunities to save money by buying elements that are similar, but not identical to the original design. Components become obsolete, are temporarily unavailable, or are counterfeited by unscrupulous distributors.

In the early to mid-2000s, personal computers from a variety of companies were failing after only a few years of normal service. Investigations suggested that defective capacitors were at fault. These small cylinders started to bulge and leak electrolyte fluid due to a buildup of hydrogen gas.

In some cases, the capacitor manufacturing process was faulted for having poor control over the amount of electrolyte dispensed into each unit. In other cases, competitors copied the primary vendor's electrolyte formula. However, they neglected to add the depolarizing agent that inhibits the formation of hydrogen during use. In all cases, the brand name computer suppliers did not monitor component quality and substitutions closely enough. The failures did not become evident until several years after delivery to the customers.

Make sure that your outsourcing partners fully understand your process needs and have the capability to consistently service the business. Beware of vendors who bid so cheaply to get the business that they cannot possibly make your product using the specified raw materials.

Managing the Relationship

It is impossible for a man to be cheated by anyone but himself.

Ralph Waldo Emerson

Toxic toy!!!

In 2007, an Australian company called Moose Enterprises introduced a toy called Bindeez. Children could assemble three-dimensional shapes using colored beads. Each bead was coated with a water-activated bonding agent, which fused them together into a rigid structure. The company outsourced production to an Asian producer with low labor costs.

In the adhesive formulation, the Australians specified a component called 1,5-pentanediol, based on its safety and low toxicity rating. However, some of the toys were produced instead with 1,4-butanediol, which is chemically similar. Both chemicals worked well in the process and the finished product. But, when ingested into the human body, 1,4-butanediol can be metabolized into gamma-hydroxybutyric acid. As children often do, some put the beads into their mouth and ended up in hospital with assorted drug-reaction symptoms.

After emergency rooms around the world reported the problems, the product was recalled globally. Afterwards, the company implemented better controls to monitor the materials and procedures of their outsourcing partners to ensure against substitutions. They also added a bitter-tasting ingredient to the formula, in an effort to prevent children from swallowing the beads. Outsourcing a process-dependent product may require unique systems to monitor performance and to prevent unexpected outcomes.

Guidelines for dealing with a toll manufacturer:

- Designate one person to be the primary commercial contact with the vendor and have him or her coordinate and monitor all communications. This individual does not need to be the sole communication channel of your organization. But, he or she must make sure that the company speaks with a common voice. Vendors find it very disconcerting when different parts of your company make conflicting demands on their time and attention.

- The organization needs to continually monitor delivered quality and end-user performance. Do not wait for complaints. Proactively look at the product the same way that your customers do.

- Eventually, some batches of defective product will slip past your controls and get into the distribution network. Suspect lots need to be recalled and inspected. Every discrete container needs code-dating that pinpoints its time and place of origin. The trace-back labeling applied by a toll supplier must be compatible with the trace-back labeling used for internally produced products. It should be in the same format and the same physical location on the package.

- The vendors should be periodically visited by someone who knows the process and speaks their language (from both a linguistic and a technology standpoint). The representative should audit compliance with the procedures that have been agreed upon. Maintain a list of red flag issues that would cause particular concern. Such inspections should involve talking to the production, packaging, and shipping personnel. The visit should be scheduled so that actual production can be observed.

- The decision to outsource must be consistent with company strategy. It is hard to implement improvement programs at the internal production plant while taking jobs away from them.
- Hire an experienced intermediary to guide the search in regions that you are not familiar with. Beware of kickback arrangements that such brokers might use to make extra money from vendors.
- Beware of making sudden changes in order levels. There is a temptation to use outside vendors to soften the spikes in demand on internal facilities. However, this can cause vendors to start and stop their process frequently as orders come and go. Some companies can deal with this. Others may have problems in retaining qualified operators for a sporadic product demand.
- Negotiate potential issues of demand variability ahead of time. Some vendors could respond quickly if they carry an inventory of parts or materials that are unique to your orders. However, they will not bear that cost unless you underwrite it.
- Follow a consistent plan on outsourcing. Avoid spreading the business to many scattered and incompatible vendors. It takes resources to teach and manage the toll suppliers. The more vendors you have, the greater the variation in procedures and problems with logistics and coordination.
- Avoid teaching the vendor your business. Choose vendors of finished products partially based on their inability or unwillingness to service your customers without you being in the loop.
- You can sign an airtight legal agreement with the most ethical people on Earth. But, if one employee leaves to work for your competitors, then you have to assume that all of your information will be compromised. Only give vendors the information that they need to service your needs.
- Avoid letting your sales organization choose whether its customers should get the outsourced product or the internal production. Its decisions rarely match up to your production capacity.
- Be ethical in all dealings.
  - If the needs are limited in duration, be truthful about that from the start.
  - Sometimes a one-time sample run is urgently needed, which will not lead to long-term business with the vendor. Be honest about the situation and pay the appropriate price for the work.
  - Treat the vendor's trade secrets with respect. Do not pass them to another company in an effort to get the same work done at a lower price. Agree in writing which information belongs to your company and which is the vendor's property.

- Beware of having your representative land in a foreign jail if your dispute turns ugly.
- Motivate the vendor to want an ongoing relationship with your company and to make good products. There is nothing worse than trying to get a short-term toll vendor to give credit for quality complaints, which you have discovered after the relationship has ended. It is even harder to go back to them for more production the following year after you have withheld payment.
- Organize the distribution system to deal with problems differently when products come from a toll vendor. The outsider's return and credit policy will be different from the internal plant's return and credit policy.
  - They may only refund the cost of product that is returned to them for inspection and is proven to be out of specification. Product failures due to undocumented requirements are not likely to be reimbursed.
  - Specifications must be well documented, explained, and agreed upon. The QC tests that verify conformance must also be clear—verify the method with comparative tests.
  - Technical terms, test methods, and procedures must be defined in the early stages.
- You may never achieve a 100% interchangeable product from both internal and external sources. Process-derived products are never exactly the same. Try to segment the various product streams into geographical or market segments, so that they do not overlap at customers.
  - The worst case is when the end user is forced to constantly adjust to unexpected performance variations in materials from multiple sources.
  - Your sales, service, and support staff must be trained to deal with sourcing differences. Make sure that the outsourcing is a good deal for them.
- Never create a situation where a vendor benefits from bad quality by getting increased orders.
- It is difficult to develop a product on an outsourced process, unless you have a very special relationship with your partner. Awkward situations occur when you overestimate your understanding of the end-user product needs or the dynamic process interactions.

The "concept map of toll processing" at the end of the chapter depicts the interactions with processing vendors and the important aspects of a relationship that is both technical and commercial in nature.

Summary

- Never suboptimize individual stages of a multistep process. Incentives and compensation should be based on the success of the entire operation, not on the efficiency of individual areas.

- Every process is unique in its own requirements. Solutions that were optimal for previous situations will not suit subsequent opportunities.

- Minimize the number of handoffs. Zero is a good target, but this is not always practical. Count every time that an individual might impact or delay the completion of a step.

- Get everyone in the plant involved with resolving process conflicts by making problems in other areas visible to them.

- Try to size sequential equipment with the same output rate, rather than running each individual unit at the maximum rate. Avoid having much, if any, in-process inventory.

- Carefully consider when to use a processing partner and how to structure the relationship. Always keep these reasons and guidelines in mind when dealing with the vendor.

- When outsourcing production to another company, every possible condition, material, attribute, and test must be documented if it might affect the product.

- Never outsource process requirements that cannot be articulated in a written specification.

- Take time to understand which knobs are available on an outsourced process compared to those available on your own process. Designate a process expert who fundamentally understands the technology, beyond just having a familiarity with how your operation works.

- Designate a primary commercial contact to be the coordinator for all communications with the vendor. Do not allow the various functions of your company to deal with the vendor in isolation. Maintain a consistent and uniform message.

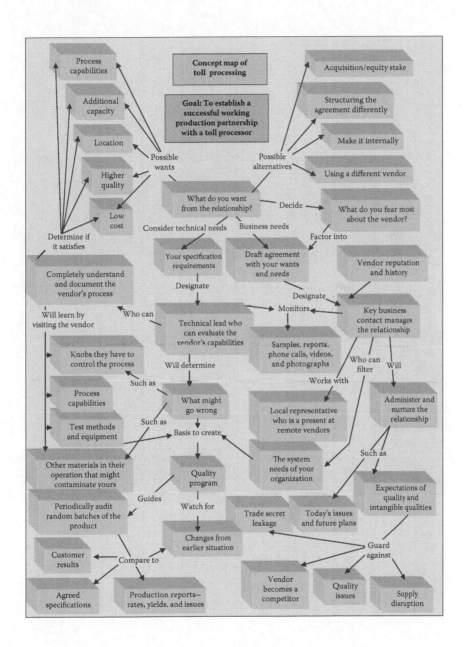

Concept map of toll processing

Goal: To establish a successful working production partnership with a toll processor

14

Human Factors

Every person interacts with the process in a different, unique, and unpredictable way. How do we maintain consistency in dynamic production systems when people are in the loop?

> Nothing is more difficult than to introduce a new order. Because the innovator has for enemies all those who have done well under the old conditions and lukewarm defenders in those who may do well under the new.
>
> **Niccolò Machiavelli**

For years, the German police hunted an unusual female serial killer. She was known only from DNA evidence collected by crime scene investigators. The suspect's profile suggested a dangerous criminal with no consistent pattern, whose widely scattered offences ranged from burglary to the murder of police officers. At one point, a reward of €300,000 was offered for information leading to her arrest.

In 2009, the manhunt was abruptly terminated. The DNA that linked the various crime scenes was found to be the result of contamination on the cotton swabs used to collect evidence. One particular female employee working on the swab production process was inadvertently depositing some of her genetic material on the product. This occurred despite hygiene procedures and controls, which should have prevented such contamination.

The crime scene investigation (CSI) experts discovered that while the sterilization measures destroyed fungal, viral, and bacterial organisms, they did not destroy all the genetic material down to the molecular level. Just because a product is double packaged and certified to the highest levels of medical safety, it does not mean that the product is DNA free.

This situation occurred because of two kinds of human factors. First, a plant employee deviated from standard operating procedures from time to time. Secondly, the criminologists made assumptions that were not supported by documentation and experimentation.

RULE #31 YOU CAN NEVER ANTICIPATE ALL HUMAN INTERACTION PROBLEMS

Scientists and engineers often overlook or underappreciate the importance of human behavior in technical situations. They may also ignore weaknesses in their own decision making. These soft factors can represent the most significant uncontrolled parameters in any dynamic process and the fitness of the resulting product.

The McDonald's fast-food chain represents an extreme example of engineering a manual process to minimize human variability. Their hamburger patties follow a strict formula and conform to exacting standards of weight and thickness. Each hamburger is cooked for a precise time to achieve the optimal temperature in the center without overcooking. Procedures dictate every activity. Despite these measures, there has historically been significant variability in the McDonald's customer experience, depending on the particular people running each franchise.

Human Input on the Plant Floor

If you dream of something worth doing and then simply go to work on it and don't think anything of personalities, or emotional conflicts, or of money, or of family distractions; . . . it is amazing how quickly you get through those 5,000 steps.

Edwin Land

One of the worst environments for worker interaction with the process occurred in the early automation of the textile industry at the start of the Industrial Revolution. In England, Luddite mobs broke into mills and destroyed machinery that threatened the livelihood of skilled weavers.

The word sabotage is derived from the sabot or shoe. Flemish weavers were said to have thrown their wooden footwear into the gears of a new class of labor-saving loom to slow the implementation of the new technology.

In both cases, the workers perceived that the benefits of process improvements would not be shared equally.

Various urban legends suggest that we should never purchase an American car that was assembled on a Monday morning or a Friday afternoon. Presumably, assembly-line workers make more mistakes and take more shortcuts while recovering from their postweekend hangovers or anticipating the coming ones.

The obvious conclusion is that the process of building a vehicle is highly dependent on the alertness, dexterity, and motivation of the individual workers involved. Anyone who performs mind numbing, physically exhausting work will understand the issues. The concept harks back to the nineteenth-century craftsmen who meticulously shaped and fit each part.

Looking deeper, it suggests that the production system fails to control routine procedural errors that will not become apparent until long after the product has been purchased. Car companies want us to believe that they have designed quality into their product through proactive engineering and precision specification of the parts. Instead, one popular perception is that quality will depend on individual worker attentiveness and morale, which cannot be monitored, controlled, or evaluated consistently.

The plant management is often most concerned about the errors committed by new employees and a few malcontents or inept bumblers. When quality problems are discovered, attention quickly turns to the usual cast of suspects. Incompetent workers may seek to conceal their mistakes and excursions. Lazy operators will seek shortcuts to ease their workload. Newly hired workers may deviate from procedures in ways that are unanticipated.

The more intricate the process is, the more likely it is that even the most trusted worker could share responsibility for quality problems and end-user dissatisfaction. The worst thing for a complex system is to introduce variability. Conscientious operators are more likely to experiment with improvements to their job.

These actions may enhance the performance of the worker's unit operation or they could make the work easier and more interesting. But, they could just as easily cause problems for downstream operations or the product's fitness-for-use (FFU). No one individual improvement or deviation is a problem in itself. The sum of all the changes and the human factors will multiply with time and the number of individuals involved.

Some workers are reluctant to share best practices with their coworkers. No one wants to be a drone. Harmonizing and enforcing the procedures might be low on the management priority list when things are running smoothly.

We need to understand that not all workers are identical. A procedure that is easy for most employees may be difficult or tiring for an employee who is different in some way. The difference could be age, flexibility, past injuries, level of dexterity, left-handedness, visual impairment, gender, strength, attention deficit, or intelligence. People are never anxious for their shortcomings to be brought to the attention of others. Workers who cannot perform the required jobs could find themselves transferred to a less desirable assignment. In subtle ways, these individuals will find ways to get through the day. Their shortcuts and work-arounds may actually be better than the written procedures. However, these different procedures do not give the same result if the product is process dependent.

When a process is highly automated, managers assume that all these differences in motor skills are a moot point. The operators just need to monitor the process at a control panel. Occasionally, they make the rounds with a clipboard to note the readouts on remote gauges. The standard operating conditions are downloaded from a master recipe. Associate degrees are required for the job. These operators are part of the knowledge economy, rather than blue-collar laborers.

But even this digital-age job description is still bedeviled by behavioral variation introduced by differences between one person and the next. Some people are more inclined to sit, while others are anxious to move around. Not all people have the same attention span or interest in the details. Some crave social contact and others are solitary. One will call maintenance at the first sign of a minor deviation, while the next shift will spend hours making minor adjustments. A third operator will stick with his or her favorite settings, regardless of the situation.

When the reasons for work instructions are not obvious, operators may choose to disregard procedures that have no obvious value. Given a lack of immediate feedback to this passive rebellion, the shop floor practices are implemented nonuniformly. Such situations develop when remotely located process experts issue advisories for which the local workers and managers have no ownership. Neither group will feel passionate about implementing these measures.

Blending operators are responsible for preparing mixtures of specialty rubber products with an assortment of additives. A measured charge of raw or synthetic rubber is dropped into the hopper of a high-intensity blender. The operator measures out the specified amounts of additives to complete

the formulation and then starts the blending cycle. The formulation expert assumes that the blending operators will carefully meter out the exact measure of each component to the limit of the accuracy of their measurement tools. Toward that end, the blending station is equipped with precision scales, which are frequently calibrated.

In the real world, human factors intervene. One operator is concerned about keeping the work area clean and orderly. Before adding the rubber, the operator brushes any dust in the hopper down into the mixer. When measuring the additive, a small amount remains in a bag at the end of the order. Not wanting to leave the nearly empty bag on a shelf, the operator empties the remaining powder into the mixer.

The other operator is concerned that components go bad if their containers sit around after opening the package. When measuring the last batch, he or she is short of the target weight. Rather than opening a fresh bag to complete the charge, the operator only adds the lesser amount. This operator wipes down the dust on all of the work surfaces with a moist cloth after each blend. The contaminated rag is disposed of.

The operators tend to make batches that have more or less of the individual components than were specified. The first operator contaminates the batch with dust left from a previous formulation. The second operator does not, but some additives are lost when he or she discards the powder adhering to the hopper. No one has told these people not to improvise their own techniques, because management never audits procedures to see exactly what they are doing. Since no one pays any attention or scrutinizes their technique, both assume that their actions are in the company's best interest.

These habits come to light when the formulator conducts management-by-walking-around (MBWA). In a more formal setting, this may be termed an audit of operation procedures. It can be very enlightening to compare the written procedures (if they exist) with the practices on the shop floor across different shifts.

Even the most detailed procedures have omissions or descriptions that leave room for interpretation. When asked for clarifications, various engineers or managers will differ on the proper techniques. We rarely issue revisions until a serious problem comes to light.

In this case, it might be helpful to provide well-labeled, reclosable storage containers. This could address the operator's reluctance to have nearly empty or almost full bags of chemicals left in the work area. These measures can be very low tech and do not require a college degree. In fact, experts may feel that mundane duties such as designing optimal work instructions are beneath their dignity. Such attitudes are a function of the organization. No one wants to listen to people who drone on about each little detail that they have addressed. "Just give me the big picture."

Ideally, the formulation expert would know and anticipate the effects of component variation. Some components can be metered out with quick and dirty techniques that speed up the process, when accuracy is not important. By some means, the operators must understand which functions are within their power to optimize and which functions need to follow a rigid procedure.

Organizational Behavior

> ... it is only your judgments that err by promising themselves effects such as are not caused by your experiments.
>
> **Leonardo da Vinci**

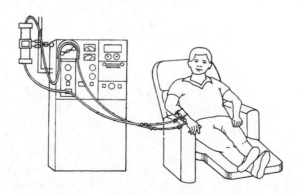

Dialysis machines prolong the lives of kidney patients by filtering toxins from their bloodstream. Baxter International manufactured filter cartridges for such a system at a plant in Ronneby, Sweden. Workers leak tested every filter pack with pressurized air. Units that failed the pressure test were examined closely to determine the point of failure. This was accomplished by wetting the surface with a 3M product called PF-5070. This volatile, nontoxic solution revealed the locations of the leaks. The defective seams were then reworked in a corrective action step. As much as 10% of Baxter's production was repaired, certified, packaged, and shipped to customers.

PF-5070 had not been tested and approved for medical applications, and certainly not for introduction into the human bloodstream. The company assumed that the solvent would fully evaporate, without affecting the end users of the product. However, some parts did retain trace amounts of the liquid, which found its way into the bloodstreams of patients. At body temperature, the solvent can vaporize, causing lethal gas bubbles. Fifty-one patients died after dialysis treatments with the Baxter filters.

A test method that originally served as a data-gathering technique became a routine part of the production process. By taking the initiative to reduce process scrap, the plant became much more efficient. This was a great success story, at least until patients started dying.

RULE #11 PROBLEM SOLVERS ARE BIASED TOWARD IMMEDIATE ACTION TO IMPROVE EFFICIENCY AND PROFITABILITY

Compensation packages rightfully give strong incentives to optimize performance. Industrial decision making often has a short attention span, wanting to address problems and move on to the next issue. The first plausible explanation is generally seized upon. The organization is directed into immediate action to resolve the situation. "Stop doing X!" or "Start doing Y!"

Real-world process problems are seldom solved so simply. The trouble rarely gives a clear signal by disappearing as soon as X is stopped. Typically, the issues have occurred sporadically in the past and may persist at a low level, even after the corrective action. Long periods of data collection may be required even to understand whether the decisive change has had a significant effect or not.

The optimal, long-term fix for the root cause will generally require patient analysis. Some organizations take immediate action, but they then ignore the important follow-up. People who advance new theories about the root cause of the problem seem to contradict the initial success. The groupthink response says, "We better not do X again, or we will look like fools if the problem comes back!" In fact, it might be wise to do X again under controlled conditions, to determine if it really triggers the problem. Some combination of X and Y might be the best choice. However, people avoid revisiting any version of option X for a long time. When asked, they can give a long, narrative account of the problem, but they rarely understand the root causes.

Quality incidents often retain the first name that people give them. This can shut down all other potential lines of inquiry because a probable cause has been identified. Suppose that a quality complaint is initially blamed on a low-cost catalyst that had recently been qualified. The "cheap catalyst" problem is immediately addressed by switching back to the higher-priced alternative. This terminology shapes the discussion and discourages people from examining the process data for alternative explanations.

A complex manufacturing operation may be mysterious to new engineers. Ideally, these individuals are intrigued and motivated by the technical challenge, rather than being intimidated. A good engineer will be anxious to dig into the process and learn to master it. This fresh perspective and openness to new approaches are strengths as well as weaknesses. However, these bright young minds are likely to be confused and demotivated by the confusing political structure of a medium to large company. Questioning the groupthink is subtly discouraged.

People generally act in ways that they perceive to be in their employer's best interest. The interactions between people grow geometrically as the organization gets larger. To deal with the complexity, political lines are drawn between the various groups. Narrowing the lines of communication reduces the background noise and limits the complexity. Scientifically trained people do not react well when subjected to such illogical behavior. They expect to write and say what they think, rather than deciphering the organization's position on every matter.

The danger to the company from such situations is that information about the technology becomes distorted and unreliable. Details concerning process behavior are the lifeblood of a technical organization. People who spin information to advance their own ideas poison and undermine the entire system.

When evaluating employees, people are most impressed by individuals who are extremely confident in themselves and their own judgment. True experts, on the other hand, often come across as being less self-confident when faced with a totally new problem. They cannot issue definitive answers until experiments are done, information is gathered, and alternatives are evaluated. When choosing between a confident scoundrel and an honest expert, we can be fooled by our own hard-wired decision-making systems.

When confronted with novel situations in complex systems, we should resist the impulse to jump at the first plausible advice offered with self-confident enthusiasm. It is better to test the various possibilities to eliminate those that are clearly impossible. With the miniaturization of communication technology, the problem gets worse. Short questions and answers are exchanged quickly, generally without detail or foundation, much faster than we can experiment to find the facts.

When making decisions, we should be mindful of cultural differences across different regions. Americans in particular can be confounded by traditional Chinese etiquette. In that culture, it may be bad form to say "no" to a superior when asked, "Do you understand?" In other places, engineers are reluctant to assume personal responsibility for important decisions, preferring group decision making. In some companies, professionals may be expected to justify and defend their conclusions in the face of adversarial questioning. Their reactions to this atmosphere will vary considerably with their background.

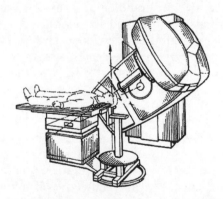

Radiation in a variety of forms is used to attack tumors in the human body. This practice can be very effective or extremely dangerous, depending on the skill of those involved. Many cases have been reported of patients exposed to high dosages, resulting in unintended tissue damage.

Medical technicians and nurses operate this equipment and monitor the patients under the supervision of a radiologist. Sometimes, they lack the tools to monitor the amount and destination of the deadly radiation during the procedure. Like sunburn, damage caused by the procedure is rarely visible in real time. Only later do burns or other symptoms reveal that mistakes have been made. In the meantime, the patients must have faith that everyone involved has done their job correctly.

Many different types of errors will cause problems. Software bugs can create a disconnect between the instrument readings and the actual power output. Calibration errors, inappropriate hardware configurations, and poor documentation of procedures have killed or injured patients. Unlike surgical procedures, it is difficult to catch the errors in real time. Fatal mistakes have been repeated over a period of years before they are detected.

Participants at every stage of a process must have immediate feedback on the effectiveness of their actions. Procedures should include ways to evaluate their performance. When errors are detected, they must be documented so that corrective action can prevent a reoccurrence.

Technical Behavior

Question everything. Learn something. Answer nothing.

Engineer's Credo

In the 1960s, the DuPont chemist Stephanie Kwolek was investigating condensation polymer synthesis. Her goal was to create an ultrastrong, stiff fiber to improve the performance properties of tires. One of her reaction products was a curious cloudy liquid. Normally, such an appearance would suggest that the solution contained solid particles in suspension, which would clog the fiber-spinning apparatus. The inquisitive researcher ran the liquid through a fine filter to remove any particulates. Still, the solution remained cloudy.

It took time and effort for Kwolek to convince a reluctant technician to process her suspicious fluid on a spinneret system to generate fibers for analysis. The technician was convinced that the solution would choke and damage the delicate apparatus. Eventually, she wore him down and got her fiber sample.

Further alarm was raised when the laboratory tests indicated absurdly high modulus values for Kwolek's samples. Rather than be embarrassed if the results proved to be wrong, she submitted multiple duplicate test requests, which confirmed the initial results. Her fibers obtained their stiffness from a rigid molecular backbone, which did not fold back on itself like other polymers. The chain strength was greatly enhanced because the individual molecules clung tightly together with hydrogen bonds, shown as dashed lines in the previous figure.

Kwolek invented Kevlar fibers by overcoming entrenched organizational preconceptions. Once she presented overwhelming evidence of their revolutionary properties, the company proceeded to scale up the product. Considerable development effort was required to do so. The laboratory techniques were not easily adapted to the existing fiber production equipment. DuPont research scientist Herb Blades developed an air-gap spinning process to generate the fiber at commercial rates. Other technical hurdles were overcome with the engineering attention of countless people. Customers were slow to adapt the product to their applications, because it was so different from their customary raw materials.

Kwolek's work demonstrates many of the essential elements needed for revolutionary product advances. These include background knowledge of the basic technology, curiosity with unexpected results, persistence in evaluating the idea, and caution to avoid overselling a concept. Kwolek's effort earned her the sole credit for Kevlar, although significant development and scale-up work was needed to implement a viable production process.

RULE #48 PROCESS AND PRODUCT KNOWLEDGE = VALUE

The value of development professionals comes from their ability to convert process knowledge into solutions for customer problems. This applies to basic research, product development, application development, and ongoing optimization. The faster a new hire comes up to speed in an assignment, the sooner the organization will benefit. This requires that process information is stored in a format easily transferable to new people. In this respect, even an old-fashioned manufacturer is a knowledge-based business.

In some companies, important intellectual property is stored in people's heads or in notes that are only decipherable by them. Rarely is all pertinent information clearly documented in ways that a new employee can absorb. If you think that your own organization is an exception to this rule, then you should make every effort to find out where the keys to the process information are really kept.

Signs of poor knowledge management:

- Only one or two people in the organization can perform certain key tasks. Consider what might happen if these individuals are hit by a bus.
- New hires take a long time to become effective. There is no written training manual or courseware. It is not clear which of the various versions of a document is the latest revision. Some copies have handwritten modifications. There is no documented chronology to indicate which one was in effect during which period of time.
- Not all specifications, recipes, and procedures are documented in consistent, standard formats.
- Nuances and subtle performance differences between one machine/line/plant and another are not documented or understood.
- When starting a project to improve an existing product, the details of the earlier development will be incomplete. The project history was never compiled into a secure central file. Bits and pieces have been removed for use in subsequent research and have never been refiled.
- It is hard to run products that have not been produced in the last few years. Process settings, configurations, and running condition records are lost or misplaced. A special person or team is required to assist.
- There are no master copies of the standard operating conditions for each different product. An operator should be able to set up and run an operation from these documents without needing to consult the expert.

- The technical staff spends more time dealing with emergencies than long-term projects. It is reactive, rather than proactive. There is never time or priority for the documentation of procedures and best practices.

It is vital that the experts are able and willing to transfer their knowledge (power) to future generations. Insecure engineers may be reluctant to share what they know and especially what they do not know. The more talkative ones may go into so much detail that the novice and the manager will be overwhelmed. It is hard to convey hard-won insight in a succinct summary. There is much to be gained by the exercise of boiling down the knowledge to its basic essence. The student will ask questions that will reveal what the tutor does not fully know.

Ways to ensure that process knowledge is well understood across the organization:

- Gradually rotate competent technical people between assignments, rather than allowing them to be the sole expert in a given area. Motivate them to be both students and teachers of relevant information.

- Create a list of the technical knowledge and skills that operators need to do certain jobs.

- Determine if those who are trained on a task can really perform the duties. This usually involves assigning them to carry out the job for at least a period of time.

- Trainers should be periodically rotated back into operations and evaluated so as to maintain their own proficiency.

- Do not allow one expert to be the white knight who rides to the rescue of a process every time there is a problem. After any such event, the expert should ensure that procedures are implemented and documented to prevent a reoccurrence.

- Compensate and reward technical people for their ability to learn new tasks and to teach others. Pay specialists based on their unique process knowledge, not the salary that the employment marketplace values them at. Do not lose valuable process knowledge because experts feel unappreciated.

- Specifications, standard operating procedures (SOPs), and standard operating conditions (SOCs) should be documented in a consistent manner with revision control. There must never be a question as to which version is in effect or where to find the master copy. These documents should contain all the details needed for a qualified operator to set up and run each product.

- Operators and the plant shift management should have decision tree guidelines telling them what to do in the event of common problems. They should be the first line of defense for 80% of the problems.
- The experienced individuals will act from their experience, without consulting the instructions. Those who are less knowledgeable should have guidelines that direct them to a similar result.
- A development project is not finished until all the paperwork is completed and reviewed. Include specifications, test methods, SOPs, and SOCs. Also, document the key findings of how the process responds to inputs, especially those not chosen for the commercial application. What approaches, materials, and theories were tested and found wanting?
- Constantly evaluate and document what is known and unknown about the process attributes.
- Known facts should be periodically tested, questioned, and communicated. It is especially important that the boundary conditions are well understood.
- Ensure that trainers have a high level of process experience. They should know and understand the logic behind best practices. A training assignment should not be a means to avoid accountability.

Marketing/Sales Behavior

The shortest and best way to make your fortune is to let people see clearly that it is in their interests to promote yours.

Jean de La Bruyere

French wine experts lament the subtle transformations of some Bordeaux wines in recent years. Reportedly, wineries have increased the temperature of the traditional maceration process to extract more color from the grape skins. This practice is driven by the dominant influence of the American wine reviewer Robert Parker. His preferences lean toward a darker color in the grand cru wines. Parker's rating points can have a significant impact on the wholesale price that vintners receive for their product.

In any market, the intangible perceptions of customers can drive process decisions, as interpreted by the sales and marketing organizations. These seemingly arbitrary decisions can be maddening for those who strive for an efficient process and clear FFU test results. Customers are notorious for raising capricious objections to soft factors in a ploy to receive price concessions. In the worst case, panicked salespeople will send the technical and manufacturing organizations into a scramble to correct a problem that never existed. A perceptive sales staff will distinguish between true customer needs and negotiating tactics.

Cynical technical people generally suspect that the market input into new product development projects is riddled with lies, exaggerations, intentional omissions, and subtle half-truths. In reality, it can take years between the time that the initial assumptions are made and the time that the new product faces the true test of value in the marketplace. In that time, the market conditions will change.

Regardless of who is right about the basic assumptions, an atmosphere of distrust can derail even a good product concept. The people who ask for a product and those who are creating it should be a cohesive unit working toward a shared objective. Everyone involved must be able to argue about the goals and requirements of a project and work out their differences. If concerns are not addressed early, they can fester into performance problems.

Every development team member should participate in some FFU tests to understand what the product needs to do and what happens in the event of failure. The team must visit customers to experience the end-use environment and see the background, deliverables, and constraints for each application.

The group should also be prepared and empowered to alter the work's direction if the market changes in the meantime. However, this must not be taken as an invitation to pursue creeping elegance or expand the project's scope.

Critical Questions: What are the unwritten customer perceptions around products like ours? How can we get a variety of product concepts in front of the decision makers to determine which is best? What trends are driving change?

One market-focused American company put tremendous effort into designing elegant packaging for a new retail product. Focus groups probed consumer opinions for the most desirable shape. Special care was taken with the color choices and the label graphics. Tooling was created to produce the optimal plastic cap and bottle combination. However, during the first production run, it was discovered that the high-speed filling systems could not consistently apply the caps onto the bottles. Some filled packages were leaking as soon as they came off the line.

The management was unwilling to retool the production process and change the award-winning package design. The plant personnel were left to their own devices to solve the problem of reliably packaging the product. Someone discovered that the problem could be mitigated by preheating the caps in an oven to make them more pliable. This procedure required extra labor. Eventually, the work-around was supplanted with a heated hopper, which fed the caps into the filling/capping machine.

RULE #59 PROCESS DECISIONS CANNOT BE MADE IN ISOLATION

A decision taken by one functional group without concern for other aspects of the process will eventually cause problems.

Unfortunately, the prototype model caught fire, spoiling a large batch of caps.

Economics of Scarcity

Everyone thinks his own burden heavy.

French Proverb

During a 2007 holiday, this author and his family stayed at a tropical beach resort. The facility had many comfortable beach chairs and shaded cabanas where guests could relax and enjoy the beautiful ocean and pool venues.

Strangely, many choice locations sat empty all day. Early risers would claim a spot with their beach towel, intending to come back later to occupy it. These people became distracted by the many other diversions, leaving the chairs to sit vacant for long periods.

This selfish and inefficient behavior is driven by the economics of scarcity. People behave differently when they expect that products and services might not be readily available. Humans are driven to hoard resources, particularly when it costs them little or nothing to do so. When this behavior is not discouraged, it forces others to do the same, aggravating a minor shortage into a serious problem. A perverse psychology can infect the development process when the compensation system predominately rewards individual accomplishment over the performance of the whole organization.

The "concept map of human interaction with the process" at the end of the chapter shows the assorted people who are affected by process dependency in some way. The "concept map of human factors in product development" describes how people influence the technical organization.

RULE #30 NEVER REWARD SELFISH BEHAVIOR

Technical professionals are prone to overbook the process equipment, technicians, and instruments that they share with other development teams. This is most likely to occur when these individuals are evaluated and compensated on the basis of their personal success. In a dysfunctional organization, the various managers will join in the arguments over scheduling, instead of organizing mutual cooperation.

Genuine resource shortages are an indication that the organization is trying to push too many projects through the development pipeline. If there is a real limitation on a very expensive piece of equipment, then creative scheduling and overtime solutions should be devised to satisfy the needs.

Management should eliminate those projects offering the least benefit so as to free-up resources for more promising projects. Engineers and technicians should be judged and rated on how efficiently they can use the available facilities to get the information that they need.

The problem is not just limited to technical departments. Salespeople are bad about encouraging their customers to order extra product during the busy season to maximize their commission income. The sales slump during the ensuing slow season is even slower because the customers have too much inventory. This scheduling annoyance may cause serious problems for the production process that has to deal with it.

Summary

- Scientists and engineers often overlook, avoid, or underappreciate human behavior. These soft factors can represent the most significant uncontrolled parameters in a dynamic process.

- Process variation can be caused by workers who are diligent and skilled.

- The variation in technique from one operator to another is best observed by audits—spending time in the operation making observations. This should be done across all locations and shifts.

- It is often necessary to take decisive action to address process problems that arise. However, managers should not allow those short-term decisions to interfere with deeper investigations and long-term solutions that may be completely different from the original fix.

- Experts must document and disseminate information to the rest of the organization.

- Rotate people between different jobs to ensure that they have a full understanding of the organization and that the process knowledge is widely communicated.

- Beware of creating a mentality for hoarding resources when individual engineers optimize their own tasks and rewards, at the expense of the larger organization.

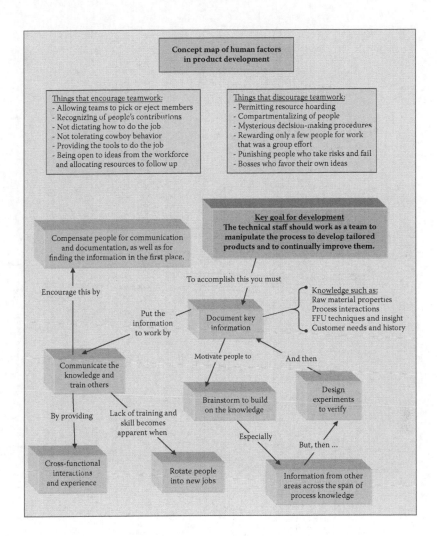

Concept map of human factors
in product development

Things that encourage teamwork:
- Allowing teams to pick or eject members
- Recognizing of people's contributions
- Not dictating how to do the job
- Not tolerating cowboy behavior
- Providing the tools to do the job
- Being open to ideas from the workforce
  and allocating resources to follow up

Things that discourage teamwork:
- Permitting resource hoarding
- Compartmentalizing of people
- Mysterious decision-making procedures
- Rewarding only a few people for work
  that was a group effort
- Punishing people who take risks and fail
- Bosses who favor their own ideas

Compensate people for communication
and documentation, as well as for
finding the information in the first place.

Key goal for development
The technical staff should work as a team to
manipulate the process to develop tailored
products and to continually improve them.

To accomplish this you must

Encourage this by

Put the
information
to work by

Document key
information

Knowledge such as:
Raw material properties
Process interactions
FFU techniques and insight
Customer needs and history

Communicate the
knowledge and
train others

Motivate people to

And then

By providing

Lack of training and
skill becomes
apparent when

Especially

Design
experiments
to verify

Brainstorm to build
on the knowledge

But, then ...

Cross-functional
interactions
and experience

Rotate people
into new jobs

Information from other
areas across the span of
process knowledge

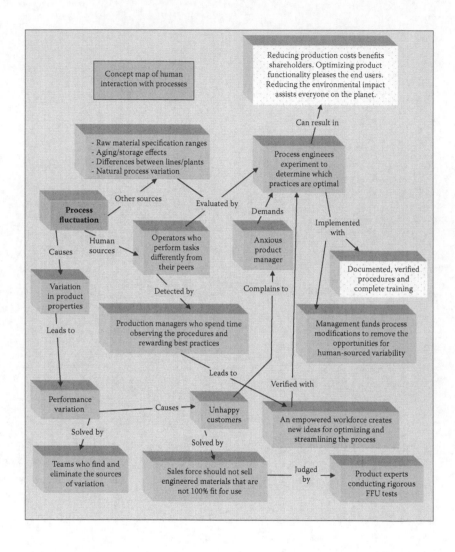

Concept map of human interaction with processes

Reducing production costs benefits shareholders. Optimizing product functionality pleases the end users. Reducing the environmental impact assists everyone on the planet.

Can result in

- Raw material specification ranges
- Aging/storage effects
- Differences between lines/plants
- Natural process variation

Other sources

Evaluated by

Process engineers experiment to determine which practices are optimal

Demands

Implemented with

Process fluctuation

Human sources

Operators who perform tasks differently from their peers

Anxious product manager

Documented, verified procedures and complete training

Causes

Variation in product properties

Detected by

Complains to

Leads to

Production managers who spend time observing the procedures and rewarding best practices

Management funds process modifications to remove the opportunities for human-sourced variability

Leads to

Verified with

Performance variation

Causes

Unhappy customers

An empowered workforce creates new ideas for optimizing and streamlining the process

Solved by

Solved by

Teams who find and eliminate the sources of variation

Sales force should not sell engineered materials that are not 100% fit for use

Judged by

Product experts conducting rigorous FFU tests

15

Managing Customer Expectations

How do we communicate the issues of process dependency to our customers? How can we turn that negative into a positive in their eyes?

> Start with the idea that you can't repeal the laws of economics. Even if they are inconvenient.
>
> **Larry Summers**

The design engineers who select engineered materials for their products rarely have all the necessary information concerning the performance of your products. The properties of traditional materials are well documented in engineering handbooks and design guides. But, a balanced perspective on new materials is hard to come by. Few suppliers are anxious to publicize the potential drawbacks of their products.

One design engineer noticed that, on paper, nylon seems to be a perfect choice for bearing surfaces. Its strength, wear properties, machinability, and low coefficient of friction (COF) suggest that nylon bearings will have a long service life. This proves to be true for those applications that operate at low speeds in dry conditions. However, in humid conditions, the situation is less favorable, because nylon components absorb moisture and swell in size. This changes its COF properties and causes the bearings to seize up. It is often necessary to disassemble the equipment and modify or replace the offending parts in order for the design to operate properly.

All machine designers eventually get burned by nylon or some other engineered material that they do not fully understand. Engineers are anxious to be the first in their field to try interesting new products. Materials that are advertised as high-tech or high-performance are the most seductive. An exciting improvement in one attribute may conceal deficiencies in other aspects of its performance.

Modern artists have also experienced problems with their choice of materials. Conservators are discovering that important artwork from the 1960s is not aging well. During that decade, artists threw off the proven, traditional methods and experimented with totally new media. Many of the choices enabled them to make bold, provocative statements. But often, these alternatives have not aged as well as the old masters art. Organic pigments have lost color, plywood substrates have delaminated, while adhesives and latex have disintegrated. Artists embraced new materials, but they failed to perform life tests.

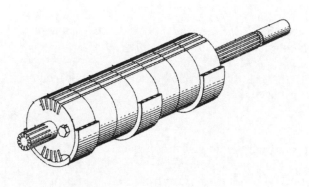

June 22, 1965   L. G. GITZENDANNER   3,191,079

HEAVY DYNAMOELECTRIC MACHINE HAVING NYLON BEARINGS

Filed Oct. 31, 1960   2 Sheets-Sheet 2

People expect that machines will eventually need repairing or replacing. However, they are not prepared for their 10 million dollar artistic master-piece to fade, sag, and crumble. Fortunately, material suppliers are rarely held responsible for the poor choices made by artists.

Your dynamic process does matter to your customers, whether they know it or not. Its shortcomings will cause them problems at the worst of times. Your ability to adjust and manipulate the process may be the key to satisfying their unmet needs. But, your process excursions, temporary work-arounds, and operator jury-rigging may cause no end of trouble.

Customer Mentality

Market research will only establish what your prospective customers were doing in the past.

The Law of Intelligence Obsolescence

In 1985, the Coca-Cola Company made a famous misjudgment of consumer expectations. The brand had been embarrassed by Pepsi-Cola advertisements touting blind taste tests in which the consumers preferred Pepsi to Coke. In response, their experts formulated a new version, which scored better in taste tests. Boldly, the marketing arm of the company introduced the improved formula as the replacement for the traditional product. Many people accepted the change without complaint. However, an angry revolt by a vocal minority and the resulting negative publicity forced an abrupt change of plans.

The issue was not a technical matter. The masters of brand marketing forgot that consumer loyalty is often driven by image and perception rather than the tangible properties of the product. Surprisingly, an earlier process change was met with much less consumer resistance. Between 1980 and 1984, Coke had gradually and quietly switched their sweetener from cane sugar to high-fructose corn syrup in the U.S. market. This change was driven by economics and it generated little reaction because it was shrouded in secrecy.

Cornstarch and cane sugar can also be used in the production of an aliphatic polyester called PLA. This polymer is considered to be environmentally friendly, due to its renewable source and biodegradability. In 2009, Frito-Lay introduced a new PLA-based package for their Sun Chips product line.

The consumers were initially receptive to the green image associated with the package, which displaced plastics made from oil and gas feedstocks. However, in late 2010, the experiment was scaled back. Many people objected to the irritating noise of the stiff PLA film when the bags were handled. They had become accustomed to the incumbent packages and were not willing to accept negative factors in their product experience.

Instant coffee has a very poor reputation among people who savor the high-quality brews served by their favorite coffeehouse. The early twentieth-century efforts in the field were judged to be of poor quality, until process improvements such as freeze drying and spray drying were introduced. These unit operations were a big improvement, but the products remain a poor imitation of freshly brewed coffee.

However, the market for soluble dry coffee is large and has been the focus of intense industrial development efforts for over a hundred years. There will always be a portion of this market who will accept marginal performance in exchange for low cost and the convenience of a quick solution for their caffeine addition.

These secondary issues can turn a new product into a "technical success/ marketing failure" situation. Businesses can be sorted into several groups based on how they choose to purchase products and services. Marketing should be geared to the needs of each customer type.

Purchasing modes of operation:

- Very price focused. These customers switch vendors frequently on the basis of price. They are willing to take off-spec, defective, or refurbished materials and will adjust their operation to deal with material variability.

RULE #21 CUSTOMER REACTIONS TO A MATERIAL WILL CONSTANTLY SURPRISE YOU

Materials always have secondary attributes that the producer is accustomed to and takes for granted. But, the customers may have negative impressions from the subtle variations in the texture, smell, or sound of a product. They might associate these properties with a product that did not work well for them in the past. Cultural, economic, geographical, and gender differences make these reactions hard to predict. An item that has found a successful niche in one region or market segment may not be as favorably received elsewhere due to differences in intangible factors.

- Somewhat oblivious or unpredictable. These customers sometimes pursue low-cost materials in one application, but they will also stick with familiar materials for long periods in other functional areas. They may buy on the basis of personal relationships with the vendor or distributor.

- Central command. These companies buy the exact same product at many different locations, based on a national or global contract. All materials must be specified by a central group. Individual plants may not have input. Prices tend to be contractual. Vendor relations can be combative, playing one supplier off against another.

- Service seekers. They expect products to be packaged and delivered in very specific ways. They may want just-in-time delivery, preassembled items, custom products, and an intensive technical service. They tend to stick with proven suppliers with which they have had a good experience.

- Quality obsessed. Customers who have process-sensitive products may be very concerned about their incoming materials. These companies want to specify every product aspect and appreciate consistency and quality from their suppliers. Small differences in secondary attributes may catch their attention and raise concerns.

- Constrained by regulations. Some food, drug, and government applications may be limited by conservative regulatory bodies. No matter how superior a product is, it may never be allowed for the application if regulators are not moved to do so. Expensive testing and lobbying may be more important to success with these customers.

- Deluded developers. Some R&D organizations are prone to making verbal promises to a vendor in return for the creation of specialized products. The manufacturing and purchasing groups may not agree to honor those terms once demand for the products ramps up.

- Tree huggers. Some customers have a business model that seeks to court the environmentally concerned consumer. They may reject a great product if it cannot be recycled or made from recycled or renewable materials. The perception about a product's environmental impact will be more important than the reality. New materials will be shunned if consumers associate them with older, less desirable products.

Developers should be very aware of what type of customer they are dealing with when customizing a product. The worst customer is one who seems to be very quality or service sensitive during the development phase and then turns out to be price focused when the product is commercialized (or vice versa). *Carefully consider which customer types you can successfully service.*

If your process is very prone to wide swings in performance, then avoid the quality-sensitive customer. Conversely, a plant that specializes in tight specifications and carefully crafted products cannot maintain a lasting relationship with price-focused end users, except as an outlet for excess inventory, scrap, or off-spec products. If suppliers operate with the mentality and staffing level of a commodity-oriented company, then they will never consistently satisfy service-seeking customers.

The central command customers will not be happy with a vendor who lacks the geographic reach to serve their far-flung locations. Failure to meet and document the letter of government requirements is always a bad thing when selling to regulated industries.

When marketing cars or other major products, it is well understood that unspoken wants and needs greatly influence purchasing decisions. Consumers might tell the auto dealer that they want reliability, value, safety, and good fuel mileage. While being shown the hybrid or the stodgy family wagon, they are drawn to the sports cars or pickup trucks. Auto companies know all the reasons why people purchase vehicles; hence, they offer a range of models that fill each emotional niche.

The sale of industrial products is less impulsive, but it can still be influenced by the heart as well as the head. IBM and AT&T traditionally sold their products and services by preying on the fears of their IT managers. "No one ever got fired for buying IBM," struck fear in the heart of a risk-averse support staff.

Have a close look at your packaging through the eyes of a customer. The price-driven customer and the tree huggers will not appreciate excessive packaging, although the quality seekers might demand it. Do you have the flexibility to package things differently, depending on who is buying it?

Do not undermine the trust of the purchasing agent who selected your product by ignoring the culture of the larger organization. Every customer has different hot-button issues. The sales staff needs to probe the emotional undercurrents and document the essential wants and needs. A company that is not prepared to meet those unwritten requirements will struggle to gain acceptance.

Customer Input

> Research is the process of going up alleys to see if they are blind.
>
> **Marston Bates**

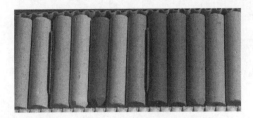

There are many examples of successful products that would never have been created without the insistence of demanding customers. These people will approach any prospective supplier that will listen to their ridiculous requirements.

In the late 1940s, Henri Sennelier ran an art supply business, directly across from the Louvre in the heart of Paris. The shop was then 60 years old and had a reputation for formulating dependable colors that some of the world's best artists needed.

Henri was asked to create an alternative art medium that addressed the shortcomings of existing paints. The requested product would be applied directly to any surface without priming or other preparation. The shop owner was highly skeptical that there was a commercial market for such a product. However, the customer is always right, especially when his name is Pablo Picasso.

Sennelier spent a year creating a line of materials, which became known as oil pastels. The product came in sticks, similar to crayons. It was a mixture of wax, pigments, and inert oils that would not harden and crack on the surface of the artwork. He produced 48 sticks, each in 40 different colors, including a range of grays that the great man had requested. Picasso bought 30 sticks of each color for himself and the remainder was snapped up by other customers.

The product was a hit and became accepted by artists the world over. Other major brands have copied the original product. These firms compete with Henri's son and granddaughter, who continue the family business to the present day.

RULE #51 DO NOT FILTER PRODUCT REQUESTS. PASS THEM TO THE PEOPLE WHO CAN EVALUATE THEM

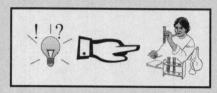

The new product request and the background details need to make their way unfiltered to the research experts who can brainstorm ways to turn them into prototype products.

Before that research goes too far, someone who understands the market should determine whether the products have the potential to become lucrative for your company. Herein lies the biggest problem in product development—determining which promising ideas merit the allocation of resources.

In Sennelier's case, he was the firm's paint formulator, the principle owner, and the friendly face behind the counter who had daily contact with the artists. With this rare combination of responsibilities, he could evaluate the ideas and their impact on his business. As the owner, he knew that satisfying the requests of the world's most famous artist was an end in itself. As a formulator, he could picture solutions to a problem and experiment with the many pigments and materials available.

In a medium or large company, there is generally no one single responsible person who has both the knowledge and the authority to make those decisions. The knowledge involves technical, market, and financial considerations. The organization must have a robust system to disseminate the appropriate information, collect opinions, and make the decision.

Researchers must look for technology that magically transforms silly ideas into viable products. The sales force should search for additional applications to provide the commercial justification for a production process that has otherwise too little volume. Some requests may be rejected immediately, based on prior experience. Others may take considerable time to evaluate.

A "no" answer should only be considered valid for that moment in time. Research and development professionals become very frustrated when their ideas are rejected and are then resurrected as someone else's idea. The effect on moral is poisonous if significant financial and professional rewards are given only to the originators of new product ideas. When only a few people are granted large rewards, then no one will be motivated to cooperate for the greater good.

From the other side, it is possible that a customer request could become the basis of a new product years after it was first proposed. In the meantime, the supplier forgets the early idea that first sparked the researchers to explore the idea. After originally being told "no," that customer may see the idea being sold to their competitors, perhaps for reasons that seemed perfectly valid to the vendor. This can result in a very unhappy customer.

In the early twentieth century, a New York tea merchant Thomas Sullivan devised a new way of packaging samples of tea. Single servings of tea leaves were packed in small cloth bags. Customers were supposed to open the sack and pour the contents into their teapot. Instead, some people dropped the entire bag into the hot water. This was faster and it allowed for clean and easy removal of the wet leaves afterwards. Customers were soon clamoring to buy their tea in this form.

Sullivan was surprised by these requests. He had accidentally solved a problem that he had not been aware of. Tea companies developed processes and materials to make large numbers of tea bags at low cost. These soon became very popular in the United States. Penetration into the English market did not come until decades later.

Critical Questions: What is our organization's reaction to customers who bring new product ideas? Are we recognized in the market as the best solution provider? Is our development group structured and equipped to jump on promising opportunities? Do we hold the customer's attention during the time needed to perfect the concept into prototype products?

Knowing Where You Stand

Everything that deceives may be said to enchant.

Plato

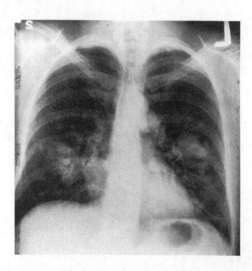

Under pressure from numerous fronts, American tobacco companies spent millions of dollars in an effort to fundamentally change cigarette design. They sought to create a product with less carcinogenic residue and less secondhand smoke. But the companies still wanted to sell an addictive product that appealed to consumer tastes. Since the 1980s, several innovative products have been put forward to deliver the pleasurable components to the smoker's lungs without exposing the rest of us to their fumes.

These products required tremendous R&D effort to mimic the experience of a regular cigarette. Philip Morris offered a nicotine-free cigarette called

Next. They used supercritical carbon dioxide to extract the nicotine from the tobacco, similar to the techniques used for decaffeinating coffee beans. However, smokers were not satisfied with a smoking experience that lacked the addictive benefits.

R. J. Reynolds devoted tremendous time and resources to developing and promoting a new cigarette that heated tobacco pellets in a tiny aluminum cylinder. It was intended to reduce the smoke, tar, and ash produced by conventional tobacco products. However, it required a special procedure to ignite the product.

In the end, no one was happy. As a group, the cigarette consumers were not satisfied with their first tastes of the novel product. Generally, they were unwilling to alter their habits to accommodate the benefits aimed at other people. The potential health benefits were not compelling. The antismoking forces did not see the innovation as a sufficient improvement and continued their push to ban all smoking in indoor settings.

Smokers have very specific expectations that are best satisfied by products that have been developed and refined to address these cravings. Customers with very specific needs and expectations can be very disappointed when product performance falls short due to batch-to-batch variation.

In the United States, the price of table sugar (known chemically as sucrose) is held artificially higher than the world price due to import restrictions and limits on domestic production. This value discontinuity has created an opportunity for alternative sweeteners in food and beverage products. One such option is high-fructose corn syrup, which is produced by converting cornstarch into sugars. The product is a blend of fructose and glucose, which is chemically very similar to sucrose. This combination circumvents the price controls on cane sugar, but it cannot be identified as sugar on package labels.

Food processes are attracted to fructose for reasons other than cost. It is sweeter than sucrose, so less of it is needed. Other than sweetness, it does not have a flavor of its own, which helps the product formulators. Finally, fructose is more attractive to water than sucrose is, so it retains moisture in products that need to be chewy, rather than dry.

Many consumers react negatively to the listing of high-fructose corn syrup on a product label. They perceive it to be less natural than chemically similar choices such as sugar and honey. A compelling process and cost advantage for one type of company does not translate into a benefit for customers, when they have other options that are emotionally attractive.

Customer perception of process dependency:

- The product might not look and perform exactly like the competition. Salespeople have to convince their customers that this is an inherent strength and not a weakness.
  - Every team member must be on message. Different functional areas cannot be telling the customer different things.
  - Focus the field people on the products, not the process. They must be selling the advantages of your materials into applications where those things matter and offer value to the end users. There is no future in defending the shortcomings of products that are not appropriate for a given application.
- There may be some inherent variation in the process from batch to batch in attributes that do not control its functionality. If the process has inherent variability in critical attributes, then you are looking at the wrong application.
- A process-driven organization may not be able to drop everything and jump into the production of a rush order from a customer.
  - Make sure that new customers know that published lead times exist for a purpose.
  - Sales needs to understand the seasonality of the customer's order patterns and communicate this to the production scheduler.
  - Significant development time may be required to meet the customer needs.
- It may not be easy to customize the product in ways that the customer requests. More importantly, it may not be possible to do so in a cost-competitive manner at the volume that the customer purchases.
- Sales staff may seem powerless to make decisions without consulting the home office.
- A process-driven supplier may not be able to meet all the customer's product needs, only those that its process is efficient at producing.
- They may feel threatened that your development team wants to learn about their operations to create accurate fitness-for-use (FFU) tests.

Process-dependent companies must act differently:

- "We aren't Burger King. You can't always have it your way. This is our product, take it or leave it." The salespeople for a process-driven company need to get that idea across to the customers without seeming to be arrogant or defensive. There must be some benefit to the relationship or they would not be looking at your product.

- A company cannot give blanket guarantees to things that have not been proven out on the process or in testing. Salespeople have to express confidence about the company's capabilities without making promises about specific requests.

  - This is especially true when FFU testing is needed. Offer to test their application, but do not guarantee good results in the initial attempt.

  - Perfect your solution with in-house FFU tests and low-volume trial customers.

  - Sell your flexibility to adjust material properties to customer applications.

  - Clarify who takes the risk when the consequences of failure are significant.

- Develop FFU tests without alarming the customers that you want to know more about their business than they do. Ideally, customers should recognize the benefits of the supplier ensuring that the products will work right.

- Monitor the product's behavior in the end use to watch for changes. Do this without making the customer nervous that the development team is constantly worried about a possible product failure.

Getting the Business

Marketing is what you do when your product is no good.

Edwin Land

In the modern world, there are many different variations of synthetic materials that can be fabricated into bottles and rigid containers for packaging consumer products. New compositions of matter are introduced every year. However, only a very few families of materials are documented in the recycling codes, which are stamped or printed on the packages. Without a code designation, consumer recycling is difficult.

Whenever a beneficial new material is commercialized, there is strong commercial pressure to force it into an existing category. This is certainly the case with polyester copolymer resins that are intended to replace polycarbonate. In recent years, concerns about bisphenol A have caused consumers to shun polycarbonate for use in demanding applications such as baby bottles. The production of new polyester grades has increased rapidly to serve this market. The new copolymers are similar to the existing homopolymer, polyethylene terephthalate (PET), which is widely used in soda and water bottles and which uses the number 1 recycling code.

However, they are not exactly the same. The new polyester grades have a higher service temperature, which allows hot water sterilization applications. These differences are not favorable for companies that recycle the number 1 waste stream back into new bottles. If the percentage of copolymer bottles in a bale of waste PET is low (less than ~5%), then mixed scrap can be reused. At high loading levels, the product properties of scrap blends will be unacceptable. Polyester producers are scrambling to modify their copolymers to make them more compatible with PET while adding functionality. Recycling operations are looking for ways to compatibilize a changing waste stream.

Customers will often hesitate to test a product if it does not offer significant cost savings or performance improvements over their current options. They may not like new suppliers that offer a range of different samples, which must be tested to fine-tune the final offering. Worse yet, the process-sensitive supplier may not even be able to set a price in advance, until experimentation reveals the best option.

This reluctance is only reasonable. The existing supplier has a known level of quality, price, and performance. Switching to a new vendor is a risky alternative that can only be justified by a major cost saving or a quantum jump in performance.

Sales strategies for process-dependent products:

- Seek out an understanding of what the customer really wants, needs, and values. The purchasing agent never knows the whole story. The people who use the product are always dissatisfied with some aspect of it.

- Sell the total value proposition: the organization and its technology, products, and service. Look for inherent advantages that will save the customer money in manpower, energy usage, production rates, and yield loss. Focus on reducing the total cost to the customer, rather than your price. Determine how to customize the offering to solve their problems.

- Document benefits that have been provided to similar customers.
- Market-focused vendors emphasize their commitment to serving the industry by offering multiple solutions to problems.
- Process-focused vendors should stress that they are basic in the technology and understand the process better than anyone else.
- Look at the customer's incumbent product and analyze whether your process has an advantage. This could be price, consistency, or performance.
- Seek to demonstrate the value of service, delivery, innovation, and quality.
- Understand the competitor's position to the customer. Are they process focused or market driven?
  - What products do they currently sell to this facility? Can all these be replaced or only a portion of them? Is the pricing picture being confused with rebates or other incentives?
  - Are they charging the customer more than other companies are in the same business? Will they drop their price to match a lower offer? Is this a lucrative market for them?
  - Are there one or two items to cherry-pick from to get a foot in the door?
- Understand who is really in control of the purchasing decision. If a material is sold to the end user by an industrial distributor (i.e., an independent sales agent or middleman), try to determine who specifies the manufacturer of the product. If the incumbent middleman and the customer's owner have a personal relationship, then it may be difficult to displace that distributor. If so, sales should be focused on getting the distributor to switch supplier allegiance.
- Consider pull-through approaches by demonstrating the benefits of your materials in the products that end users buy from your customer.
  - This approach can anger customers if it is not done in a tactful and cooperative fashion.
  - Expect a long sales cycle and more support requirements.
- Consumers can be persuaded through advertising to buy products made with your materials.
  - This approach may take a long time and may only work for patient, committed companies.
  - Make sure that other competitors cannot ride your coat tails and achieve the same benefits with a similar-looking product. Your brand must carry through to the end user.

Keeping the Business

When you stop talking, you've lost your customer. When you turn your back, you've lost her.

Estée Lauder

Large companies in the processed food business have multiple cooking laboratories to evaluate and demonstrate products in a variety of applications. Some have a kitchen layout for each major class of customer. A few are exact duplicates of the cooking facilities in specific chains, such as McDonalds. These test kitchens permit the FFU evaluation of products in a realistic customer setting. Development teams can understand the environment that the products will experience. When the products are ready, full-scale demonstrations of the new concepts are done in the test kitchen. Customers can see exactly how the products will affect the efficiency and workflow of their operations. Additionally, video presentations and training materials are produced using these facilities. This approach provides a service for smaller chains that lack the facilities to evaluate new equipment. It is the optimal means to demonstrate a systems solution that combines consumable products, equipment, and services.

Market-focused companies optimize performance for customers:

- Use an incumbent position with customers to continuously improve the value of your offering. Constantly generate data, linking process parameters to performance in each application. A gradual process evolution steadily improves product functionality, rather than being a cost-cutting exercise.

- What are the shortcomings of this product in each area of the customer's operation?
- Use your sales and service presence to report back specific results on each batch, especially if the FFU is variable. Adjust the process to zero on conditions that consistently perform best. Lock them in.
- Customize and differentiate products to optimize each particular subapplication.
- Find similar applications at other companies where you can do comparable improvements.
- Market-focused and technology-focused companies develop new products that will cannibalize existing ones to improve their market position.

- Brainstorm how a competitor could get this business. What are the product's strengths and weaknesses in this application?
  - Do not limit the equation to product attributes. Consider distribution channels, payment terms, service, customized packaging, and delivery.
  - Do not drop your price just because the purchasing agent complains. Examine how well the alternatives work versus your product and the comparative value.
  - Demonstrate how you can justify your price without appearing arrogant.
- Model the total value proposition to understand how it changes in new situations.
  - Do rising fuel and labor costs improve or damage your competitiveness? How can you improve efficiency before the energy market changes and you lose the business?
  - Can you proactively improve your solution before someone takes the business away?
- Use intimate customer access to find and possibly test ideas for additional products. It can be far easier to do so with an existing relationship rather than an unfamiliar one. This role is not restricted to the sales and marketing staff. Send technical people in there as well.
  - This contact will spark ideas for new products and solution opportunities.
  - Provide advice and services to help the customer even if it does not immediately result in more business. Make your experts an asset that the customers want in their facility.

- Test your latest innovations on customers with whom you have a good relationship. They will consistently give better cooperation and feedback than people who do not know you.

The more employees that you put in contact with customers, the wider your span of knowledge will be.

Critical Questions: What services could we offer to encourage customers to choose our products? Should we sell these services or provide them for free?

Summary

- Customers may not appreciate that your lengthy product development cycle is an evolution, generally requiring experimentation in their facility.
- Innovation means that customers will constantly experiment with new and unexpected uses for your products.
- Classify potential customers based on their operating philosophy before starting to develop a customized product for them. Do not go too far with the project before determining their requirements.
- Customers may spout a corporate philosophy to express their needs for a new product, but then act very differently when their time and money are required.
- Do not employ the same sales strategies for process-dependent materials as those that are used for conventional products.
- Consider how all the customers' functional areas will perceive the product and its inherent variability, not just the purchasing department. What do they want and need?
- Customer requests and their background details need to make their way unfiltered to the research experts who can brainstorm ways to turn them into prototype products. Before that research goes too far, someone who understands the market and the company strategy should determine whether the prototypes have the potential to become lucrative products. If not, promptly communicate that decision back to the customer.
- Be open and honest with the customers. Tell them what they can expect. Do not undertake a development project, hoping that your

inherent process shortcomings can be worked around, if the customer really requires a flawless solution.

- Sell a total solution, including your ability to tailor the process to get the optimal properties.
- Use your incumbent position with existing customers to constantly refine and improve your product. Consider how your competitors might offer an even better solution.

The "concept map of customer expectations" summarizes the love/hate relationship between customers and process-dependent suppliers. Their relationship depends on a clever manipulation of a dynamic process.

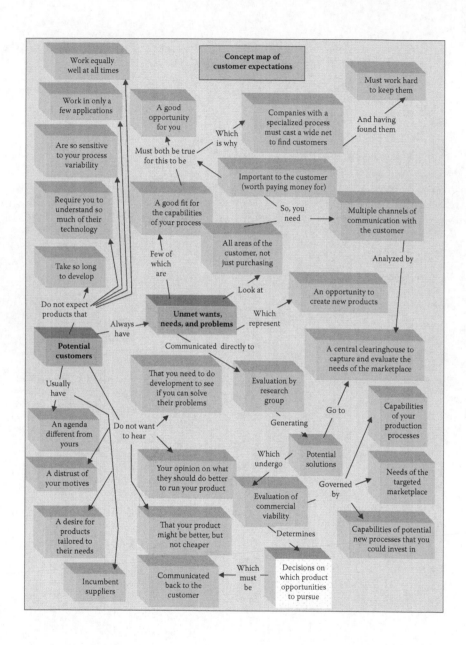

16

Proprietary Systems

How do we integrate product offerings to provide complete solutions? How must the process change to interface with a proprietary system?

> Clothes make the man. Naked people have little or no influence on society.
>
> **Mark Twain**

The classic proprietary product combination is the safety razor system. King Camp Gillette conceived of the idea while shaving when he was a traveling salesman in the 1890s. From the beginning, he dreamed of selling cheap, disposable blades, which would fit into a reusable holder. The realization of his vision required 6 years of experimentation before a production process could be constructed. The process to stamp, harden, temper, and sharpen the thin metal sheets into uniform blades was more challenging than he anticipated.

The system generated income by selling profitable replacement blades. The handle was more expensive than strictly necessary, because adjustability was designed into it. This allowed consumers to change the spacing between the cutting edge and a guard surface to accommodate different application situations, with only one blade design. This greatly simplified the logistics of making, distributing, and explaining the product, while accommodating a range of consumer preferences.

Modest production was initiated in 1903, years later than Gillette's original expectations. Sales rose quickly over the following decade, allowing him to retire as a wealthy man. This success attracted competitors, sparking battles over intellectual property and forcing Gillette to constantly innovate and improve.

In 1947, Edwin Land introduced his proprietary instant photography system with a similar business model. Customized cameras were designed to accommodate special film packs to bypass the lengthy development step of conventional photography. In doing so, he took George Eastman's convenience factor to a new level. Polaroid designed an intricate production process to combine the photosensitive film, the paper backing, and the chemicals needed to develop a picture within minutes.

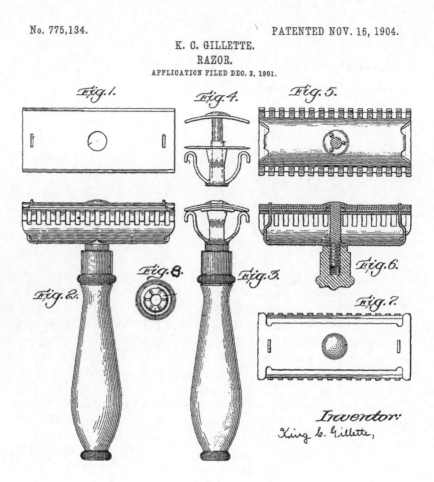

No. 775,134.   PATENTED NOV. 15, 1904.
K. C. GILLETTE.
RAZOR.
APPLICATION FILED DEC. 3, 1901.

Inventor:
King C. Gillette,

On the downside, the system was considerably more expensive and had poor image quality compared to darkroom processing techniques. But, it created a market niche that had been overlooked by the dominant Eastman Kodak company. Many consumers were willing to pay more for instant results. Over the years, Polaroid refined the system to offer more features and improve quality. However, it was always deficient in terms of unit cost and resolution compared to the less convenient alternatives.

On the commercial side, robotic assembly systems, programmable logic controllers, and laboratory spectrometers may require software customization, replacement parts, and periodic maintenance that are provided under the terms of an annual service contract. Contracts of this nature sometimes produce a better margin for the vendor than the original equipment sale. The idea is that they form the basis of an ongoing relationship, which will generate additional business in future years.

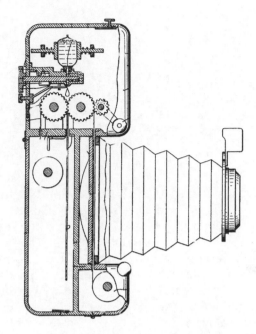

Many vendors bundle products to offer one-stop solutions. In return for the free use of a copying machine, a business might commit to buying all its paper needs from a given distributor. Such commercial arrangements, while useful to both parties, are not proprietary systems. The copier is capable of using reams of standard office paper, which are readily available from a host of different suppliers.

RULE #41 IDENTIFY AT LEAST ONE CRITICAL ATTRIBUTE THAT TIES YOUR CONSUMABLES INTO YOUR PROPRIETARY SYSTEM

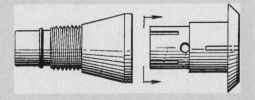

In a truly proprietary system, there is some attribute that only allows one type of consumable product to be used on the equipment. This could be a unique mechanical, electrical, or pneumatic coupling that is incompatible with the industry standard.

Around the seventh century AD, the Byzantine Empire developed an effective weapons system known as "Greek fire." This early flamethrower was highly destructive and frightening. Its use in battle was credited with breaking the two sieges of Constantinople. Much is made of the lost recipe, which was probably based on a petroleum derivative, similar to naphtha. In fact, the secret was not just the combustible agent, but also a proprietary system of pressurized vessels and brass nozzles to deliver the flaming liquid. The secrets of the technology were so closely held that they were lost to the Byzantines after only a few years.

In the 1940s, Harvard chemists developed a modern equivalent, which used the aluminum salts of naphthenic and palmitic acids as a thickening agent for gasoline. This gel is useful for military applications because it burns more slowly than pure gasoline and it does not dissipate into the air as quickly. As with Greek fire, napalm requires specialized equipment for it to be delivered effectively to its wartime target.

Ideally, the marriage of specialized materials and equipment brings unique benefits to the customers, prompting them to choose that system, rather than alternative solutions. It is highly desirable to have some factor, such as patent protection, which prevents competitors from offering one or more knockoff components to the system. This is critically important when the pricing structure requires the supplier to take a loss on one of the system components.

Benefits of proprietary systems:

- You control all aspects of the customer interface. Compatibility problems between your material and third-party equipment are less likely.

- The barrier to entry for competitors is higher.

- Customers are reassured that your complete solution to their problem will not be complicated by integration issues between multiple vendors.

- Sweeping innovation can be implemented quickly when you control the supporting infrastructure.
- System elements can be designed to accommodate the process variability of your materials.
- Material changes can be implemented smoothly because you control the design of the entire system.
- The systems approach is a good way to supply turnkey solutions to a series of similar end users. It requires your company to be market focused.

System Trade-offs

Wisdom consists in being able to distinguish among dangers and make a choice of the least harmful.

Niccolo Machiavelli

This product may contain up to 10% ethanol by volume.

Automotive gasoline and a car's catalytic convertor are two components of a system. This is not a proprietary system, because the components are not sold together for the exclusive benefit of one vendor. Standard fuel is expected to function in the gasoline engines of all the different suppliers. However, the two products must work together. Lead additives for gasoline had to be phased out when catalyst systems were mandated for cars to reduce emissions.

More recently, the U.S. Congress has contemplated doubling the ethanol content of gasoline from 10% up to 20%. This is intended to reduce the use of fossil fuel and cut greenhouse gas emissions. Some environmentalists question whether ethanol derived from corn fermentation will reduce the overall carbon dioxide output. Industry experts worry that increased levels of ethanol will increase the operating temperatures of catalytic converters to dangerous levels. This could significantly reduce the service life of the costly catalyst element in cars already on the road.

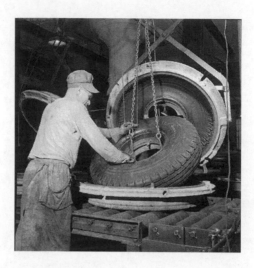

The combination of automobiles and tires is another example of open systems. The two products are always made by different companies. In most cases, engineers from two or more companies will cooperate to create specifications to ensure the safety and reliability of the final product. This was the case for the Ford Motor Company and the Bridgestone Corporation in the 1990s. Engineers collaborated to specify a Firestone radial tire for the Ford Explorer.

Ford designers created fitness-for-use (FFU) issues by selecting a tire that was slightly undersized for the heavy sport utility vehicle (SUV). They compounded the problem by recommending a low inflation pressure in the owner's manual. The underinflation was intended to address complaints that the Explorer had an uncomfortably rough ride. This low tire inflation pressure provided a smoother ride for the vehicle, but increased tire stress. The stress on the tires increased at high speeds and temperatures. If the tires were made correctly, these severe operating conditions would only result in excessive tread wear. But, some Firestone radial tires also had defect points of poor bond strength. The defective product suffered from ply separation and catastrophic failures under high-stress operating conditions.

Within several years, there were alarming reports of rollover accidents with the Explorer that were triggered by tire blowouts while running at highway speeds. Due to their high center of gravity, SUVs are more prone to rolling over in the event of tire failures. Published accounts blamed the two companies for over 200 deaths and 700 injuries. Both sides could point to mistakes made by the other. The sum of all the errors created a serious safety problem.

RULE #28 THOROUGHLY EVALUATE ALL ASPECTS OF SYSTEM COMPONENTS BEFORE IMPLEMENTING THEM

There are advantages and disadvantages to offering a truly proprietary system as opposed to conforming to an industry standard, whether it is formal or informal. Conduct extensive tests to ensure that the system can handle the full range of property variations and end-user conditions. Once the system design is deployed in large numbers, it can be very difficult and expensive to correct errors.

Downsides of proprietary systems:

- Customers may not see the benefits of a closed system over an industry standard. The supplier must offer a compelling advantage in either price or performance to induce customers to lock in a single-source vendor. Small suppliers will find this to be a difficult sale.
- Third-party suppliers may find it attractive to invest in process equipment to sell compatible products for this system. It is especially embarrassing when these knockoffs function better than the originals or are tremendously cheaper.
- Customer problems with component incompatibility become your problem.
- Customers can be clever at introducing competitive consumables into a closed system. Witness ink syringes to replenish the disposable ink cartridges in home printers. Creating special cartridge/printer combinations to prevent reuse requires significant development spending.

- Vendors of closed systems are limited in their ability to source additional consumables from outside vendors in the event of a shortage or problems in internal facilities. Very specialized equipment is generally needed to produce the product.

- Makers of system components may not be willing to work with you on other opportunities if you are their competitor with your proprietary system.

- It may be difficult to take advantage of process improvements that were developed by the original equipment manufacturers for the rest of the industry.

- In large-volume applications, the open standard may have a considerable price advantage over proprietary systems.

- Obsolete or defective inventory may be more difficult to dispose of due to the proprietary interface features. Selling off-spec components at a deep discount to your customers will cannibalize your sales.

- If your company is not a major player in the market, it is hard to convince customers to lock into your proprietary system.

- This is not a technique geared to process-oriented companies.

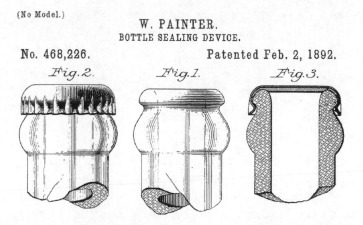

(No Model.)

W. PAINTER.
BOTTLE SEALING DEVICE.

No. 468,226.   Patented Feb. 2, 1892.

Fig.2.   Fig.1.   Fig.3.

In 1891, William Painter solved the problem of sealing bottles containing beer and carbonated soft drinks. The incumbent solutions tended to either be expensive or prone to leakage. Metal caps were not desirable because they could deteriorate when in contact with the liquids and could impart off-tastes to the product.

Painter's solution was a composite cap that combined a crimped metal body with a cork-sealing surface. The cork provided a hermetic seal while isolating the metal from the liquid inside. This approach required changes to the bottle design and the filling/capping machines that Painter patented.

Painter's complete solution included a bottle opener that allowed consumers to easily remove the cap. Despite the investment needed in new equipment, the advantages were so compelling that crown caps remain the dominant closure system for beer bottles. The materials have changed since the original product was introduced, but the concept is fundamentally the same.

Developing Products for Proprietary Systems

> Great minds must be ready not only to take opportunities, but to make them.
>
> **Charles Caleb Colton**

One of the early product systems was the combination of the safety match and the special striking surface that is needed to ignite it. This combination was commercialized by John Edvard Lundström, who made his native Sweden the leading nineteenth-century producer of matches.

Lundström's product was safer than the incumbent for two reasons. First, the ignition was not caused by friction alone. Rather, it was the act of rubbing the head of the match (containing potassium chlorate) against the striking surface (containing red phosphorus), that triggered the reaction. When stored apart, there was little chance that the product could ignite spontaneously. Secondly, the red phosphorus was considerably less toxic than the white version that was used in earlier match designs. This innovation greatly improved the health conditions for the factory workers. However, significant experimental work was needed to develop a reliable and consistent match/striker combination that could be manufactured economically.

In the 1930s, the National Cash Register (NCR) Company of Dayton, Ohio, specialized in using electromechanical technology to automate financial transactions at the point of purchase in retail shops. Their products became ever more complex to solve customer problems. Multiple copies of each receipt were desired. However, the poor quality of the existing carbon paper technology was holding up progress in this field. The company recognized the need for a new technology to make multiple copies of the printed output of their machines. The ideal solution would be less messy and more reliable. Toward this end, they employed Barry Green to develop a new consumable for use on their systems.

Green developed a process to encapsulate pigments so that their color would only be transferred at the desired point. The technique involved dispersing the pigment into tiny droplets in a gelatin emulsion. By then changing the acidity, the gelatin chains could be forced to precipitate out of the solution, coating the pigment droplets. A treatment with formaldehyde would then cross-link the gelatin coating into a durable, protective shell. Once dried, these tiny spheres were coated onto paper. A clay coating on the opposite surface facilitated the permanent transfer of the dye.

The resulting product was cleaner and more convenient than separate sheets of carbon paper, greatly enhancing the performance of NCR's systems. This same technology later enabled Chester Carlton's photocopiers (see Chapter 7) by allowing toner powder to be free flowing.

Until the late nineteenth century, all the world's national armies employed black powder as the propellant in their firearms. Gunpowder was largely unchanged since the Middle Ages. Despite widespread use, it had many deficiencies. The dense smoke revealed the position of the gunners while blocking their own view of the battlefield. Its residue fouled the weapons and necessitated frequent cleaning. These problems only worsened as mechanical innovations increased the rate of fire of new weapons.

In 1884, the French chemist Paul Vieille overcame these shortcomings with a new approach. Nitrocellulose was known to be an excellent propellant; however, it was unstable and tended to explode in storage. Vieille found a way to stabilize the material into a gel, which could be produced

on an industrial scale. The French tried to keep the secret to themselves, but within a few years, every national army had its own version of smokeless powder.

A gun and its ammunition work together as a system, so changes to one element require accommodation in the other. The increased power of the new propellant upsets the delicate balance of system design. Muzzle velocity and internal pressure went up dramatically. A longer range and better accuracy were achieved, but the weapons had to be completely redesigned to handle the increased power.

The existing weapon designs were scrapped and the entire arsenal was completely replaced in a matter of a few years. Infantry rifles transitioned to a smaller bore and higher rate of fire. The effective range of artillery increased dramatically, requiring new targeting techniques. A new family of machine guns became viable because the clean-burning powder did not foul the mechanism. In short, improvements to a consumable component of a system forced a complete redesign of the reusable components and enabled new designs that had not been possible before.

Military leaders did not fully anticipate how this revolution would alter their way of doing business. The enhanced firepower made it very dangerous to advance large troop formations across open ground. Mobile warfare became untenable as men hid in trenches to stay alive during World War I.

The product development of components for a closed system is different from creating stand-alone products. Decisions must be considered holistically and choices can only be made on the basis of extensive testing.

Johannes Gutenberg is credited with developing history's most influential integrated systems approach in the middle of the fifteenth century—the movable-type printing press. In fact, the innovation was more than just a printing machine; it was a complete process, incorporating the press equipment, a new ink that was tailored to the application, and a process for making a large volume of interchangeable type elements.

Individual characters were impressed into a soft substrate—first clay and later copper—which was filled with molten metal to create the type pieces. Gutenberg formulated an alloy of lead, tin, and antimony that could be easily cast and could stand up to repeated cycles in the press.

The traditional water-based inks were not suitable for press operations, because they tended to run and smear. Gutenberg invented a new ink family based on lampblack and linseed oil that was heated in order to partially polymerize the oil. The paper substrate needed to be damp, in order to prevent the ink from spreading as it soaked into the paper.

Finally, the technology of wine and oil processing was adapted to create the press itself. Pressure was needed to create a uniform impression across the paper. A sliding drawer was added to allow access to the typeface for inking. In Gutenberg's process, multiple colors could be applied to create more decorative results.

Like most technology pioneers, Gutenberg had great difficulty in generating cash flow with his innovations. Eventually, he lost the rights to his new process and possession of the equipment. The financial backer took over the operation, and continued to improve its functionality. The system was not proprietary, because a wide range of craftsmen seemed to have contributed to and copied from Gutenberg's work. Within 50 years of the first Gutenberg Bible, the new publishing industry had produced thousands of titles.

Fritz Pfleumer invented magnetic tape for audio recording in the 1920s. His proof of concept was made of thin strips of paper that he coated with iron(III) oxide using lacquer as an adhesive. BASF refined this prototype into a commercial product based on cellulose acetate tape. The German electronics firm AEG built tape recorders similar to that shown below. Together, BASF's coated tape and AEG's equipment was a proprietary, integrated system. Later, this became an open standard as the tape products of many companies could be used on most reel-to-reel recording equipment.

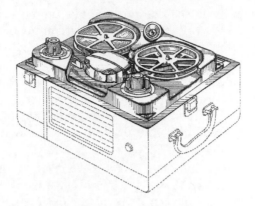

Developing products for a proprietary system:

- The supplier must assume total responsibility for the FFU testing.
  - The variation of your material must be well within the tolerance needs of the system.
  - This forces you to maintain backward compatibility of your consumable product as you update the system. Alternately, the system supplier must continue to offer all old versions for a long period of time.
- There must be good communication and cooperation between the various groups that develop and support the different parts for the system.
  - If one factor absolutely must drive everything else in that system, then make sure that all groups accept this principle. This could be due to cost, safety, or performance issues.
  - The different development groups for the various system components should periodically meet to discuss the opportunities and problems. If some important aspect of one component is causing difficulties in the system, then all the functional areas should jointly brainstorm ways to smooth out the problems.
  - Always remember that the total system profitability is the important thing. Do not allow one arm of the organization to suboptimize a segment at the expense of the total business.
  - A different management mentality is generally needed to make process-dependent materials and equipment systems. The two products cannot both be administered with the same production philosophy.

- All the variability in the one process must allow that component to mate with all the variability in the other components. Customers cannot be expected to pick and choose the ones that work together.

- Patents are an important way to protect the sale of your consumable product from low-cost competition. Consider patenting the combination of all the system aspects working together to perform some function at the customer's end. The downside of patenting end-use applications is that it may require you to sue your own customers to prevent infringement.

- Build adjustability into the prototype/field-test equipment and product so as not to lock into a tight corner—avoid committing to permanent tooling, molds, or high-volume designs until the testing validates the equipment/product compatibility.

- Look for opportunities to extend the system approach to related applications to existing customers.

  - Be prepared to add more components to the platform and to continuously improve it. Consider offering a toothbrush or dental floss head that fits onto the razor handle.

  - Highly functional systems that solve multiple problems for a customer are very difficult for competition to displace.

- The system approach needs a compelling reason to justify its existence.

  - For instance, the one-piece disposable razor cannot be cheaper and work as well as your special razor plus replaceable blade model.

  - Do not make up bogus benefits to lock the business into a closed system. The customers will figure it out eventually.

When a proprietary system's consumable material is process sensitive, then special care must be taken to ensure business continuity. Confidence is required that the system can be supported even if raw material or process changes are needed in production. It is absurdly dangerous to build a system approach around a single-sourced raw material, for instance. If one supplier's plant is shutdown by a fire, hurricane, or labor action, then your entire business model is at risk.

Dealing with risk in the systems approach:

- Always qualify alternate materials, even if they cost more or are harder to run in your process. Be prepared to switch to the backup on short notice.

- When a material is a system component, the process consistency must be very tight. Do not assume that even minor process changes will not affect the system. Test everything.

- If an irreplaceable minority component in the recipe is single sourced, consider stockpiling a large amount, in the event of a failure in the supply chain. Do not implement just-in-time inventory management on essential materials.
- Split your production operation into redundant sites, if possible. Do not let a localized problem become a complete disaster.

A system should be designed to allow for the natural variability of sensitive components. Imagine a material that will change size in the extremes of temperature and humidity in shipment and storage. Do not design a proprietary system that requires a component made with that material to always have exactly the right measurements to function correctly. This may seem obvious; however, system designers may not appreciate the problem until they encounter performance problems at the extreme edges of the envelope. Specifications should be created to document the complete range of ambient conditions that the system components may encounter. Performance should be verified across the entire spectrum before making guarantees to the customers.

The total system might require a very tight control on a particular process condition on your production equipment. Competitors will have a hard time making a knockoff product to work with your system until they learn what the vital attribute is. The customers might not even realize that the system is proprietary until they sample products from your competitors, in an effort to save a little money. But, such a system represents a 100% check on the stability of your process.

In the most extreme cases, it may not be possible to use each batch of consumables without adjustments to the system. With such a sensitive product, you could encode information that will allow the system to compensate. A consumable component would be quality control tested to measure the properties that govern performance. Your system would then read that information and compensate for the unique nature of each batch or unit, without the customer even knowing what is going on in the background.

The key factors for system components can be found in the "concept map of proprietary systems" that follows. You must maintain a delicate balance between the needs of the primary and the secondary products in your system through constant testing and comparison.

Summary

- The ideal system is a unique combination of components that offers the best solution for the customer's problem.

- The worst system offerings have poor synergy between the components and they are obviously crafted to force customers to buy extra things, which they neither want nor need.

- Customers seek simple solutions to complex problems. Proprietary systems should provide a solution that is free of compatibility issues from multiple vendors.

- The supplier assumes total responsibility for FFU when it provides a system.

- Each production location must have access to all system components to verify that their parts function. Specifications must detail every aspect of the interface requirements.

- The systems approach often dictates that your product be made to more exacting requirements or with special interface features. It is harder to implement changes or to source product from outsiders during supply-chain disruptions.

- Proprietary systems tend to block direct competition. But it can be very embarrassing when competitors introduce components that work better in your system.

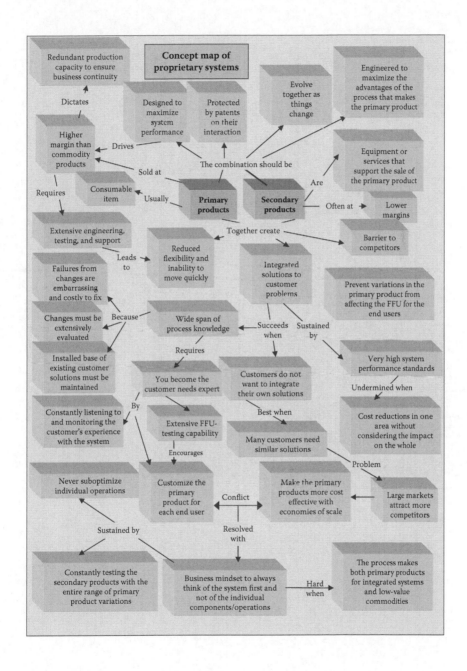

Concept map of proprietary systems

Index